EMIS DATA REVIEWS SERIES 8

Series Editor: Prof. B.L. Weiss

Properties of Lattice-Matched and Strained

Indium Gallium Arsenide

ELECTRONIC MATERIALS INFORMATION SERVICE
Other books in the EMIS Data Reviews Series from INSPEC:

No. 6	**Properties of Indium Phosphide** (1991)
No. 7	**Properties of Aluminium Gallium Arsenide** (1993)
No. 8	**Properties of Lattice-Matched and Strained Indium Gallium Arsenide** (1993)
No. 10	**Properties of Narrow Gap Cadmium-based Compounds** (1994)
No. 13	**Properties of Silicon Carbide** (1995)
No. 14	**Properties of Metal Silicides** (1995)
No. 15	**Properties of III-V Quantum Wells and Superlattices** (1996)
No. 17	**Properties of Wide Bandgap II-VI Semiconductors** (1997)
No. 18	**Properties of Porous Silicon** (1997)
No. 19	**Properties of Amorphous Silicon and its Alloys** (1998)
No. 20	**Properties of Crystalline Silicon** (1999)
No. 21	**Properties, Processing and Applications of Indium Phosphide** (1999)
No. 22	**Properties, Processing and Applications of Glass and Rare-Earth-Doped Glasses for Optical Fibres** (1998)
No. 23	**Properties, Processing and Applications of Gallium Nitride and Related Semi-Conductors** (1999)
No. 24	**Properties of Silicon Germanium and SiGe:Carbon** (2000)
No. 25	**Physical Properties of Liquid Crystals, Volume 1: Nematics** (2000)
No. 26	**Properties, Growth and Applications of Diamond** (2000)
No. 28	**Properties of Lithium Niobate** (2002)
No. 29	**Properties of Amorphous Carbon** (2002)

Details of books in the EMIS Data Reviews Series are available at www.theiet.org/publications/booksandconferences

Properties of Lattice-Matched and Strained

Indium Gallium Arsenide

Edited by Pallab Bhattacharya

University of Michigan, USA

The Institution of Engineering and Technology

Published by The Institution of Engineering and Technology, London, United Kingdom

First published 1993 (0 85296 865 5)
Paperback edition 2006

The Institution of Engineering and Technology
Michael Faraday House
Six Hills Way, Stevenage
Herts, SG1 2AY, United Kingdom

www.theiet.org

British Library Cataloguing in Publication Data
A catalogue record for this product is available from the British Library

ISBN (10 digit) 0 86341 662 4
ISBN (13 digit) 978-0-86341-662-0

Printed in the UK by Hobbs the Printers Ltd, Hampshire
Reprinted in the UK by Lightning Source UK Ltd, Milton Keynes

Contents

Foreword

During the past two decades, considerable research has been devoted to the epitaxial growth techniques, material properties, and the device behaviour of the ternary solid solution $Ga_xIn_{1-x}As$. This system is one of the best examples of a semiconductor material where the research and scientific interest are driven by significant technology needs and opportunities.

In the early 1970s, $Ga_xIn_{1-x}As$ was investigated for optoelectronic devices. One application was for light-emitting diode sources at the 1.06 µm emission wavelength of the Nd:YAG laser. This structure used vapour-phase epitaxy and composition grading from the GaAs substrate to the desired composition. Also at this time, the minimum in the optical fibre attenuation was near 1.05 µm which made it desirable to develop semiconductor sources and detectors for this wavelength. The other application was for photoemissive cathodes for the detection of Nd:YAG emission. These studies led to the determination of the room-temperature energy gap and lattice constant for the quaternary $Ga_xIn_{1-x}As_yP_{1-y}$.

By the mid 1970s, liquid-phase epitaxial growth of $Ga_xIn_{1-x}As_yP_{1-y}$ on InP had been demonstrated and improvements in optical fibres had resulted in the transmission attenuation minimum at a wavelength of 1.3 µm. The activity due to the success of the $InP/Ga_xIn_{1-x}As_yP_{1-y}/InP$ double heterostructure laser made InP substrates readily available from commercial suppliers. Near the end of the 1970s, $Ga_{0.47}In_{0.53}As/InP$ photodiodes with a spectral dependence between wavelengths of 1.0 and 1.7 µm were demonstrated.

During the two year period of 1977 to 1978, several innovations occurred which would ensure the technological importance of $Ga_xIn_{1-x}As$. The first was the prediction from transport calculations that the saturated drift velocities of $Ga_{0.47}In_{0.53}As$, which lattice matched to InP, would be higher than for GaAs, and, therefore, would be a promising material for microwave semiconductor devices. The next important result was the demonstration that high quality $InP/Ga_{0.47}In_{0.53}As/InP$ double-heterostructure lasers could be fabricated. The third innovation was the demonstration of modulation-doped heterostructures to provide enhanced carrier velocities for field-effect transistors. The heterostructure $Al_xGa_{1-x}As/GaAs$ was used for this demonstration because this system was the most mature technology. Modulation-doped heterostructures proved to be ideally suited to take advantage of the high electron saturated drift velocity of $Ga_xIn_{1-x}As$.

Another innovation in the early 1980s which provided many possible applications for $Ga_xIn_{1-x}As$ was the development of strained-layer superlattices which became known as pseudomorphic layers. The lattice-matched heterostructure of $Ga_{0.47}In_{0.53}As/Al_{0.48}In_{0.52}As$ also became useful in a variety of applications. By the mid 1980s, low threshold quantum well lasers had been demonstrated. A few years later, the erbium-doped fibre amplifier was shown to be well suited as postamplifiers, preamplifiers and in-line repeaters for optical fibre communication systems. This optical amplifier required a laser pump at a wavelength of 0.98 µm. The best laser source was shown to be heterostructure lasers with $Ga_xIn_{1-x}As$ strained layer quantum well active layers. Also, the epitaxial growth technologies for molecular beam epitaxy and organometallic chemical vapour deposition matured to become suitable growth technologies for preparation of the required multiple thin layers for a variety of $Ga_xIn_{1-x}As$ heterostructures. From this point on, the publications on the growth, properties and devices made with $Ga_xIn_{1-x}As$ grew almost exponentially with time. These devices included photodiodes, metal-semiconductor-metal photodetectors, heterostructure lasers, surface-emitting vertical-cavity lasers, resonant tunnelling structures, quantum wires and dots, modulation doped field-effect transistors (or HEMTs), and heterojunction-bipolar transistors as well as optoelectronic integrated circuits. Threshold current for $Ga_xIn_{1-x}As$ heterostructure lasers reached the mA range and current gain cutoff frequencies, f_T, of transistors exceeded 100 GHz.

With this rapid growth in the research and technology related to $Ga_xIn_{1-x}As$, it is very timely to collect the widely dispersed data describing this important semiconductor. This book, with material contributed by those whose work has made the field successful, will provide the information for further advances. I look forward to having this resource among my reference books.

H. Craig Casey Jr.
Department of Electrical Engineering
Duke University
Durham, NC 27708-0291
USA

July 1993

Introduction

The $In_xGa_{1-x}As$ (or GaInAs) ternary alloys span the range of lattice constants between the binary compounds GaAs (x = 0, bandgap 1.43 eV) and InAs (x = 1, bandgap 0.35 eV). These alloys and their heterostructures with wider bandgap semiconductors are playing a pivotal role in the study of quantum systems and in the development of high-performance electronic and optoelectronic devices. Except the alloy with composition $In_{0.53}Ga_{0.47}As$, which is lattice-matched to InP, all other compositions are usually grown epitaxially on GaAs or InP substrates, and are therefore 'mismatched' materials. It is important to realise that almost all work - epitaxial growth, characterisation and device applications - with InGaAs is being done with thin (~1 μm) single layers or heterostructures formed with other compositionally different lattice-matched or mismatched semiconductors. Because of this, and due to the lack of 'bulk' InGaAs, this ternary compound has not been investigated and used in the same way as GaAs, InP, or Si. As a consequence, its thermal and mechanical properties are not known as well as those of GaAs or InP. Nonetheless, this has not hindered research and development of this compound and some of its more exciting composition- and strain-dependent material properties are being fully exploited to enhance device properties or to realise novel concepts.

$In_xGa_{1-x}As$ is a direct bandgap semiconductor throughout the entire composition range. $In_{0.53}Ga_{0.47}As$ lattice-matched to InP, with a bandgap of 0.74 eV at room temperature, is an important material for device applications. Very high electron mobility, high electron velocity, and a large intervalley separation in the conduction band are some of its favourable properties. Heterostructure devices using this compound, such as modulation doped field effect transistors, have demonstrated record DC, high-frequency, and noise performance at room and cryogenic temperatures. Lattice-matched heterostructures include $In_{0.53}Ga_{0.47}As/InP$, $In_{0.53}Ga_{0.47}As/In_{0.52}Al_{0.48}As$, $In_{0.53}Ga_{0.47}As/In_{1-x}Ga_xAs_yP_{1-y}$ and $In_{0.53}Ga_{0.47}As/In_xGa_yAl_{(1-x-y)}As$. Its bandgap also corresponds to the spectral range where optical fibres have very low loss and dispersion. Therefore $In_{0.53}Ga_{0.47}As$ is used for the development of sources, modulators and detectors for optical communication. It is not often that one finds a material suitable for both electronic and optoelectronic applications. Due to this fortuitous coincidence, InGaAs and its heterostructures are being used for the development of optoelectronic integrated circuits (OEIC).

The development and applications of mismatched $In_xGa_{1-x}As$ alloys are fascinating. Strained semiconductors were initially developed as a tool for bandgap engineering. It was subsequently realised that biaxial strain in the material, incorporated during epitaxy, can be exploited for band structure engineering. Thus, there are two types of mismatched materials: relaxed, or strained-relieved, which can only be used for bandgap tailoring, and pseudomorphic, which can be used for bandgap and band structure engineering. The change in the band structure with biaxial strain uniquely alters the electronic and optical properties of semiconductors. These changes have been exploited to enhance the performance of devices such as FETs and quantum well lasers to unprecedented levels. It is also important to note that while $In_xGa_{1-x}As$ grown on GaAs is only compressively strained, $In_{0.53\pm x}Ga_{0.47\mp x}As$ grown on InP substrates can be both compressively or tensilely strained.

It is amply evident that InGaAs and related heterostructures have established themselves as important materials for enabling technologies in microelectronics and optoelectronics. It is befitting, therefore, to dedicate a volume of the EMIS series to these materials and heterostructures. This volume describes their most important properties in the form of Datareviews written by renowned experts in the field. Their research contributions, and those of many others, have helped in the firm establishment of these heterostructures as technologically important materials. The volume is conveniently classified into 9 groups: (1) Structural Properties of InGaAs; (2) Thermal, Mechanical and Lattice Vibrational Properties of InGaAs; (3) Band Structure of Lattice-Matched and Strained InGaAs Alloys; (4) Transport Properties; (5) Surface Properties; (6) Radiative and Non-Radiative Recombination; (7) Optical and Electro-Optic Properties of InGaAs; (8) InGaAs Technology; and (9) Device Application of InGaAs and Related Heterostructures. It was felt that the volume would be incomplete without a chapter on device applications of InGaAs. Amongst the host of devices made with these materials, only the most important ones are described. Since heterostructures of InGaAs are often made with other lattice-matched quaternaries, it was deemed useful to devote an article specifically to these lattice-matched compounds.

While many aspects of the information contained in this book are available in publications and review articles, this volume should evolve as a condensed and valuable resource for students, researchers and practising engineers.

Before concluding, it is appropriate to comment on the chemical formula InGaAs itself. In accordance with the position of the constituent group III atoms in the periodic table, a more accurate description and formula is GaInAs. However, InGaAs is more often used at the present time and this nomenclature has been retained. Finally, I would like to thank John L. Sears, Managing Editor of the EMIS series, for his support and patience and the authors and reviewers for their contributions.

Pallab Bhattacharya
Department of Electrical Engineering and Computer Science
University of Michigan
Ann Arbor, Michigan 48109-2122
USA

July 1993

Contributing Authors

Dr S. Adachi	Gunma University, Department of Electronic Engineering, 1-5-1 Tenjin-cho, Kiryu 376, Gunma, Japan	2.1, 2.4, 3.4
Dr A.R. Adams	University of Surrey, Department of Physics, Guildford, Surrey, GU2 5XH, UK	3.2
Professor I. Adesida	University of Illinois, Microelectronics Laboratory, 208 North Wright Street, Urbana, IL 61801, USA	8.3
Dr S.A. Alterovitz	NASA-Lewis, 21000 Brookpark Road, M/S 54-5, Cleveland, OH 44135, USA	7.1
Professor P.K. Bhattacharya	University of Michigan, Department of Electrical Engineering and Computer Science, 1301 Beal Avenue, Ann Arbor, MI 48109-2122, USA	4.3, 6.3, 8.2
Dr D. Bimberg	Technische Universitat Berlin, Institut fur Festkorperphysik, Hardenbergstr. 36, 100 Berlin 12, Germany	6.1, 6.2
Dr N. Chand	AT&T Bell Laboratories, Room 7C-304, 600 Mountain Avenue, Murray Hill, NJ 07974, USA	4.4, 8.1
Dr Y.C. Chen	University of Michigan, Department of Electrical Engineering and Computer Science, 1301 Beal Avenue, Ann Arbor, MI 48109-2122, USA	8.2
Dr D.J. Dunstan	University of Surrey, Department of Physics, Guildford, Surrey, GU2 5XH, UK	2.3
Dr M. Dutta	U.S. Army Research Laboratory, Electronics and Power Sources Directorate, AMSRL-EP-EF, Fort Monmouth, NJ 07703-5601, USA	7.3, 7.4
Dr N.K. Dutta	AT&T Bell Laboratories, Optical Materials Research, Room 6E-414, 600 Mountain Avenue, Murray Hill, NJ 07974-0636, USA	9.2
Professor L.F. Eastman	Cornell University, School of Electrical Engineering, 424 Phillips Hall, Ithaca, NY 14853, USA	4.1

Dr P.N. Fawcett	Imperial College of Science, Technology and Medicine, Interdisciplinary Research Centre, Prince Consort Road, London, SW7 2BZ, UK	5.1
Dr E.A. Fitzgerald	AT&T Bell Laboratories, Room 1E-447, 600 Mountain Avenue, Murray Hill, NJ 07974, USA	1.2
Professor M. Ilegems	Ecole Polytechnique Federale de Lausanne, Microelectronics and Optoelectronics Institute, CH-1015 Lausanne, Switzerland	1.3
Dr R.F. Karlicek Jr.	AT&T Bell Laboratories, Room 7C-304, 600 Mountain Avenue, Murray Hill, NJ 07974, USA	8.1
Dr K.W. Kim	North Carolina State University, College of Engineering, 213 Page Hall, Raleigh, NC 27695-7901, USA	4.2
Dr P. Kordos	Institut fur Schicht- und Ionentechnik, Forschungszentrum Julich, W-5170 Julich, Germany	5.3
Professor M.A. Littlejohn	North Carolina State University, College of Engineering, 213 Page Hall, Raleigh, NC 27695-7901, USA	4.2
Professor S. Mahajan	Carnegie-Mellon University, Department of Metallurgical Engineering and Materials Science, Pittsburgh, PA 15213, USA	1.4
Dr M. Marso	Institut fur Schicht- und Ionentechnik, Forschungszentrum Julich, W-5170 Julich, Germany	5.3
Dr M. Matsuura	Keio University, Department of Electrical Engineering, Faculty of Science and Technology, 3-14-1 Hiyoshi, Yokohama 223, Japan	2.2
Professor R. Merlin	University of Michigan, College of Literature, Science and the Arts, 3035C Randall Laboratory, Ann Arbor, MI 48109-1120, USA	7.2
Dr J. Pamulapati	U.S. Army-ETDL, SLCET-ED, Fort Monmouth, NJ 07703-5000, USA	7.3, 7.4

Professor D. Pavlidis	University of Michigan, Department of Electrical Engineering and Computer Science, 1301 Beal Avenue, Ann Arbor, MI 48109-2122, USA	9.3
Professor T.P. Pearsall	University of Washington, Department of Electrical Engineering, 101 Wilson Hall, M/S FB-10, Seattle, WA 98195, USA	9.1
Dr A.D. Prins	University of Surrey, Department of Physics, Guildford, Surrey, GU2 5XH, UK	2.3
Dr J. Shah	AT&T Bell Laboratories, Room 4D-415, Holmdel, NJ 07733, USA	6.4
Professor J. Singh	University of Michigan, Department of Electrical Engineering and Computer Science, 1301 Beal Avenue, Ann Arbor, MI 48109-2122, USA	3.1, 3.5
Dr B. Srocka	Technische Universitat Berlin, Institut fur Festkorperphysik, Hardenbergstr. 36, 100 Berlin 12, Germany	6.1, 6.2
Professor G.E. Stillman	University of Illinois at Urbana-Champaign, Center for Compound Semiconductor Microelectronics, 151 Microelectronics Laboratory, 208 North Wright Street, Urbana, IL 61801, USA	3.3
Professor G.B. Stringfellow	University of Utah, Department of Materials Science, 304 EMRO, Salt Lake City, UT 84112, USA	1.1
Professor N.S. Takahashi	Keio University, Department of Electrical Engineering, Faculty of Science and Technology, 3-14-1 Hiyoshi, Yokohama 223, Japan	2.2
Dr H. Tian	North Carolina State University, College of Engineering, 213 Page Hall, Raleigh, NC 27695-7901, USA	4.2
Professor C.R. Wie	State University of New York at Buffalo, Department of Electrical and Computer Engineering, Bonner Hall, Amherst, NY 14260, USA	8.4

Professor H.H. Wieder	University of California at San Diego, Department of Electrical and Computer Engineering, Mail Code 0407, La Jolla, CA 92093-0407, USA	5.2
Dr V.A. Wilkinson	University of Surrey, Department of Physics, Guildford, Surrey, GU2 5XH, UK	3.2
Professor C.W. Wilmsen	Colorado State University, Department of Electrical Engineering, Fort Collins, CO 80523, USA	5.4
Dr J.E. Zucker	AT&T Bell Laboratories, Room 4F-319, Crawfords Corner Road, Holmdel, NJ 07733, USA	7.5

Acknowledgements

It is a pleasure to acknowledge the work both of the contributing authors named on the previous pages and of the following experts in the field who have also participated in the production of this book:

Dr S. Adachi	Gunma University, Department of Electronic Engineering, 1-5-1 Tenjin-cho, Kiryu 376, Gunma, Japan
Dr J.M. Ballingall	Martin Marietta, Electronics Laboratory, Building 3, Room 120, PO Box 4840, Syracuse, NY 13221-4840, USA
Professor S.M. Bedair	North Carolina State University, Department of Electrical and Computer Engineering, Box 7911, Raleigh, NC 27695-7911, USA
Professor P. Berger	University of Delaware, Department of Electrical Engineering, 140 Evans Hall, Newark, Delaware 19716-3130, USA
Dr A. Brown	Hughes Research Laboratory, RL-61, 3011 Malibu Canyon Road, Malibu, CA 90265, USA
Professor A. Gossard	University of California at Santa Barbara, Department of Electrical and Computer Engineering, Santa Barbara, CA 93106-9560, USA
Professor G. Guillot	INSA, Physique de la matiere - 502, 69621 Villeurbanne Cedex, France
Dr J. Hayes	Bellcore, Room 3X-373, 331 Newman Springs Road, Red Bank, NJ 07701-7040, USA
Dr B. Hong	Bellcore, Room 3Y-368, 331 Newman Springs Road, Red Bank, NJ 07701-7040, USA
Professor J. Laskar	University of Hawaii at Manoa, Department of Electrical Engineering, 2540 Dole Street, Homes Hall 483, Honolulu, Hawaii 96822, USA
Dr J. Loehr	Wright Laboratories, Solid-State Electronic Directorate, WL-ELR Building 620, WPAFB, OH 45433, USA
Dr M. Ludowise	Hewlett-Packard Laboratories, 3500 Deer Creek Road, Palo Alto, CA 94304, USA
Professor A. Majerfeld	University of Colorado at Boulder, Department of Electrical Engineering, Campus Box 425, Boulder, CO 80309, USA

Dr T. Masselink	IBM T.J. Watson Research Center, Room 20-230, PO Box 218, Yorktown Heights, NY 10598, USA
Dr M. Mondry	University of California at Santa Barbara, Department of Electrical and Computer Engineering, Santa Barbara, CA 93106-9560, USA
Dr D. Nichols	AT&T Bell Laboratories, Room 6E-401, PO Box 636, 600 Mountain Avenue, Murray Hill, NJ 07974, USA
Professor B. Orr	University of Michigan, Department of Physics, 2071 Randall Laboratory, Ann Arbor, MI 48109-1120, USA
Professor S. Pang	University of Michigan, Department of Electrical Engineering and Computer Science, 1301 Beal Avenue, Ann Arbor, MI 48109-2122, USA
Professor R.V. Ramaswamy	University of Florida, Department of Electrical Engineering, 133 Larsen Hall, Gainesville, FL 32611, USA
Dr M. Stroscio	U.S. Army Research Office, Electronics Division, 4300 South Miami Boulevard, Research Triangle Park, NC 27709-2211, USA
Dr S. Tiwari	IBM T.J. Watson Research Center, Room 30-144, PO Box 218, Route 154, Ketchuwan Road, Yorktown Heights, NY 10598, USA
Dr R. Tober	Harry Diamond Laboratories, 2800 Powder Mill Road, SLCHD-ST-AP, Adelphi, MD 20783-1197, USA
Professor C. Tu	University of California at San Diego, Department of Electrical and Computer Engineering, 9500 Gilman Drive, La Jolla, CA 92093-0407, USA
Professor J. Wager	Oregon State University, Department of Electrical and Computer Engineering, Corvallis, OR 97331-3211, USA
Professor H. Winful	University of Michigan, Department of Electrical Engineering and Computer Science, 1301 Beal Avenue, Ann Arbor, MI 48109-2122, USA
Professor M.C. Wu	University of California at Los Angeles, Department of Electrical Engineering, 405 Hilgard Avenue, Los Angeles, CA 90024-1594, USA

Professor S. Yalisove	University of Michigan, Department of Materials Science and Engineering, 2118 H.H. Dow Building, Ann Arbor, Michigan 48109-2136, USA
Professor H.D. Yao	University of Nebraska at Lincoln, Department of Electrical Engineering, Lincoln, NE 68588-0511, USA

Abbreviations

The following abbreviations are used in this book.

AC	alternating current
AES	Auger electron spectroscopy
ALE	atomic layer epitaxy
APD	avalanche photodiode
AS	admittance spectroscopy
B	binary
BEP	beam equivalent pressure
BH	buried heterostructure
BT	bipolar transistor
CBE	chemical beam epitaxy
CBM	conduction band minimum
CHINT	charge injection transistor
C-V	capacitance-voltage
CVD	chemical vapour deposition
CW	continuous wave
DA	disorder-activated
DALA	disorder-activated longitudinal acoustic
DATO	disorder-activated transverse optic
DC	direct current
DCHEMT	doped channel high electron mobility transistor
2DEG	2-dimensional electron gas
DH	double heterostructure
DHHEMT	double heterojunction high electron mobility transistor
DL	dielectric level
DLOS	deep level optical spectroscopy
DLTS	deep level transient spectroscopy
DSA	dimethyl succinic acid
EBER	electron beam electroreflectance
ER	electroreflectance
EXAFS	extended X-ray absorption fine structure
FCC	face centred cubic
FET	field effect transistor
FK	Franz-Keldysh
FWHM	full width at half maximum
GRIN	graded index
GRIN-SCH	graded index separate confinement heterostructure
GSMBE	gas source molecular beam epitaxy

HBT	heterojunction bipolar transistor
HEMT	high electron mobility transistor
HFET	heterostructure field effect transistor
HH	heavy hole
HIGFET	heterostructure insulated gate field effect transistor
HWHM	half width at half maximum
IMPATT	impact avalanche transit time
IP	internal photoemission
IR	infrared
I-V	current-voltage
JFET	junction field effect transistor
LA	longitudinal acoustic
LDA	local-density functional approach
LEC	liquid encapsulated Czochralski
LED	light emitting diode
LEED	low energy electron diffraction
LF	low frequency
LH	light hole
LO	longitudinal optic
LPE	liquid phase epitaxy
LS	light scattering
MB	Matthews and Blakeslee (model)
MBE	molecular beam epitaxy
MD	modulation doped
MEDFET	metal-doped channel field effect transistor
MESFET	metal-semiconductor field effect transistor
MIS	metal insulator semiconductor
MISFET	metal insulator semiconductor field effect transistor
MMIC	monolithic microwave integrated circuit
MO	metalorganic
MOCVD	metalorganic chemical vapour deposition
MODFET	modulation doped field effect transistor
MOMBE	metalorganic molecular beam epitaxy
MOSFET	metal oxide semiconductor field effect transistor
MOVPE	metalorganic vapour phase epitaxy
MQW	multiple quantum well
MS	model-solid approach
MSM	metal-semiconductor-metal
NDR	negative differential resistance
NL	neutrality level
OA	optical absorption
OEIC	optoelectronic integrated circuit
OMCVD	organometallic chemical vapour deposition

OMVPE	organometallic vapour phase epitaxy
PAE	power added efficiency
PC	photoconductive response
PDP	phonon deformation potential
PECVD	plasma enhanced chemical vapour deposition
PHEMT	pseudomorphic high electron mobility transistor
PIN	p-type intrinsic n-type
PL	photoluminescence
PLE	photoluminescence excitation spectroscopy
PP	pseudopotential
PPC	persistent photoconductivity
PR	photoreflectance
Q	quaternary
QW	quantum well
RC	resistance capacitance
RF	radio frequency
RHEED	reflection high energy electron diffraction
RIE	reactive ion etching
RT	resonant tunnelling
RTA	rapid thermal annealing
SA	succinic acid
SAM	separate absorption and multiplication
SCR	space charge region
SDH	Shubnikov-de Haas
SE	semi-empirical estimation
SHHEMT	single heterojunction high electron mobility transistor
SI	semi-insulating
SIMS	secondary ion mass spectroscopy
SL	superlattice
SRV	surface recombination velocity
STM	scanning tunnelling microscopy
T	ternary
TA	transverse acoustic
TBP	tertiary butyl phosphine
TEM	transmission electron microscopy
TLM	transmission line method
TM	transition metal
TO	transverse optic
TP	transport property
UPS	ultraviolet photoemission spectroscopy
UV	ultraviolet

VPE	vapour phase epitaxy
VRBE	vacuum referred binding energy
XPS	X-ray photoemission spectroscopy

CHAPTER 1

STRUCTURAL PROPERTIES

1.1 Lattice parameters and crystal structure of InGaAs
1.2 Lattice-mismatch and dislocations in InGaAs/GaAs strained heterostructures
1.3 InP-based lattice-matched heterostructures
1.4 Microstructure of InGaAs epitaxial layers

1.1 Lattice parameters and crystal structure of InGaAs

G.B. Stringfellow

February 1993

A INTRODUCTION

The crystal structures of the III-V alloys were determined many years ago. The lattice constants of the pure compounds were also accurately determined. Nevertheless, due to the importance of these materials quantities, they are reviewed briefly here. Special attention is paid to the lattice parameter of InGaAs layers grown epitaxially on InP substrates.

B CRYSTAL STRUCTURE

GaAs and InAs are well known from X-ray diffraction studies to always have the zinc-blende structure (space group T_d^2-F43m). These two compounds are known to be completely miscible [1,2]. For a disordered alloy, the zinc-blende crystal structure persists throughout the entire composition range. However, in recent years, ternary III-V alloys, especially those with compositions near the centre of the composition range, x ~ 0.5 for an $A_xB_{1-x}C$ alloy, have been found to order. The most common ordered structure consists of a doubling of the periodicity in one of the 4 {111} directions in the lattice, forming a structure with R3m symmetry (the Cu-Pt structure). Thus, in some regions of the alloy the crystal structure is not zinc-blende. For $In_{1-x}Ga_xAs$ (with x near 0.5) the Cu-Pt ordered structure has been observed in alloys grown on (001)-oriented substrates by vapour phase epitaxy [3]. On the other hand, $Ga_{0.5}In_{0.5}As$ grown by molecular beam epitaxy on (110)-oriented InP substrates is found to order in the Cu-Au-I structure with a doubling of the periodicity in one of the {100} directions, giving P4m2 symmetry [4,5].

In bulk III-V alloys, the Cu-Au-I structure is calculated to have a lower energy than the Cu-Pt structure [6]. However, ordering occurs at the surface during epitaxial growth [7]. For vapour phase epitaxial growth on the (001) plane, surface reconstruction is believed to alter the relative energies of the two phases making the Cu-Pt structure more stable [8]. Kinetic effects also play a role. It is postulated that layer-by-layer growth via the motion of [110]-oriented steps on the reconstructed (001) surface also favours formation of the Cu-Pt structure [7].

C LATTICE PARAMETERS

The lattice parameter is known to be a function of pressure, temperature, and the addition of dopants. For pure materials measured at atmospheric pressure and room temperature the lattice constants of the end compounds GaAs and InAs have been precisely determined. GaAs has a lattice constant of 5.65325 Å at 300 K [3]. InAs has a lattice constant of 6.0583 Å at 298.15 K [10].

As mentioned above, GaAs and InAs are completely miscible. The lattice constant of bulk single crystals of InGaAs was studied by Hockings et al [11]. The lattice constant was determined from X-ray diffraction results from homogeneous samples. The solid composition was determined from chemical analysis and X-ray fluorescence measurements. An error limit of $\pm$ 0.005 in composition was given. The error limit for the X-ray measurements is estimated to be $\pm$ 0.001 Å. The lattice parameter was found to vary linearly with solid composition for $In_{1-x}Ga_xAs$ alloys. The best experimental data give the relationship

$$a = 6.0583 - 0.4050x \qquad (1)$$

An extremely important value of solid composition is that having the same lattice parameter as InP. This allows the growth of extremely perfect epitaxial layers on InP substrates without strain and mismatch dislocations. The value of x giving a lattice constant equal to that of the InP at 300 K is x = 0.468, since the lattice constant of InP is reported to be 5.8687 Å at this temperature [12].

Epitaxial layers of $Ga_{0.47}In_{0.53}As$ lattice matched to InP substrates have been grown by a number of techniques. This presents an opportunity to verify the results obtained for bulk crystals as represented by Eqn (1).

For example, Nakajima et al [13] grew layers by liquid phase epitaxy on InP substrates having the (100), (111)B and (111)A orientations. They reported that the lattice parameter is exactly that of InP for x = 0.47 $\pm$ 0.01 for both (100) and (111)B substrate orientations. The composition was determined by electron microprobe analysis (accuracy of approximately $\pm$ 0.01) and lattice constant from X-ray analysis.

A potential problem with such experiments is that the X-ray diffraction pattern of the exactly lattice matched layer is indistinguishable from that of the InP substrate. When slightly lattice mismatched, the epilayer is strained to match the interatomic spacing of the substrate in the growth plane. This causes an exaggeration of the lattice parameter difference between the epilayer and substrate from X-ray diffraction measurements. However, this can be corrected for and the results are judged to be reliable within the accuracy limits quoted above.

Cole et al [14] measured the strain due to lattice mismatch directly using ion beam crystallography. Layers of InGaAs were grown by organometallic vapour phase epitaxy on InP substrates. Composition was determined by Rutherford backscattering analysis. They found that a composition x = 0.47($\pm$ 0.01) gives an epitaxial layer with no measurable strain, with a lattice constant of 5.868($\pm$ 0.004) Å.

D CONCLUSION

The crystal structure and lattice parameters of the $In_{1-x}Ga_xAs$ ternary compounds are reviewed. The constituent binaries GaAs and InAs are completely miscible. The ternaries are lattice-matched to GaAs while the compound with x = 0.47 is lattice-matched to InP.

REFERENCES

[1] M.B. Panish, Ilegems [in *Progress in Solid State Chemistry* Eds H. Reiss, J.O. McCaldin (Pergamon, Oxford, 1972) vol.7 p.66-9]

[2] G.B. Stringfellow [*J. Cryst. Growth (Netherlands)* vol.27 (1974) p.21-34]

[3] M.A. Shahid, S. Mahajan, D.E. Laughlin, H.M. Cox [*Phys. Rev. Lett. (USA)* vol.58 (1987) p.2567-70]

[4] T.S. Kuan, W.I. Wang, E.L. Wilkie [*Appl. Phys. Lett. (USA)* vol.51 (1987) p.51-3]

[5] O. Ueda, Y. Nakata, T. Fujii [*Appl. Phys. Lett. (USA)* vol.58 (1991) p.705-7]

[6] S.H. Wei, Z. Zunger [*Phys. Rev. B (USA)* vol.39 (1989) p.3279]

[7] G.B. Stringfellow, G.S. Chen [*J. Vac. Sci. Technol. B (USA)* vol.9 (1991) p.2182-8]

[8] S. Froyen, A. Zunger [*Phys. Rev. Lett. (USA)* vol.66 (1991) p.2132-4]

[9] J.B. Mullin, B.W. Straughan, C.M.H. Driscoll, A.F.W. Willoughby [in *Inst. Phys. Conf. Ser. (UK)* vol.24 (1975) p.275]

[10] J.V. Ozolinsh, G.K. Averkieva, A.F. Ilvinsh, N.A. Goryunova [*Sov. Phys. Cryst. (USA)* vol.7 (1963) p.691]

[11] E.F. Hockings, I. Kudman, T.E. Seidel, C.M. Schmelz, E.F. Steigmeir [*J. Appl. Phys. (USA)* vol.37 (1966) p.2879-87]

[12] G. Giesecke, H. Pfister [*Acta Crystallogr. (Denmark)* vol.11 (1958) p.369]

[13] K. Nakajima, T. Tanahashi, K. Akita, T. Yamaoka [*J. Appl. Phys. (USA)* vol.50 (1979) p.4975]

[14] J.M. Cole, L.G. Earwalker, A.G. Cullis, N.G. Chew, S.J. Bass [*J. Appl. Phys. (USA)* vol.60 (1986) p.2639]

1.2 Lattice mismatch and dislocations in InGaAs/GaAs strained heterostructures

E.A. Fitzgerald

February 1992

A INTRODUCTION

The ternary compound $In_xGa_{1-x}As$ is of technical interest for its electronic and optical properties. InGaAs is potentially useful for optoelectronic applications in three forms:

(a) Bulk InGaAs, or lattice-matched $In_{0.53}Ga_{0.47}As$ on InP substrates
(b) Strained InGaAs on GaAs substrates
(c) Relaxed InGaAs on GaAs substrates

The first category has dominated applications since the 1.55 μm wavelength emission of $In_{0.53}Ga_{0.47}As$ can be transmitted through optical fibre with low signal loss for long distance telecommunications.

B RELAXED, BULK InGaAs

The advantageous properties of InGaAs lie in the change in band structure as InAs is mixed with GaAs. The electron mobility, the bandgap size, and the lattice constant of GaAs and InAs are drastically different as shown in TABLE 1.

TABLE 1 The bandgap, electron mobility, and lattice constant of GaAs [1,2] and InAs [1,3].

Material	Mobility μ (300 K, cm^2/Vs)	Bandgap E_g (eV)	Lattice constant a (Å)
GaAs	9200	1.424	5.6533
InAs	30,000	0.354	6.0584

By mixing InAs with GaAs, one can increase the electron mobility and decrease the bandgap, if the alloy is totally miscible. Fortunately, most III-V alloy solutions are miscible due to a large enthalpy of mixing, forming completely soluble ternaries. Using a regular solution model, the enthalpy of mixing (or heat of mixing) is [4]

$$\Delta H_M = \alpha X_{GaAs} X_{InAs} \tag{1}$$

where α is a constant independent of temperature and X_{GaAs} and X_{InAs} are the mole fractions of GaAs and InAs, respectively. α for InGaAs alloys is ~ -2650 cal/mole [5,6]. The benefit of this negative heat of mixing is that the alloy is completely miscible, and therefore the desired properties can be attained by adjusting the InAs mole fraction. The bandgap does not follow a linear behaviour with InAs mole fraction (x) [7]

$$E_g = 0.324 + 0.7(1\text{-}x) + 0.4(1\text{-}x)^2 \qquad (2)$$

or [1]

$$E_g = 0.36 + 0.505(1\text{-}x) + 0.555(1\text{-}x)^2 \qquad (3)$$

whereas the lattice constant does [7]

$$a = 6.058 - 0.405(1\text{-}x) \qquad (4)$$

The change in lattice constant is not a deleterious effect if the InGaAs is a bulk substrate. Such material would be useful for discrete 1.3 and 1.55 μm devices. However, one adverse effect of a completely miscible solution is that if a crystal is grown from a melt via the Czochralski method, the liquid composition becomes much different than the composition of the solid, creating an inhomogeneous boule. Thus, it is usually necessary to grow epitaxial layers of $In_xGa_{1-x}As$ on substrates which can be grown in bulk, such as InP or GaAs. If only lattice-matched $In_xGa_{1-x}As$ is considered, $In_{0.53}Ga_{0.47}As$ is the only composition which can be grown (lattice-matched to InP).

C STRAINED $In_xGa_{1-x}As$ ON GaAs SUBSTRATES

The advantages of growing strained InGaAs on GaAs substrates are the ability to:

(a) Implement In compositions which are not lattice-matched to InP
(b) Utilise the effects of the strain
(c) Utilise the band discontinuities between InGaAs and GaAs
(d) Integrate InGaAs-based devices with GaAs devices

InGaAs compositions with greater In concentrations will have a higher electron mobility, as shown in TABLE 1. Thus, strained layers without defects can utilise the superior properties of InGaAs. The strain itself can be useful, since in-plane mobilities can be enhanced in two-dimensional electron gases confined in a strained InGaAs layer between GaAs. The band discontinuities between the InGaAs and GaAs allow confinement of electrons and holes to the InGaAs layer, creating the two-dimensional electron gas. A field effect transistor has been fabricated utilising these properties [8]. The same confinement of electrons and holes in the InGaAs which allows the formation of the two-dimensional electron gas can be used to make efficient quantum well emitters. Since the recombination occurs in the InGaAs, the emission wavelength will be longer than the GaAs emission wavelength. Any such optical and electrical device can be integrated with GaAs devices on the substrate.

The successful implementation of the applications above requires that the material is of high quality, i.e. that the strain in the InGaAs layer does not create defects. If an $In_xGa_{1-x}As$ layer is grown on GaAs, the layer will stay totally strained and defect-free if the mismatch strain is small enough, and/or the thickness is small enough. Early work by Frank and van der Merwe [9-12] used a one-dimensional lattice to determine the critical lattice-mismatch for defect formation, and a two-dimensional model was formulated by van der Merwe [13-14]. However, a much more simple and more easily utilised mismatch epitaxy theory was developed by Matthews using continuum elasticity [15-16]. We have also performed calculations based on Matthews' models to clarify some discrepancies in the literature, and we summarise those calculations below [17].

As a strained layer is deposited, strain energy increases, and, at some thickness termed the critical thickness, it will be energetically advantageous to introduce misfit dislocations. By either balancing the forces on an assumed, pre-existing threading dislocation, or by minimising the total energy consisting of strain and dislocation energy, we can arrive at an expression for this critical thickness (h_{c60}),

$$h_{c60} = \frac{D(1 - \nu\cos^2\Theta)(\ln(\frac{h_{c60}}{b}) + 1)}{Yf} \tag{5}$$

$$D = \frac{G_{GaAs}G_{InGaAs}b}{\pi\,(G_{GaAs} + G_{InGaAs})(1 - \nu)} \tag{6}$$

$$G = C_{44} - \frac{1}{3}\,(2C_{44} + C_{12} - C_{11}) \tag{7}$$

$$b = \frac{\sqrt{2}}{2}\,a_{InGaAs} \tag{8}$$

$$\nu = \frac{C_{12}}{C_{12} + C_{11}} \tag{9}$$

$$Y = C_{11} + C_{12} - 2\frac{C_{12}^2}{C_{11}} \tag{10}$$

$$f = \frac{a_{InGaAs} - a_{GaAs}}{a_{InGaAs}} \tag{11}$$

ν is the Poisson ratio. In Eqn (5), we have assumed 60° dislocations are present, and therefore $\Theta = 60°$. The dislocation types are discussed in Section E. Eqn (5) also approximates the anisotropic crystal structure by using the anisotropic factor, $2C_{44} + C_{12} - C_{11}$ in the definition of G in Eqn (7), and by using the definition for the Young's modulus in Eqn (10). If edge dislocations were assumed, then the critical thickness expression would be

$$h_{c90} = \frac{D(1 - \nu\cos^2\Theta)(\ln(\frac{h_{c90}}{b}) + 1)}{2Yf} \qquad (12)$$

The two curves for both the 60° and 90° (edge) dislocations are plotted in FIGURE 1. The 60° dislocation curve is the physical curve, since edge dislocations cannot glide in zinc-blende crystals. To generate FIGURE 1, all the elastic constants were varied with x by assuming a linear extrapolation from GaAs to InAs. The elastic constants are given in TABLE 2 [1,2,3].

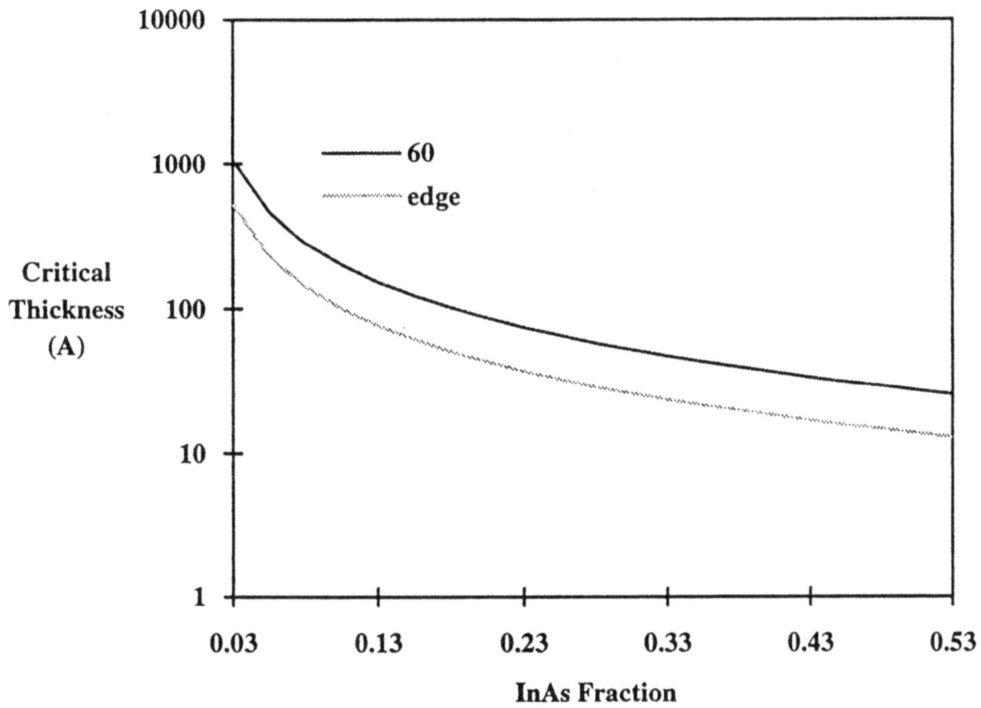

FIGURE 1 Critical thickness for defect introduction for $In_xGa_{1-x}As$ on GaAs.

TABLE 2 C_{11}, C_{12} and C_{44} for GaAs and InAs.

Material	C_{11} (dynes/cm^2)	C_{12} (dynes/cm^2)	C_{44} (dynes/cm^2)
GaAs	11.88×10^{11}	5.38×10^{11}	5.94×10^{11}
InAs	8.329×10^{11}	4.526×10^{11}	3.959×10^{11}

If the elastic constants are not varied with In fraction, the effect on the critical thickness curves is small. Even when comparing critical thickness curves from different materials systems, such as Ge_xSi_{1-x}/Si, there is little effect on the critical thickness curves [17]. Critical thickness curves can be derived assuming an isotropic crystal. If Eqn (5) is derived for an isotropic crystal, a more simple expression is produced.

$$h_{c60} = \frac{b(1 - \nu\cos^2\Theta)(\ln(\frac{h_{c60}}{b}) + 1)}{4\pi(1+\nu)f} \qquad (13)$$

FIGURE 2 is a plot for the $In_xGa_{1-x}As/GaAs$ system comparing the critical thickness curves derived using isotropic and anisotropic elastic constants. Some difference is seen, but Eqn (13) remains a good approximation.

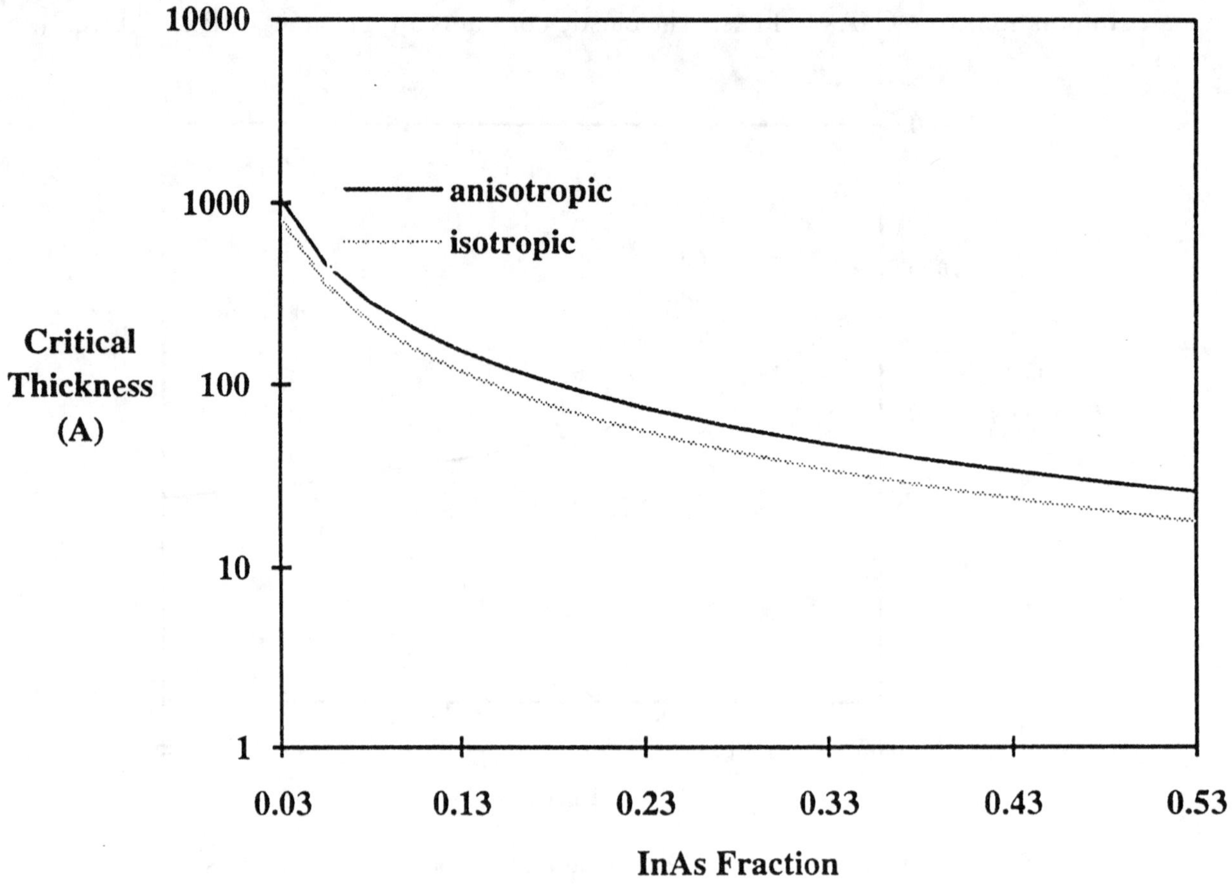

FIGURE 2 Comparison of critical thickness curves using isotropic elastic constants (Eqn (13)) vs anisotropic elastic constants (Eqn (5)).

D RELAXED InGaAs/GaAs

We can also calculate the equilibrium amount of strain in a structure where the thickness has exceeded the critical layer thickness by interchanging f and h_{c60} in Eqn (5):

$$\varepsilon = \frac{D(1 - \nu\cos^2\Theta)(\ln(\frac{h_{c60}}{b}) + 1)}{Yh_{c60}} \qquad (14)$$

where ε is the residual elastic strain in the epilayer, and therefore $(f - \varepsilon)$ is the amount of mismatch strain relieved by the formation of 60° misfit dislocations. FIGURE 3 is a plot of ε vs film thickness for different x. As one can see, once h_{c60} is exceeded, the residual elastic strain varies little with x.

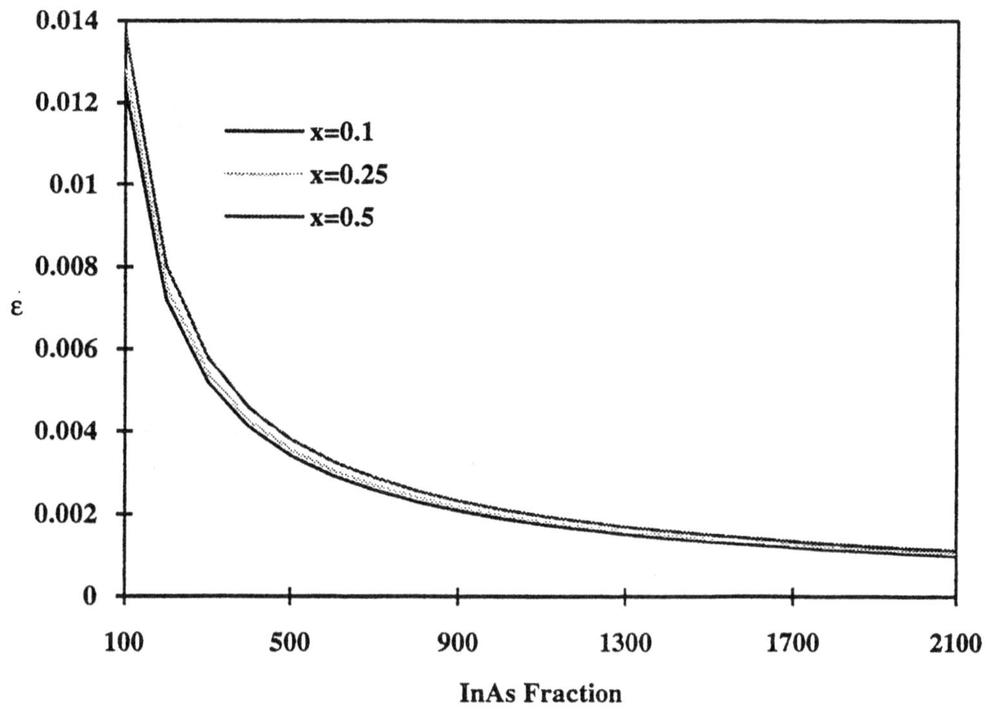

FIGURE 3 Equilibrium residual strain in an epitaxial layer of $In_xGa_{1-x}As$ on GaAs.

E KINETIC FACTORS

It is well known that non-equilibrium, strained structures can be achieved with epitaxial growth techniques. This non-equilibrium behaviour is related to the types of dislocations and their ability to nucleate and propagate through the structure.

E1 Dislocation Types

It has been generally observed, in many (100) epitaxial zinc-blende or diamond systems, that for low misfit systems (f <~ 1.5%), most of the misfit dislocations at the interface are 60° dislocations, and very few threading segments (i.e. misfit dislocations which thread through the epilayer to reach the top surface) are present. High misfit systems possess many edge dislocations at the interface, and many threading segments in the epilayer as well. Transmission electron microscopy is one method of observing the dislocation morphology in the transition from low to high misfit in InGaAs/GaAs [18-20]. The 60° dislocations are able to glide in cubic crystals, because the primary slip system is <110>{111}, i.e. the slip direction is <110>, and the glide plane is {111}. The terminology of the 60° dislocation is derived from the angle between the Burgers vector (slip direction, <110>) and the line direction, which is often another <110> which is 60 or 120° away. Detailed atomic arrangements of the 60° dislocation have been studied by Hornstra [21]. One important consideration in the zinc-blende structure is that the dislocation core can terminate with a row of either Ga or As atoms. Thus, two dislocation types are possible, termed α and β dislocations.

FIGURE 1 clearly shows that because the edge dislocation is more efficient at relieving misfit strain than the 60° dislocation, the critical thickness is less. However, for low misfit systems, a misfitted interface composed entirely of edge dislocations is impossible, since the edge dislocation glide plane is {100} and the slip system glide plane is {111}. Thus, in low misfit systems in which the dislocations must propagate to the interface, 60° dislocations will dominate, despite their relative inefficiency. High misfit systems do contain many edge misfit dislocations, but the exact nature of their formation is unclear. With such a high misfit, InGaAs growth is typically three-dimensional and the critical thickness is small. The edge dislocations are likely introduced via climb at the edge of the islands.

E2 Dislocation Propagation

The equilibrium calculations determine the ground state of the system, but the dislocations flow through the material at a finite velocity. Thus, very low temperatures are expected to inhibit dislocation flow, and therefore may induce a degree of metastability. The empirical expression which best describes dislocation velocities in crystals is:

$$v = B\tau_{eff}^{m} e^{-\frac{U}{kT}} \tag{15}$$

where τ is the effective shear stress (proportional to the mismatch strain), m is typically near unity, B is a materials-related fitting parameter, and U is an activation energy. Experiments on GaAs crystals have determined these parameters for both α and β dislocations under different doping, temperature and stress conditions [22,23]. The velocities of dislocations in InGaAs will certainly be affected by the addition of In. Small amounts of In appear to harden the lattice at high temperature [24], suggesting lower dislocation velocities, whereas significant In concentrations should result in higher dislocation velocities expected for the softer InAs.

E3 Nucleation of Dislocations

The inability to nucleate the desired dislocations when above the critical thickness may also introduce metastability. If we consider the homogeneous nucleation of a dislocation loop, we can derive the critical radius and activation energy for that process [16,17]. However, the activation energy to nucleate a dislocation half-loop is quite large, implying in the InGaAs/GaAs system that dislocations would not be introduced below $x = 0.60$, clearly in disagreement with experiment. Thus, it is likely that dislocations in low mismatch systems are introduced via inhomogeneities, i.e. wafer edges, pre-existing threading dislocations, particulates, and oval defects introduced during growth [25]. The best indication that InGaAs/GaAs depends on heterogeneous nucleation sources has occurred from experiments involving pre-growth substrate patterning [26,27]. In these experiments, by isolating heterogeneous nucleation sources on mesas, other mesas free of nucleation sources can be defect-free. The dominant source of the misfit dislocations in these experiments was threading dislocations from the substrates. Subsequent experiments on non-patterned substrates also showed that the threading dislocation density in the GaAs substrate was a dominant misfit dislocation source [28].

E4 Adding Kinetics to Equilibrium Expressions

Matthews first tried to incorporate kinetics into the equilibrium equations to describe relaxation in the Ge/GaAs system [15]. Eqn (15) was not accepted as a universal empirical description at that time, and a diffusive dislocation core model for dislocation velocity was used. Dodson and Tsao later used Eqn (15) in their model and included a multiplication term analogous to bulk crystal growth to arrive at a differential equation describing relaxation [29,17]:

$$\frac{\partial \delta}{\partial t} = C(\delta_\infty - \delta)^2(\delta - \delta_0) \qquad (16)$$

C is a constant, δ is the strain relieved by dislocations ($\delta = f - \varepsilon$), and δ_0 is a 'dislocation source density' which was necessary to fit experimental data.. The realistic velocity expression results in a more realistic modelling of δ with time. However, the δ_0 and C vary for different compositions and materials systems, so the utility of Eqn (16) in predicting relaxation in InGaAs/GaAs is questionable. A universal analytical description of relaxation employing kinetics is difficult due to different types of nucleation sources and dislocation interactions [30] in real systems.

F DESIGN OF MISMATCHED STRUCTURES

The materials engineer must first decide the strain state of the mismatched heterostructure:

(a) totally strained and in equilibrium
(b) totally strained and in metastable equilibrium
(c) partially relaxed
(d) totally relaxed

The two easiest regimes in which to engineer heterostructures are (a) and (d). For example, if a thin strained quantum well is desired for a laser or an FET, regime (a) is desired and the composition and thickness of the InGaAs layer should be determined using Eqn (5) and FIGURE 1. If the InGaAs is below the critical thickness curve, no dislocations can be formed at the interface, even if processing temperatures are raised; interdiffusion and dopant diffusion will determine the processing temperatures. Regime (d) is useful for forming relaxed InGaAs devices, such as long wavelength detectors and emitters, on GaAs substrates. If uniform In composition layers are deposited on GaAs, it is likely the InGaAs will be partially relaxed, falling into category (c). For a particular materials and growth system, the proper parameters in Eqn (16) can be found and used as a guide to achieve the desired lattice constant and strain. However, recent results on compositional grading, i.e. increasing the composition with thickness, results in fully relaxed layers with few threading dislocations [31-34]. Thus, these layers will also be stable with respect to processing since the structures are grown to achieve equilibrium. It is well known that the critical thickness can be exceeded without relaxation, and thus metastable strained layers can be fabricated (b). However, these layers will be susceptible to relaxation during processing, and therefore fabricating metastable device structures with high yield is challenging.

G CONCLUSION

The mismatched epitaxy issues for the InGaAs/GaAs system have been summarised. When designing the device heterostructure, the materials engineer must decide the strain regime of the InGaAs/GaAs heterostructure. If strained, stable InGaAs is desired, the equilibrium critical thickness should be used as a design rule. If fully relaxed structures are desired, graded In composition should be used to insure full relaxation and low threading defect density. Kinetics dominate the metastable strained and partially relaxed InGaAs/GaAs growth regimes, and engineering devices in these regimes is difficult at best.

REFERENCES

[1] S. Adachi [*J. Appl. Phys. (USA)* vol.53 (1982) p.8775-92]

[2] O. Madelung [*Landolt-Bornstein Numerical Data and Functional Relationships in Science and Technology* vol.22 (Springer-Verlag, 1987) p.82-97]

[3] O. Madelung [*Landolt-Bornstein Numerical Data and Functional Relationships in Science and Technology* vol.22 (Springer-Verlag, 1987) p.117-123]

[4] D.R. Gaskell [*Introduction to Metallurgical Thermodynamics* (McGraw-Hill, 1981) p.382]

[5] E.Kuphal [*J. Cryst. Growth (Netherlands)* vol.67 (1984) p.441-57]

[6] T.P. Pearsall, M. Quillec [*Appl. Phys. Lett. (USA)* vol.35 (1979) p.342]

[7] O. Madelung [*Landolt-Bornstein Numerical Data and Functional Relationships in Science and Technology* vol.22 (Springer-Verlag, 1987) p.142-4]

[8] J.J. Rosenberg, M. Benlarmi, P.D. Kirchner, J.M. Woodall, G.D. Pettit [*IEEE Electron Device Lett. (USA)* vol.EDL-6 (1985) p.491]

[9] F.C. Frank, J.H. van der Merwe [*Proc. R. Soc. A (UK)* vol.198 (1949) p.205]

[10] F.C. Frank, J.H. van der Merwe [*Proc. R. Soc. A (UK)* vol.198 (1949) p.216]

[11] F.C. Frank, J.H. van der Merwe [*Proc. R. Soc. A (UK)* vol.200 (1949) p.125]

[12] F.C. Frank, J.H. van der Merwe [*Proc. R. Soc. A (UK)* vol.201 (1949) p.261]

[13] J.H. van der Merwe [*J. Appl. Phys. (USA)* vol.34 (1963) p.117]

[14] J.H. van der Merwe [*J. Appl. Phys. (USA)* vol.34 (1963) p.123]

[15] J.W. Matthews, S. Mader, T.B. Light [*J. Appl. Phys. (USA)* vol.41 (1970) p.3800]

[16] J.W. Matthews [*Epitaxial Growth* vol.B (Academic Press, 1975)]

[17] E.A. Fitzgerald [*Mater. Sci. Rep. (Netherlands)* vol.7 (1991)]

[18] K.H. Chang, P.K. Bhattacharya, R. Gibala [*J. Appl. Phys. (USA)* vol.66 (1989) p.2993]

[19] K. Breen [*J. Electron. Mater. (USA)* vol.21 (1992) p.409]

[20] V. Krishnamoorthy, P. Ribas, R.M. Park [*Appl. Phys. Lett. (USA)* vol.58 (1991) p.2000]

[21] J. Hornstra [*J. Phys. Chem. Solids (UK)* vol.5 (1958) p.129]

[22] H. Steinhardt, P. Haasen [*Phys. Status Solidi A (Germany)* vol.49 (1978) p.93]

[23] S.K. Choi, M. Mihara, T. Ninomiya [*Jpn. J. Appl. Phys. (Japan)* vol.16 (1977) p.737]

[24] S. Guruswamy, J.P. Hirth, K.T. Faber [*J. Appl. Phys. (USA)* vol.60 (1986) p.4136]

[25] E.A. Fitzgerald [*J. Vac. Sci. Technol. B (USA)* vol.7 (1989) p.782]

[26] E.A. Fitzgerald, P.D. Kirchner, R. Proano, G.D. Pettit, J.M. Woodall, D.G. Ast [*Appl. Phys Lett. (USA)* vol.52 (1988) p.1496]

[27] E.A. Fitzgerald et al [*J. Appl. Phys. (USA)* vol.65 (1989) p.2220]

[28] J.F. Klem, W.S. Fu, P.L. Gourley, E.D. Jones, T.M. Brennan, J.A. Lott [*Appl. Phys. Lett. (USA)* vol.56 (1990) p.1350]
[29] B.W. Dodson, J.Y. Tsao [*Appl. Phys. Lett. (USA)* vol.51 (1987) p.1325, vol.52 (1988) p.852]
[30] L.B. Freund [*J. Appl. Phys. (USA)* vol.68 (1990) p.2073]
[31] E.A. Fitzgerald et al [*Appl. Phys. Lett. (USA)* vol.59 (1991) p.811]
[32] E.A. Fitzgerald et al [*J. Vac. Sci. Technol. B (USA)* vol.10 (1992) p.1807]
[33] P. Ribas, V. Krishnamoorthy, R.M. Park [*Appl. Phys. Lett. (USA)* vol.57 (1990) p.1040]
[34] K.L. Kavanagh, J.C.P. Chang, J. Chen, J.M. Fernandez, H.H. Weider [*J. Vac. Sci. Technol. B (USA)* vol.10 (1992) p.1820]

1.3 InP-based lattice-matched heterostructures

M. Ilegems

June 1993

A INTRODUCTION

Amongst the III-V compounds crystallising in the zinc-blende structure, four compounds, AlP, GaP, AlAs and GaAs, have lattice constants smaller than InP, while four others, InAs, AlSb, GaSb and InSb, have lattice constants larger than that of InP. In principle, 6 ternary alloys ($Ga_xIn_{1-x}As$, $Al_xIn_{1-x}As$, $GaAs_xSb_{1-x}$, $AlAs_xSb_{1-x}$, GaP_xSb_{1-x} and AlP_xSb_{1-x}) and an infinite number of quaternary and higher order alloys can be formed with lattice constants matching that of InP.

Of these, the alloys formed by mixing P or As with Sb on the group V sublattice exhibit extended miscibility gaps which prevent the growth of ternary mixed crystals lattice matched to InP under normal conditions of temperature encountered in epitaxy. The data here will thus be restricted to the $Ga_xIn_{1-x}As$ and $Al_xIn_{1-x}As$ ternary and to the $Ga_xIn_{1-x}As_yP_{1-y}$ and $Al_xGa_yIn_{1-x-y}As$ quaternary alloys which are of primary interest for applications in micro and optoelectronic devices.

B INTERPOLATION METHODS

Material parameters of ternary (T) and quaternary (Q) alloys can be estimated by interpolating the parameters of the binary (B) or ternary alloys depending on the information available. In the absence of ordering, the expressions used are of the form [1,2]

- for ternary alloys of the type $A_xB_{1-x}C$:

$$T_{ABC}(x) = x.B_{AC} + (1-x).B_{BC} - x(1-x).C_{ABC} \quad (1)$$

- for quaternary alloys of the type $A_xB_yC_{1-x-y}D$

$$Q(x, y) = x.B_{AD} + y.B_{BD} + (1-x-y).B_{CD} - xy.C_{ABD} - x(1-x-y).C_{ACD} - y(1-x-y).C_{BCD} \quad (2)$$

- for quaternary alloys of the type $A_xB_{1-x}C_yD_{1-y}$

$$Q(x, y) = \{x(1-x).[yT_{ABC}(x) + (1-y)T_{ABD}(x)] + y(1-y).[xT_{ACD}(y) + (1-x)T_{BCD}(y)]\}/[x(1-x)+y(1-y)] \quad (3)$$

In the latter case, the Q values reduce to the T values on the quaternary plane boundaries and to the average of the T values at the midpoint composition (x = y = 0.5). The constant C in Eqn (1) is known as the bowing parameter. In practice, the results predicted by Eqns (2) and (3) are presented by an expansion of Q in a product polynomial in x and y of the form

$$Q(x,y) = \Sigma\, C_{ij}.x^i.y^j \quad (4)$$

where the coefficients are obtained by fitting Eqn (4) to Eqn (3).

C LATTICE CONSTANT MATCHING COMPOSITIONS

The lattice constants of the (Ga,In)As and (Al,In)As alloys are found to vary linearly with composition according to Vegard's law [3]. Using the binary lattice constant values [4] $a_0 = 5.451$ Å for AlP, $a_0 = 5.4512$ Å for GaP, $a_0 = 5.6622$ Å for AlAs, $a_0 = 5.6533$ Å for GaAs, $a_0 = 5.8688$ Å for InP and $a_0 = 6.0584$ Å for InAs, the lattice matching composition values to InP are

$x = 0.468$ for $Ga_xIn_{1-x}As$
$x = 0.476$ for $Al_xIn_{1-x}As$
$x = 0.1896y/(0.4176 - 0.0125y) \sim 0.47y$ for $Ga_xIn_{1-x}As_yP_{1-y}$ $(0 \leq y \leq 1)$
$y = 0.468 - 0.983x$ for $Al_xGa_yIn_{1-x-y}As$ $(0 \leq x \leq 0.476,\ 0 \leq y \leq 0.468)$

D NEAREST NEIGHBOUR BOND DISTANCE

In the binary compounds, each group III or group V element has four nearest neighbours of the opposite type situated at a distance of $\sqrt{3}\, a_o/4$ at the corners of a regular tetrahedron. In the ternary and quaternary alloys considered here, it has been found that the cation-anion distances remain close to their values in the pure binary crystals, as shown in FIGURES 1 and 2 for the (Ga,In)As ternary and (Ga,In)(As,P) quaternary alloy.

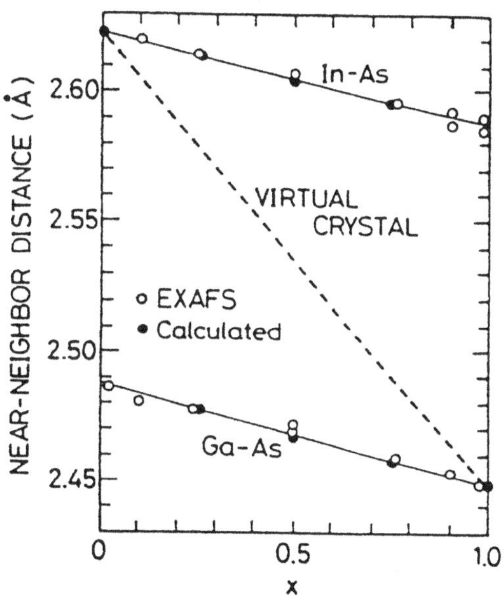

FIGURE 1 Ga-As and In-As nearest-neighbour distances versus InAs mole fraction in the (Ga,In)As ternary alloy as measured by EXAFS [5]. The dashed line represents the average anion-cation spacing (Vegard's law). Figure after refs [1,6].

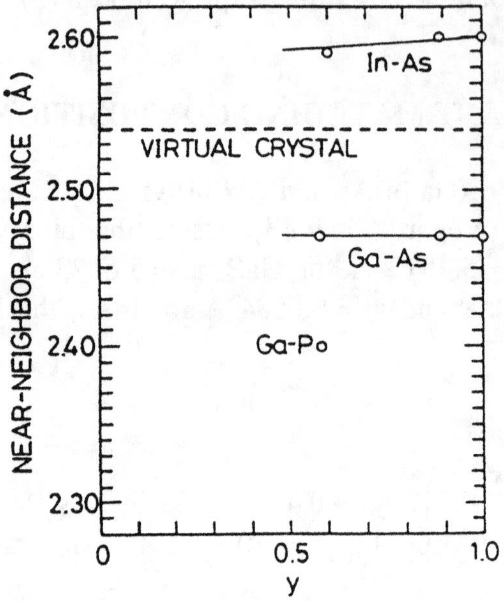

FIGURE 2 Ga-P, Ga-As and In-As nearest-neighbour distances versus composition y in the $Ga_xIn_{1-x}As_yP_{1-y}$ quaternary alloy lattice matched to InP as measured by EXAFS [7]. The dashed line represents the average anion-cation spacing (In-P nearest neighbour distance). Figure after refs [1,6].

In a related way, the long wavelength optical phonon spectra in (Ga,In)(As,P) lattice matched to InP show a four mode behaviour with independent GaAs-, InAs-, InP- and GaP-like modes observable over most of the composition range (FIGURES 3 and 4). Similarly, a three mode behaviour with InAs-like, GaAs-like and AlAs-like modes is observed in the (Al,Ga,In)As system [8].

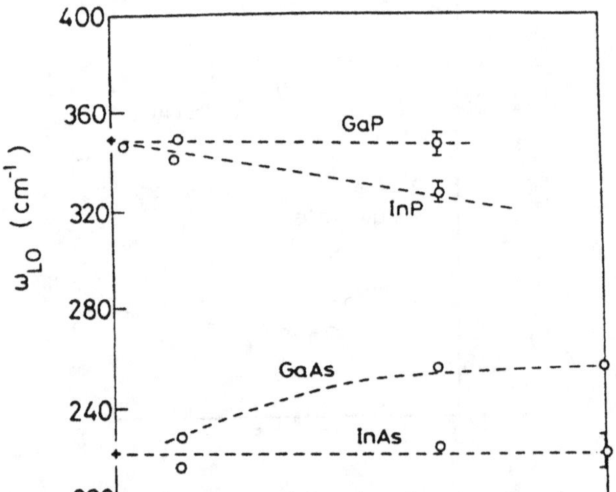

FIGURE 3 Longitudinal optical (LO) phonon frequency versus composition y for $Ga_xIn_{1-x}As_yP_{1-y}$ lattice matched to InP. Experimental data from ref [9]. Figure after ref [1].

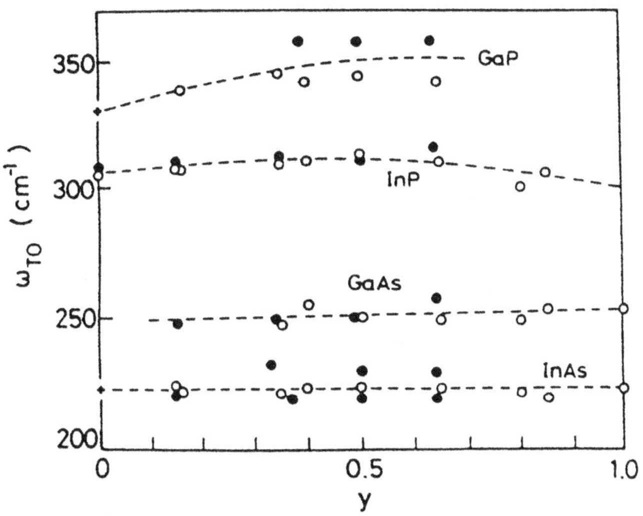

FIGURE 4 Transverse optical (TO) phonon frequency versus composition y for $Ga_xIn_{1-x}As_yP_{1-y}$ lattice matched to InP. Experimental data from refs [10, 11]. Figure after ref [1].

Thermal, mechanical, dielectric and transport properties of the ternary and quaternary alloys have been reviewed recently by Adachi [1] and are presented in further sections of the present Datareview.

E DIRECT AND INDIRECT ENERGY BANDGAPS

E1 $Ga_xIn_{1-x}As_yP_{1-y}$

The electronic bandgap structure of the quaternary $Ga_xIn_{1-x}As_yP_{1-y}$ alloy has been reviewed by Adachi [1] and Madelung [12] amongst others. The band structure is direct over the entire range of compositions lattice matched to InP. Generally, it is found that the bandgap variation for the lattice matched quaternary alloy may be expressed with good accuracy by a quadratic function of the composition y :

$$E_g(y) = a + by + cy^2$$

where a and b are determined by the values at the end points and where the bowing parameter c measures the deviation from linearity.

Along the lattice matched composition line, the following relations hold for the direct (Γ) and indirect (X, L) energy gaps (in eV) , as obtained from the literature data for the constituents and interpolation to experimental photoluminescence, electroreflectance, and optical absorption and transmission measurements :

$$E^{\Gamma}_{g,\ direct}(y) = 1.35 - 0.775y + 0.149y^2 \qquad (298\ K,\ [13])$$

$$E^{\Gamma}_{g,\ direct}(y) = 1.425 - 0.7668y + 0.149y^2 \qquad (4.2\ K,\ [13])$$

$$E^{X}_{g,\ indirect}(y) = 2.21 - 0.88y \qquad (298\ K,\ [1])$$

$$E^{L}_{g,\ indirect}(y) = 2.05 - 0.85y \qquad (298\ K,\ [1])$$

The relation for the direct Γ minimum is plotted in FIGURE 5.

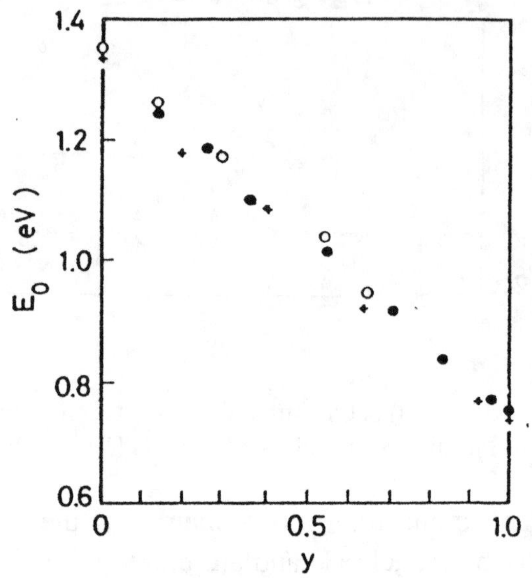

FIGURE 5 Values of the lowest direct bandgap energy at 300 K versus composition y for $Ga_xIn_{1-x}As_yP_{1-y}$ lattice matched to InP. Experimental data from refs [14-16]. Figure after ref [1].

E2 $Al_xGa_yIn_{1-x-y}As$

The band structure for the quaternary $Al_xGa_yIn_{1-x-y}As$ alloy is direct over the entire range of compositions lattice matched to InP. The room temperature fundamental bandgap is given as a function of x along the lattice matched range by [17]

$$E^{\Gamma}_{g,\ direct}(x) = 0.75 + 1.0496x + 1.0645x^2 - 0.033x^3 \qquad (0 \le x \le 0.476)$$

The resulting variation is plotted in FIGURE 6.

E3 $Ga_xIn_{1-x}As$ and $Al_xIn_{1-x}As$ Matched Ternary Compositions

Along the ternary GaInAs and AlInAs boundaries, the following values are generally accepted:

$$E^{\Gamma}_{g,\ direct}(GaInAs) = 0.74 \pm 0.01\ eV \qquad [1,12,20]$$

$$E^{\Gamma}_{g,\ direct}(AlInAs) = 1.49 \pm 0.1\ eV \qquad [1,12]$$

Corresponding low temperature values are

$$E^{\Gamma}_{g,\ direct}(GaInAs,\ 7\ K) = 0.802\ eV \qquad [21]$$

$$E^{\Gamma}_{g,\ direct}(AlInAs,\ 77\ K) = 1.513\ eV \qquad [22]$$

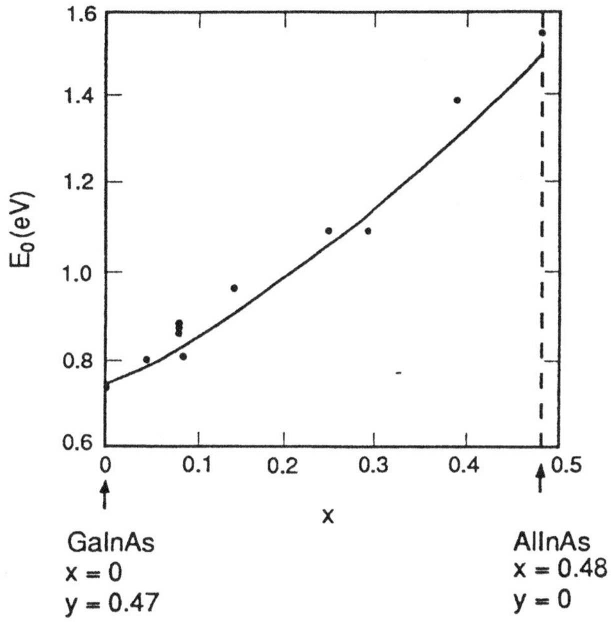

FIGURE 6 Values of the lowest 300 K direct bandgap energy versus composition x for $Al_xGa_yIn_{1-x-y}As$ lattice matched to InP. Experimental data from refs [18,19].

Somewhat lower values for the bandgap of AlInAs lattice matched to InP, i.e. 1.511 eV and 1.439 eV at 1.5 and 300 K, respectively, are given in a recent study [23].

Conduction and valence band offsets are given in a subsequent section of this Datareview.

F ELECTRON AND HOLE EFFECTIVE MASSES

F1 $Ga_xIn_{1-x}As_yP_{1-y}$

The dependence on composition of the electron and hole effective masses in the lattice matched quaternary $Ga_xIn_{1-x}As_yP_{1-y}$ system have been studied experimentally and have been reviewed by Adachi [1]. The results are plotted in FIGURES 7, 8 and 9.

The interpolation relationships for electron and hole effective masses can thus be written as

electron effective mass	$m_e/m_0 = 0.079 - 0.038y$	[1,26]
light hole effective mass	$m_{lh}/m_0 = 0.120 - 0.099y + 0.030y^2$	[1]
heavy hole effective mass	$m_{hh}/m_0 = 0.46$	[35]

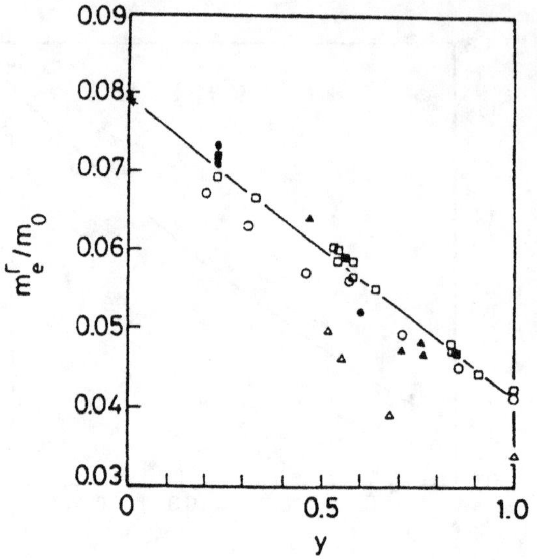

FIGURE 7 Electron effective mass as a function of composition y for $Ga_xIn_{1-x}As_yP_{1-y}$ lattice matched to InP. The solid line is a linear interpolation between the values of InP (m_e/m_0 = 0.079, ref [24]) and $Ga_{0.47}In_{0.53}As$ (m_e/m_0 = 0.041, ref [25]). Experimental data from refs [26-31]. Figure after ref [1].

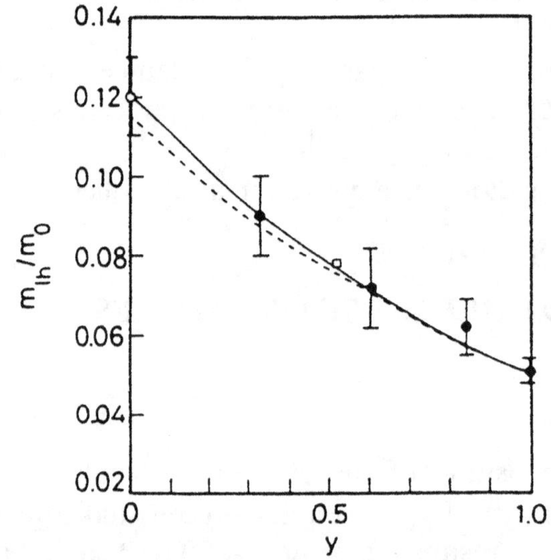

FIGURE 8 Light hole effective mass as a function of composition y for $Ga_xIn_{1-x}As_yP_{1-y}$ lattice matched to InP. The solid line is calculated from the experimental data assuming a quadratic relationship between effective mass and composition. Experimental data from refs [24,32-34]. End point values are m_{lh} = 0.12 m_0 for InP [24], m_{lh} = 0.051 m_0 for GaInAs [34]. Figure after ref [1].

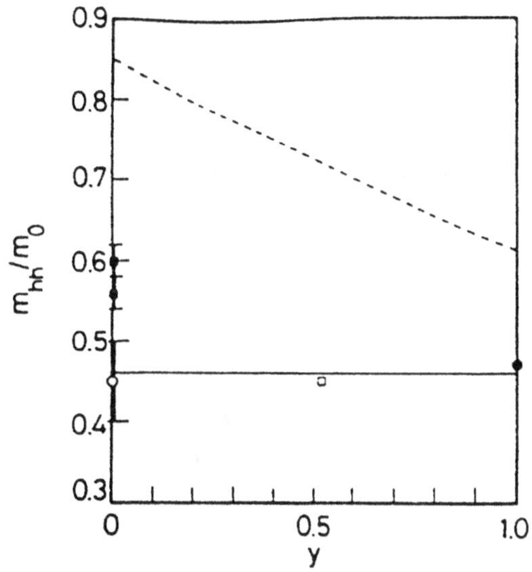

FIGURE 9 Heavy hole effective mass as a function of composition y for $Ga_xIn_{1-x}As_yP_{1-y}$ lattice matched to InP. The solid line is drawn assuming that the heavy hole mass is independent of composition. Experimental data from refs [24,32,33,35]. Figure after ref [1].

F2 $Al_xGa_yIn_{1-x-y}As$

Few data are available on the effective mass variations in this quaternary system. In the absence of experimental data, theoretical values for the effective mass of a ternary compound of the type $A_xB_{1-x}C$ are interpolated from the masses of the constituent binaries.

F3 $Ga_xIn_{1-x}As$ and $Al_xIn_{1-x}As$ Matched Ternary Compositions

The following values apply at the lattice matched compositions (room temperature values or estimates) :

- for $Ga_xIn_{1-x}As$

electron effective mass	$m_e/m_0 = 0.041$	[25]
light hole effective mass	$m_{lh}/m_0 = 0.051$	[34]
heavy hole effective mass	$m_{hh}/m_0 = 0.46$	[1], FIGURE 9

- for $Al_xIn_{1-x}As$

electron effective mass $m_e/m_0 = 0.075$
(Eqn (5), using reduced mass values of 0.13 for AlAs [36] and 0.0236 for InAs [37])

light hole effective mass $m_{lh}/m_0 = 0.096$
(Eqn (5), using reduced mass values of 0.14 for AlAs [36] and 0.055 for InAs [37])

heavy hole effective mass $m_{hh}/m_0 = 0.41$
(Eqn (5), using reduced mass values of 0.44 for AlAs [36] and 0.39 for InAs [37])

G CONCLUSION

The basic material parameters needed for the design of heterostructures lattice matched to InP are generally known. Because of the difficulty of separating effects due to residual deviations from lattice matched composition and residual strain, the precision of the ternary and especially quaternary data is expected to be further improved as more detailed studies are carried out.

REFERENCES

[1] S. Adachi [*Physical Properties of III-V Semiconducting Compounds : InP, InAs, GaAs, GaP, InGaAs, and InGaAsP* (Wiley-Interscience, New York, 1992)]. Figures from this reference are copyright © by John Wiley & Sons, Inc. and reprinted with their permission.

[2] S. Adachi [*J. Appl. Phys (USA)* vol.61 (1987) p.4869]

[3] L. Vegard [*Z. Phys. (Germany)* vol.5 (1921) p.17]

[4] G. Giesecke [*Semicond. Semimet.* vol.2, Eds R.K. Willardson, A.C. Beer (Academic Press, New York, 1966) p.63]

[5] J.C. Mikkelsen, J.B. Boyce [*Phys. Rev. Lett. (USA)* vol.49 (1982) p.1412 and *Phys. Rev. B (USA)* vol.28 (1983) p.7130]

[6] T. Fukui [*Jpn. J. Appl. Phys. (Japan)* vol.23 (1984) p.L208 and *J. Appl. Phys. (USA)* vol.57 (1985) p.5188]

[7] H. Oyanagi, Y. Takeda, T. Matsushita, T. Ishiguro, A. Sasaki [in *Gallium Arsenide and Related Compounds* (Institute of Physics, Bristol, 1986) p.295]

[8] R. Borroff, R. Merlin, A. Chin, P.K. Bhattacharya [*Appl. Phys. Lett. (USA)* vol.53 (1988) p.1652]

[9] R.K. Soni, S.C. Abbi, K.P. Jain, M. Balkanski, S. Slempkes, J.L. Benchimol [*J. Appl. Phys. (USA)* vol.59 (1986) p.2184]

[10] A. Pinczuk, J.M. Worlock, R.E. Nahory, M.A. Pollack [*Appl. Phys. Lett. (USA)* vol.33 (1978) p.461]

[11] C. Pickering [*J. Electron. Mater. (USA)* vol.10 (1981) p.901]

[12] O. Madelung (Ed.) [*Semiconductors: Group IV Elements and III-V Compounds, in Data in Science and Technology*, R. Poerschke, editor in chief (Springer-Verlag, New York, 1991)]

[13] T.P. Pearsall [*GaInAsP Alloy Semiconductors*, T.P. Pearsall (Ed.) (Wiley & Sons, New York, 1982) p.295]

[14] K. Nakajima, A. Yamaguchi, K. Akita, T. Kotani [*J. Appl. Phys. (USA)* vol.49 (1978) p.5944]

[15] P.M. Laufer, F.H. Pollak, R.E. Nahory, M.A. Pollack [*Solid State Commun. (USA)* vol.36 (1980) p.419]

[16] J.A. Lahtinen, T. Tuomi [*Phys. Status Solidi B (Germany)* vol.130 (1985) p.637]

[17] D. Olego, T.Y. Chang, E. Silberg, E.A. Caridi, A. Pinczuki [*Gallium Arsenide and Related Compounds* (Institute of Physics, Bristol, 1982) p.195]

[18] K. Mazu, T. Mishima, S. Hiroi, M. Konagai, K. Takahashi [*J. Appl. Phys. (USA)* vol.53 (1982) p.7558]

[19] P. Parayanthal, C.S. Ro, F.H. Pollak, C.R. Stanley, G.W. Wicks, L.F. Eastman [*Appl. Phys. Lett. (USA)* vol.43 (1983) p.109]

[20] J.Y. Marzin, M.N. Charasse, B. Sermage [*Phys. Rev. B (USA)* vol.31 (1985) p.8298]

[21] I.C. Bassignana, C.J. Miner, N. Puetz [*J. Appl. Phys. (USA)* vol.65 (1989) p.4299]

[22] R. Houdré, F. Gueissaz, M. Gailhanou, J.D. Ganière, A. Rudra, M. Ilegems [*J. Cryst. Growth (Netherlands)* vol.111 (1991) p.456]

[23] D. Oertel, D. Bimberg, R.K. Bauer, K.W. Carey [*Appl. Phys. Lett. (USA)* vol.55 (1989) p.140]

[24] P. Rochon, E. Fortin [*Phys. Rev. B (USA)* vol.12 (1975) p.5803]

[25] T.P. Pearsall et al [*Gallium Arsenide and Related Compounds* (Institute of Physics, Bristol, 1979) p.94]

[26] R.J. Nicholas, J.C. Portal, C. Houlbert, P. Perrier, T.P. Pearsall [*Appl. Phys. Lett. (USA)* vol.34 (1979) p.492]

[27] H. Brendecke, H.L. Störmer, R.J. Nelson [*Appl. Phys. Lett. (USA)* vol.35 (1979) p.772]

[28] E.H. Perea, E.E. Mendez, C.G. Fonstad [*J. Electron. Mater. (USA)* vol.9 (1980) p.459]

[29] J.B. Restorff, B. Houston, R.S. Allgaier, M.A. Littlejohn, S.B. Phatak [*J. Appl. Phys. (USA)* vol.51 (1980) p.2277]

[30] R.J. Nicholas, S.J. Sessions, J.C. Portal [*Appl. Phys. Lett. (USA)* vol.37 (1980) p.178]

[31] J.M. Chamberlain, A.A. Reeder, R.J. Turner, E. Kuphal, J.L. Benchimol [*Solid-State Electron. (UK)* vol.30 (1987) p.217]

[32] K. Alavi, R.L. Aggerwal, S.H. Groves [*J. Magn. Magn. Mater. (Netherlands)* vol.11 (1979) p.136]

[33] J. Leotin, R. Barbaste, S. Askenazy, M.S. Skolnick, R. Stradling, J. Tuchendler [*Solid State Commun. (USA)* vol.15 (1974) p.693]

[34] C. Hermann, T.P. Pearsall [*Appl. Phys. Lett. (USA)* vol.38 (1981) p.450]

[35] K. Alavi, R.L. Aggerwal, S.H. Groves [*Phys. Rev. B (USA)* vol.21 (1980) p.1311]

[36] M.P.C. Krijn, [*Semicond. Sci. Technol. (UK)* vol.6 (1991) p.27]

[37] P.D. Wang, S.N. Holmes, T. Le, R.A. Stradling, I.T. Ferguson, A.G. de Oliviera, [*Semicond. Sci. Technol. (UK)* vol.7 (1992) p.767]

1.4 Microstructure of InGaAs epitaxial layers

S. Mahajan

March 1993

A INTRODUCTION

InGaAs is a scientifically interesting and technologically important compound semiconductor. It finds applications in state-of-the-art lightwave communication systems and high speed devices.

InGaAs epitaxial layers, emitting at 1.67 μm, can be grown lattice-matched to InP substrates. A variety of epitaxial growth techniques, such as liquid phase epitaxy (LPE), organometallic vapour phase epitaxy (OMVPE) and molecular beam epitaxy (MBE), have been used to grow these layers.

InGaAs layers crystallise in the zinc-blende structure which consists of two interpenetrating FCC units. One of the units is displaced with respect to the other by $^a/_4$ <111>, where a is the lattice parameter of the layer. One unit is occupied by group III atoms, whereas group V atoms reside on the second unit. Conceptually, InGaAs can be obtained from GaAs by replacing some of the Ga atoms of group III sub-lattice by In atoms. An interesting question is whether or not these substitutions occur in a random fashion? A number of investigators have addressed this issue. A consensus has emerged that In and Ga atoms are not distributed at random within group III sub-lattice. Two types of deviation from randomness are observed: (i) phase separation and (ii) atomic ordering. These two features constitute the microstructure and their characteristics are discussed in this article. Also their influence on electronic and optical properties is considered.

B PHASE SEPARATION AND ATOMIC ORDERING

FIGURE 1 shows a series of dark-field images of a (001) lattice-matched InGaAs epitaxial layer grown by MBE at 500°C. The modulated structure is clearly evident (McDevitt et al [1]). It is difficult to discern the alignment of modulations in FIGURES 1(a) and 1(b). However, it is apparent from FIGURES 1(c) and 1(d) that the speckle microstructure is composed of two orthogonal modulations which lie along the two in-plane <100> directions. Also, the principal strains associated with the modulations are along their propagation direction; an identical result is obtained for the layers grown by LPE. The period of the modulation is ~ 5 - 6 nm as measured from the <400> images. Furthermore, modulations along the growth direction were not observed.

That phase separation is two-dimensional in nature and that it occurs along the soft directions lying in the growth plane are borne out by the recent work of McDevitt et al [1]. The implication is that phase separation occurs at the surface while the layer is being grown. If this were not so, the directions along which modulations evolve would not depend on orientation

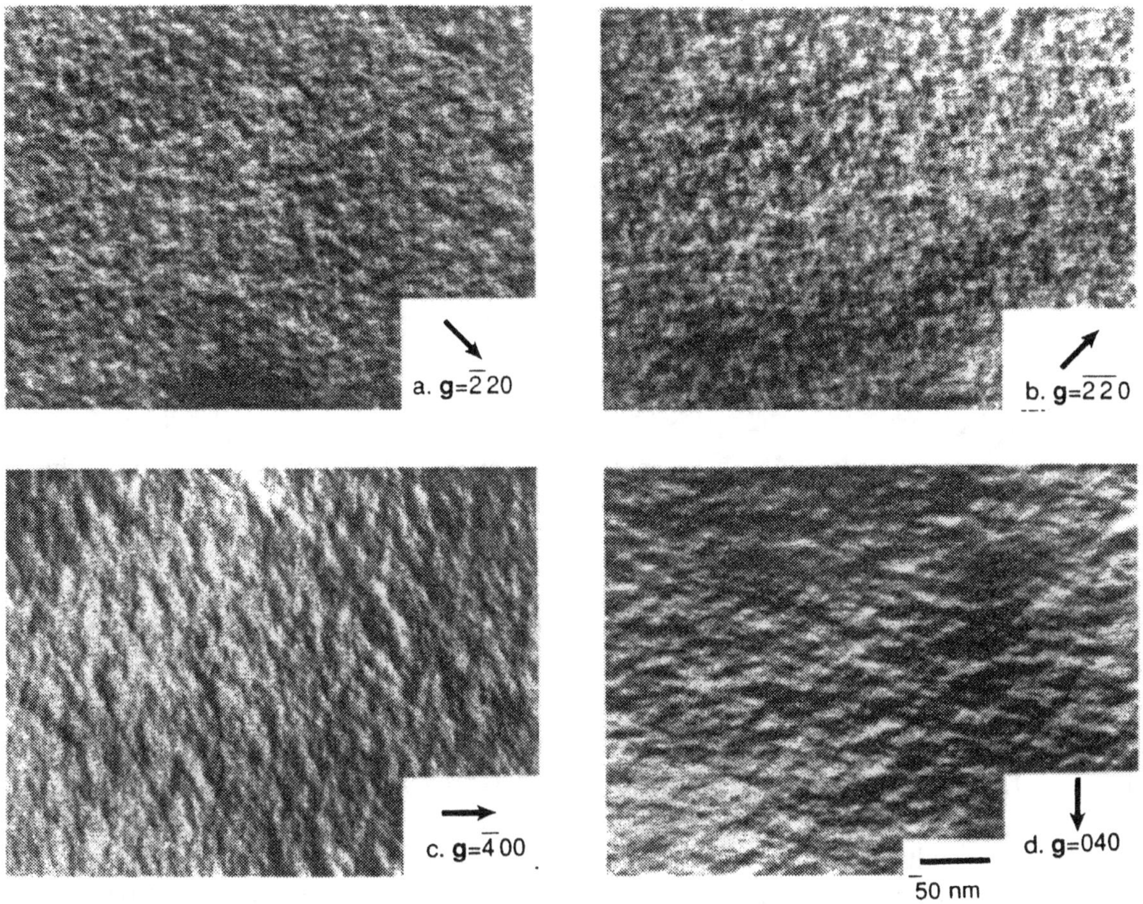

FIGURE 1 Series of dark-field images obtained from a (001) InGaAs layer grown by MBE; its emission wavelength is 1.67 μm. After McDevitt et al (1992).

of the growth surface. Instead, they would develop along the three <100> directions, the softest directions in the zinc-blende structure, in each case.

Using a field-ion-microscope, equipped with an atom probe, MacKenzie et al [2] have shown that distinct deviations from compositional uniformity exist in InGaAs layers grown by OMVPE. The composition varies locally, on a scale typically 10 - 20 nm, from the mean composition by up to 5 at %. This observation suggests that the observed modulations may represent In- and Ga-rich regions.

McDevitt et al [1] have investigated the temperature dependence of the period of modulation in InGaAs layers grown by MBE, the temperature range investigated being 400 to 500°C. The dominant effect of the reduced-growth temperature is to decrease the period of the modulations. The respective periods in the layers grown at 400, 450 and 500°C were 3, 3.5 and 7 nm.

An attempt has been made to analyse quantitatively the temperature dependence of the wavelength discussed above [1]. Since the composition modulations must develop by surface diffusion, the wavelength of the modulation (λ) may be related to the surface diffusion coefficient (D_S) by the following relation:

$$\lambda^2 = AD_s t \qquad (1)$$

where A is a constant of proportionality and t is the time required for the deposition of a bi-layer on the (001) surface. It is implicitly assumed in Eqn (1) that the atomic arrangement at the surface is frozen in the bulk because bulk diffusion is extremely slow in these materials. It can be shown that for a constant growth rate, i.e. constant t, λ is related to temperature (T) via the following equation:

$$\lambda^2 = B\exp(-E_s/kT) \qquad (2)$$

where B is a constant and E_S is the activation energy for surface diffusion of the atomic constituent that is rate controlling. On plotting ln λ vs $^1/_T$ McDevitt et al [1] obtained a linear relation and estimated E_S to be ~ 0.35 eV. They have argued that this value could represent the activation energy for the diffusion of As atoms on the (001) surface consisting of Ga and In atoms. This correlation implies that the evolution of phase-separated regions in InGaAs layers is controlled by the surface diffusion of As atoms.

Under appropriate growth conditions, atomic ordering coexists with phase separation in layers grown by a variety of vapour phase techniques on {001} substrates [3]. This ordering is CuPt type in which alternate {111} layers in InGaAs are either enriched or depleted in In or Ga atoms. As a result of ordering real space periodicity along the <111> direction is doubled. This can be assessed from FIGURE 2 that shows diffraction patterns obtained from $[1\bar{1}0]$ and [110] cross-sections. In FIGURE 2(a) only the pattern corresponding to the zinc-blende structure is observed, whereas in FIGURE 2(b) extra diffraction spots at $\frac{\bar{1}}{2}, \frac{1}{2}, \frac{1}{2}, \frac{1}{2}, \frac{\bar{1}}{2}, \frac{1}{2}$ and other equivalent positions are seen. The atomic arrangement compatible with ordering on the $(\bar{1}11)$ is shown schematically in FIGURE 3. The preceding observation implies that ordering occurs only on two planes, i.e. $(\bar{1}11)$ and $(1\bar{1}1)$, out of the possible four {111} planes. Since in the bulk the four {111} planes are equivalent, it can be inferred that ordering like phase separation may occur at the surface while the layer is growing.

The influence of growth conditions, i.e. growth temperature and growth rate, on ordering in GaInP layers has been extensively investigated [4-7]. Such extensive studies have not been carried out on InGaAs layers, but the effects of growth conditions are going to be qualitatively similar. For a fixed V/III ratio, ordering is optimal at an intermediate growth temperature, whereas it is poor at lower and higher growth temperatures.

The effects of growth rates on atomic ordering have been investigated by [3,7,8]. A common feature of these studies is that the increase in growth rate impairs the perfection of ordering. This effect is manifested most dramatically in electron diffraction patterns. As the growth rate is increased, diffuse intensity spikes develop through superlattice reflections and they become more pronounced with increasing growth rate.

The growth on vicinal (001) surfaces can bias an ordering variant [9]. This implies that surface steps play a role in the development of the ordered structure.

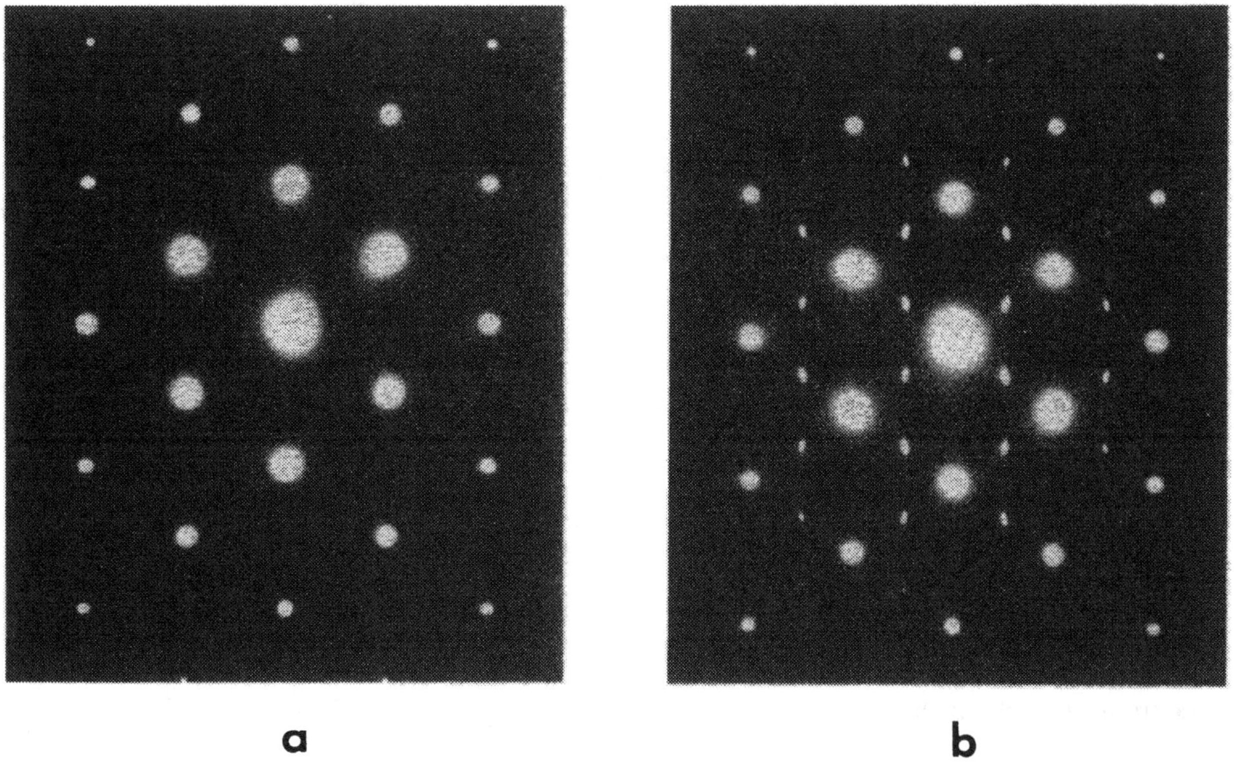

FIGURE 2 Electron diffraction patterns obtained from (a) $[1\bar{1}0]$ and (b) [110] cross-sections. Note the presence of superlattice spots in (b).

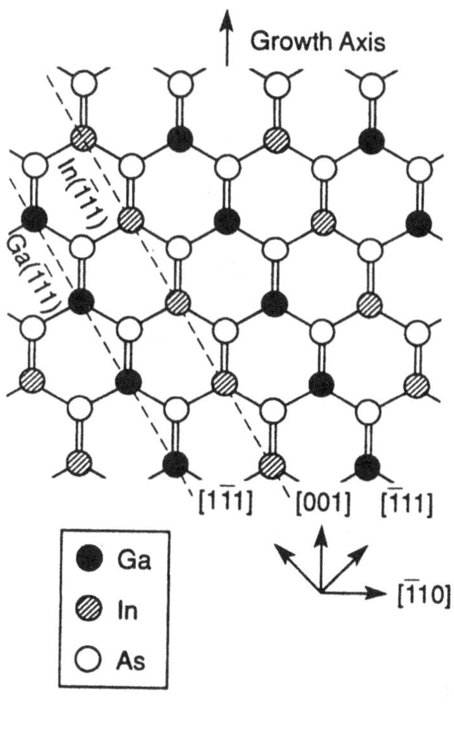

FIGURE 3 Schematic illustrating the atomic arrangement within InGaAs that is ordered on the $(\bar{1}11)$ plane.

Several models have been proposed to rationalise the evolution of CuPt-type ordered structures. However, only one model can satisfactorily explain the observed ordering characteristics [10-12]. They have invoked that the occurrence of 2 x surface reconstruction on group V terminated (001) surfaces produces sub-surface strains. As a result, dilated and compressed regions are produced in sub-surface layers between and underneath the dimers, respectively, along the $[1\bar{1}0]$ direction. When such a surface is subjected to a flux of In, Ga and As, the species required to grow an InGaAs layer, In atoms will segregate in the dilated regions, whereas Ga atoms would occupy compressed regions. This is inferred because the covalent tetrahedral radius of In atoms is larger than that of Ga atoms. As a result, strain energy is lower. If it is assumed that reconstruction always occurs at the growing surface, the occurrence of CuPt type ordering can be rationalised in a satisfactory manner.

An interesting question concerns the chronology of phase separation versus that of atomic ordering. Since phase separated microstructures reflect the crystallography of the growth surface and since phase separation is two-dimensional in nature, it is envisaged that phase separation occurs first. This suggestion is quite reasonable because the heat of mixing is positive in mixed III-V epitaxial layers. When an additional bilayer is grown, phase separated regions experience the influence of surface reconstruction. As a result atomic rearrangement occurs in the sub-surface, phase separated layer, leading to the evolution of CuPt ordering. It is visualised that perfect long range order will never be realised because the time available for rearrangement is too short.

C INFLUENCE OF PHASE SEPARATION AND ATOMIC ORDERING ON ELECTRONIC AND OPTICAL PROPERTIES

The decrease in carrier mobility has been reported for all compositions lying within miscibility gaps [13,14]. The detailed study by Kuphal [13] reports a dip in electron mobility of (001) InGaAsP layers grown at 630°C, whereas this is not observed in the layer deposited at 740°C. Bhattacharya and Wu [14] have attributed the decrease to a large difference in the values of alloy scattering potential (ΔU). The estimated values of ΔU are in the range of 0.6 to 0.8 eV; such high values of ΔU could result only from phase separation.

The photoluminescence (PL) measurements on phase separated layers have produced conflicting results. Most investigators have reported broadening of near bandgap PL peaks for layers whose compositions lie within the miscibility gap. On the other hand, several studies show very narrow PL peaks. At present the origin of discrepancy between the two sets of observations is not well understood.

The recent study of Ueda et al [15] has demonstrated convincingly that the occurrence of long range order in InGaAs layers, deposited on (110) InP substrates by MBE, enhances electronic mobility. The observed mobility is much higher than that calculated due to alloy scattering, implying that the observed enhancement results from atomic ordering.

Gomyo et al [16] and Kondow et al [17] have shown that the bandgap of (001) GaInP layers, grown by OMVPE, is lowered due to the occurrence of atomic ordering. The decrease in bandgap may be enhanced further using atomic layer epitaxy [18].

D CONCLUSION

In summary, it is now well recognised that In and Ga atoms within (001) InGaAs layers are not distributed at random within group III sub-lattice. Two types of deviation from randomness are observed: (i) phase separation and (ii) atomic ordering. The presence of phase separation reduces electronic mobility and appears to broaden PL peaks, whereas atomic ordering enhances mobility and reduces bandgaps.

REFERENCES

[1] T.L. McDevitt, S. Mahajan, D.E. Laughlin, W.A. Bonner, V.G. Keramidas [*Phys. Rev. B (USA)* vol.45 (1992) p.6614-22]
[2] R.A.D. MacKenzie, J.A. Liddle, C.R.M. Grovenor [*J. Appl. Phys. (USA)* vol.69 (1991) p.250-6]
[3] S. Mahajan, M.A. Shahid, D.E. Laughlin [*Inst. Phys. Conf. Ser. (UK)* no.100 (1989) p.143-53]
[4] A. Gomyo, K. Kobayashi, S. Kawata, I. Hino, T. Suzuki, T. Yuasa [*J. Cryst. Growth (Netherlands)* vol.77 (1986) p.367-73]
[5] T. Suzuki, A. Gomyo, S. Iijima [*J. Cryst. Growth (Netherlands)* vol.93 (1988) p.396-405]
[6] M. Kondow, H. Kakibayashi, S. Minagawa [*J. Cryst. Growth (Netherlands)* vol.88 (1988) p.291-6]
[7] S.R. Kurtz, J.M. Olson, A. Kibbler [*Appl. Phys. Lett. (USA)* vol.57 (1990) p.1922-4]
[8] B.T. McDermott et al [*Appl. Phys. Lett. (USA)* vol.56 (1990) p.1172-4]
[9] P. Bellon, J.P. Chevalier, E. Augarde, J.P. Andre, G.P. Martin [*J. Appl. Phys. (USA)* vol.66 (1989) p.2388-94]
[10] S. Mahajan [in *Semiconductor Physics* Eds A. Fazzio, J.R. Leite, A.S. Chaves (World Scientific, Singapore, 1992) p.79-98]
[11] S. Mahajan, B.A. Philips [in *Physical Metallurgy* Eds C.T. Liu, R.W. Cahn, G. Sauthoff (Academic Publishers, Dordrecht, 1992) p.93-106]
[12] J.E. Bernard, S. Froyen, A. Zunger [*Phys. Rev. B (USA)* vol.44 (1991) p.11178-95]
[13] E. Kuphal [*J. Cryst. Growth (Netherlands)* vol.67 (1984) p.441-57]
[14] P.K. Bhattacharya, J.W. Wu [*J. Appl. Phys. (USA)* vol.58 (1985) p.1410-14]
[15] O. Ueda, Y. Nakata [*Mat. Sci. Forum (Switzerland)* vol.83-87 (1992) p.1297-1302]
[16] A. Gomyo, T. Suzuki, K. Kobayashi, S. Kawata, I. Hino, T. Yuasa [*Appl. Phys. Lett. (USA)* vol.50 (1987) p.673-5]
[17] M. Kondow, H. Kakibayashi, S. Minagawa, Y. Inoue, T. Nishino, Y. Hamakawa [*J. Cryst. Growth (Netherlands)* vol.93 (1988) p.412-17]
[18] B.T. McDermott, N.A. El-Masry, B.L. Jiang, F. Hyuga, S.M. Bedair, W.M. Duncan [*J. Cryst. Growth (Netherlands)* vol.107 (1991) p.96-101]

CHAPTER 2

THERMAL, MECHANICAL AND LATTICE VIBRATIONAL PROPERTIES OF InGaAs

2.1 Thermal conductivity of InGaAs

S. Adachi

December 1992

A INTRODUCTION

Lattice thermal conductivity, or thermal resistivity, results from interactions between phonons and from the scattering of phonons by crystalline imperfections. Knowledge of the thermal conductivity of semiconductors forms an important part in the design of power-dissipating devices, such as diodes, transistors and optoelectronic devices (laser and light-emitting diodes). The thermal conductivity value is also necessary in calculating the figure of merit for thermoelectric devices (e.g. Peltier devices). A useful description of the theoretical and practical aspects of the thermal conductivity in semiconductors is given by Steigmeier and Kudman [1], Holland [2], Maycock [3], Bhandari and Rowe [4] and Srivastava [5] who review some of the work done on III-V compounds, including some of the ternary and quaternary alloys.

B THEORETICAL CONSIDERATIONS

An exact calculation of lattice thermal conductivity for the III-V binaries is possible in principle, but lack of knowledge of various parameters (e.g. anharmonic forces and lattice vibration spectra) and the difficulty of obtaining exact solutions of phonon-phonon interactions are formidable barriers to progress. In the case of alloy compounds, an additional contribution being a result of a random distribution of constituent atoms in sublattice sites should be taken into consideration [6-11]. A phenomenological model of the thermal conductivity for ternary alloys was proposed by Abeles [6]. The lattice thermal conductivity was expressed in terms of the lattice parameters and mean atomic weights of the alloy and its constituents. Abeles' model has been successfully used for many III-V ternaries and quaternaries by Adachi [8] and Nakwaski [9]. In the ternary $In_{1-x}Ga_xAs$, only the In-Ga disorder due to the random distribution of the In and Ga atoms in the cation sublattice should be taken into consideration.

C EXPERIMENTAL DATA

C1 Room Temperature Values

Various experimental data on the III-V binaries, ternaries and quaternaries have led to an increased understanding of the thermal conductivity phenomena in these materials. (Because of the many dependencies of the thermal conductivity K, which in this review is always given in units of W/cm K, it is not possible to give a general account and the only data presented is taken at room temperature for an undoped or a lightly doped sample.) Measured K values for the $In_{1-x}Ga_xAs$ end-point binaries are in the ranges 0.25 - 0.30 W/cm K for InAs and 0.37 - 0.54 W/cm K for GaAs (see TABLE 1).

It is important to point out that when large numbers of foreign atoms are added to the host lattice as in alloying, the thermal conductivity may decrease significantly. The experimental data on III-V ternaries (InGaAs [3,12], AlGaAs [13], InGaP [7], InAsP [3,14] and GaAsP [3,15]) and quaternaries (InGaAsP [16], InGaAsSb [17,18] and AlGaAsSb [17,18]), exhibited strong nonlinearity with respect to the alloy composition. Such compositional variations can be successfully explained by the quadratic function with respect to the alloy composition x [8]:

$$K(x) = A + Bx + Cx^2 \qquad (1)$$

where C is the so-called nonlinear (bowing) parameter.

Room temperature thermal conductivity for the $In_{1-x}Ga_xAs$ ternary system has been measured by Abrahams et al [12] by employing the technique of placing the specimen between a heater and a sink. The ternary alloys were prepared by reacting InAs with a predetermined amount of GaAs, either by gradient freezing or by zone-levelling. The ingots were polycrystalline with very coarse grains elongated parallel to the growth direction.

The thermal conductivity K for $In_{1-x}Ga_xAs$ alloy lattice-matched to InP (x = 0.47) has been reported by Both et al [16] using a dynamic technique. The sample used was grown by liquid-phase epitaxy. They also measured the K values for epitaxial InGaAsP quaternaries lattice matched to InP. The thicknesses of these layers were about 2 - 3.5 μm.

It is well known that dilute alloying of GaAs with InAs is a very effective technique for reducing the dislocation density of semi-insulating crystals grown by the liquid-encapsulated Czochralski (LEC) method. Ohmer et al [19] have reported the effect of this small addition of In on the thermal properties of GaAs (thermal diffusivity, specific heat, and thermal resistivity). All of the dilute $In_{1-x}Ga_xAs$ alloys ($0.987 \leq x \leq 1.00$) used in this experiment were grown by the high pressure LEC method from high-purity GaAs melts contained in pyrolytic boron nitride crucibles. Alloying was achieved by adding In to the melt with an excess of As to maintain a stoichiometric balance between group III and V components. They have found that for x = 0.005 the thermal diffusivity and conductivity are reduced to 60% and 50% of the values for GaAs, respectively.

In TABLE 1 we list the measured K values (and thermal resistivity 1/K) for $In_{1-x}Ga_xAs$ alloy along with those for GaAs and InAs binaries. FIGURE 1 also compares the calculated results of Eqn (1) (solid line) to the experimental data of the $In_{1-x}Ga_xAs$ alloy system. The corresponding alloy-disorder bowing C for 1/K was found to be -72 cm K/W (A = 3.70 cm K/W; B = 70.57 cm K/W) [8].

It is clear from FIGURE 1 that Eqn (1) shows good agreement with the experimental data. However, the expression cannot be made to successfully fit the rapid increasing region found by Ohmer et al [19]. This region is in accordance with theory if the size difference is considered and combined coherently with the mass difference [11].

TABLE 1 Room-temperature thermal conductivity (K) and resistivity (1/K) of $In_{1-x}Ga_xAs$ alloy.

x	K (W/cm K)	1/K (cm K/W)	Ref
1.00 (GaAs)	0.44 ± 0.04	2.27 ± 0.2	[3]
	0.37	2.7	[12]
	0.45	2.2	[13]
	0.54 ± 0.03	1.85 ± 0.1	[15]
	0.395	2.53	[19]
	0.445	2.25	[20]
	0.419	2.39	[21]
0.99987	0.317	3.15	[19]
0.9996	0.336	2.98	[19]
0.9995	0.401	2.49	[19]
0.999	0.245	4.08	[19]
0.9987	0.346	2.89	[19]
0.9984	0.228	4.39	[19]
0.9983	0.218	4.59	[19]
0.9947	0.197	5.08	[19]
0.9938	0.161	6.21	[19]
0.9931	0.151	6.62	[19]
0.99	0.24	4.17	[22]
0.9872	0.241	4.15	[19]
0.987	0.181	5.52	[19]
0.98	0.215	4.65	[22]
0.96	0.18	5.56	[22]
0.92	0.15	6.67	[22]
0.88	0.082	12.20	[12]
0.78	0.068	14.71	[12]
0.66	0.056	17.86	[12]
0.56	0.048	20.83	[12]
0.47	0.043 - 0.061	16.4 - 23.4	[16]
0.36	0.055	18.18	[12]
0.31	0.061	16.39	[12]
0.24	0.074	13.51	[12]
0.16	0.095	10.53	[12]
0.00 (InAs)	0.27	3.70	[3]
	0.289	3.46	[12]
	0.30	3.33	[22]
	0.27	3.70	[23]
	0.273	3.66	[24]
	0.25	4.00	[25]

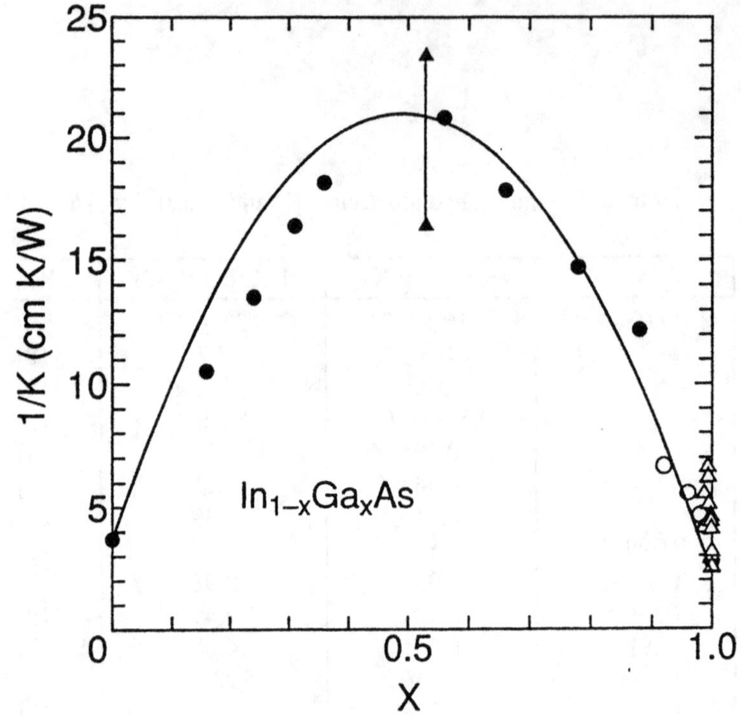

FIGURE 1 Lattice thermal resistivity 1/K as a function of composition x for $In_{1-x}Ga_xAs$ alloy. The solid line is calculated from Eqn (1). The experimental data are taken from Abrahams et al (solid circles) [12], Both et al (solid triangles) [16], Ohmer et al (open triangles) [19] and Arasly et al (open circles) [22].

C2 Effects of Temperature and Doping

We limited our discussion to room temperature. Data at other temperatures are scarce but we do know that many III-V binaries obey a rule [26]

$$K = A/T^n \tag{2}$$

with

$$1.20 \le n \le 1.55 \tag{3}$$

These are stronger temperature dependences than that predicted for three-phonon processes, and probably indicate the presence of higher order processes. It seems reasonable to expect Eqn (2) to apply for all values of x over a limited temperature range, say 150 to 600 K. If we take n = 1.375 for $In_{1-x}Ga_xAs$ ternary and Eqn (3) applies then over this range, the maximum error introduced is 13% which may be less than the other uncertainties.

The value of A in Eqn (2) is known to be dependent on the free-carrier concentration. The data listed in TABLE 1 were for undoped or lightly doped materials. Very heavily doped samples (N_a, $N_d > 10^{19}$ cm^{-3}) could have significantly lower thermal conductivities (i.e. smaller A values) [27].

D CONCLUSION

The room temperature thermal conductivity of $In_{1-x}Ga_xAs$ alloy decreases markedly with alloying and exhibits a minimum value of about 0.05 W/cm K at alloying composition of x ~ 0.5. This value is about 5 ~ 10 times as small as those of related binaries (i.e. GaAs and InAs). An approximate analytic expression provides quite good agreement with the experimental data.

REFERENCES

[1] E.F. Steigmeier, I. Kudman [*Phys. Rev. (USA)* vol.141 (1966) p.767-74]

[2] M.G. Holland [*Semicond. Semimet. (USA)* vol.2 (Physics of III-V Compounds, Academic Press, New York, 1967) ch.1 p.3]

[3] P.D. Maycock [*Solid-State Electron. (UK)* vol.10 (1967) p.161-8]

[4] C.M. Bhandari, D.M. Rowe [*Thermal Conduction in Semiconductors* (Wiley, New York, 1988)]

[5] G.P. Srivastava [*The Physics of Phonons* (Adam Hilger, Bristol, 1990)]

[6] B. Abeles [*Phys. Rev. (USA)* vol.131 (1963) p.1906-11]

[7] W. Both, F.P. Herrmann [*Cryst. Res. Technol. (Germany)* vol.17 (1982) p.K117-22]

[8] S. Adachi [*J. Appl. Phys. (USA)* vol.54 (1983) p.1844-8]

[9] W. Nakwaski [*J. Appl. Phys. (USA)* vol.64 (1988) p.159-66]

[10] F.L. Madarasz, P.G. Klemens [*Int. J. Thermophys. (USA)* vol.8 (1987) p.257-62]

[11] F. Szmulowicz, F.L. Madarasz, P.G. Klemens, J. Diller [*J. Appl. Phys. (USA)* vol.66 (1989) p.252-5]

[12] M.S. Abrahams, R. Braunstein, F.D. Rosi [*J. Phys. Chem. Solids (UK)* vol.10 (1959) p.204-10]

[13] M.A. Afromowitz [*J. Appl. Phys. (USA)* vol.44 (1973) p.1292-4]

[14] R. Bowers, J.E. Bauerle, A.J. Cornish [*J. Appl. Phys. (USA)* vol.30 (1959) p.1050-4]

[15] R.O. Carlson, G.A. Slack, S.J. Silverman [*J. Appl. Phys. (USA)* vol.36 (1965) p.505-7]

[16] W. Both, V. Gottschalch, G. Wagner [*Cryst. Res. Technol. (Germany)* vol.21 (1986) p.K85-7]

[17] W. Both, A.E. Bochkarev, A. Drakin, B.N. Sverdlov [*Cryst. Res. Technol. (Germany)* vol.24 (1989) p.K161-6]

[18] W. Both, A. Bochkarev, A. Drakin, B. Sverdlov [*Electron. Lett. (UK)* vol.26 (1990) p.418-19]

[19] M.C. Ohmer, W.C. Mitchel, G.A. Graves, D.E. Holmes, H. Kuwamoto, P.W. Yu [*J. Appl. Phys. (USA)* vol.64 (1988) p.2775-7]

[20] A. Amith, I. Kudman, E.F. Steigmeier [*Phys. Rev. (USA)* vol.138 p.A1270-6]

[21] T. Yao [*Appl. Phys. Lett. (USA)* vol.51 (1987) p.1798-800]

[22] D.G. Arasly, R.N. Ragimov, M.I. Aliev [*Sov. Phys.-Semicond. (USA)* vol.24 (1990) p.225-7]

[23] R. Bowers, R.W. Ure Jr., J.E. Bauerle, A.J. Cornish [*J. Appl. Phys. (USA)* vol.30 (1959) p.930-4]

[24] E.F. Steigmeier, I. Kudman [*Phys. Rev. (USA)* vol.132 (1963) p.508-12]

[25] G. Le Guillou, H.J. Albany [*Phys. Rev. B (USA)* vol.5 (1972) p.2301-8]

[26] S. Adachi [*Physical Properties of III-V Semiconductor Compounds: InP, InAs, GaAs, GaP, InGaAs, and InGaAsP* (Wiley-Interscience, New York, 1992)]

[27] J.C. Brice [in *Properties of Gallium Arsenide, 2nd Edition*, EMIS Datareviews Series no.2 (INSPEC, IEE, 1990) ch.1 p.21-3]

2.2 Thermal expansion coefficient of InGaAs

N.S. Takahashi and M. Matsuura

December 1992

A INTRODUCTION

Ternary $In_{0.53}Ga_{0.47}As$ perfectly lattice-matched to an InP substrate at room temperature is a candidate for application to optoelectronic devices such as photo-detectors for fibre optic communication systems. Much attention has also been given to strained-layer superlattices and quantum well heterostructures of lattice-mismatched $In_{1-x}Ga_xAs$ grown on GaAs and InP substrates for achieving high-performance electronic and optoelectronic devices.

Thermal expansion coefficients have been measured only in $In_{0.53}Ga_{0.47}As$ thin films prepared by heteroepitaxial growth techniques such as MBE, MOCVD and LPE. We review here the linear thermal expansion coefficients of $In_{0.53}Ga_{0.47}As$. The numerical expression for the linear thermal expansion coefficient of $In_{1-x}Ga_xAs$ using the linear interpolation scheme is presented.

B $In_{0.53}Ga_{0.47}As$ VALUE

The experimental values of the linear expansion coefficient α for $In_{0.53}Ga_{0.47}As$ are listed in TABLE 1. The result given by Bert et al is calculated with the value for InP (α = 4.56 x 10^{-6}/K) [1].

TABLE 1 Values for $In_{0.53}Ga_{0.47}As$.

α (10^{-6}/K)	Ref
5.66	Bisaro et al [2]
5.64	Bert et al [3]
5.72	Krasil'nikov et al [4]
5.72	Bak-Misiuk et al [5]
5.72	Pietsch and Marlow [6]
mean 5.692 $\pm$ 0.035	

C $In_{1-x}Ga_xAs$ VALUE

No linear expansion coefficients are known for the entire range of $In_{1-x}Ga_xAs$ alloy composition. This necessitates the use of linear interpolation (i.e. Vegard's law) for providing useful and reliable parameters. For lattice parameters and thermal expansion coefficients of most ternary compounds, Vegard's law is obeyed. Therefore, given α for GaAs and InAs, a linear relationship for the linear expansion coefficient of $In_{1-x}Ga_xAs$ can be obtained. Refs [7-10] give data about the linear thermal expansion coefficient of InAs crystals. A mean value of 5.20 x 10^{-6}/K is recommended. Brice [11] has given a recommended value for GaAs

($\alpha = 6.03 \times 10^{-6}$/K). From these values we obtain the linear thermal expansion coefficient of $In_{1-x}Ga_xAs$:

$$\alpha(x) = 5.20 + 0.83x \ (10^{-6}/K) \qquad (1)$$

The validity of Eqn (1) has been checked by comparison with the above mentioned experimental value for $In_{0.53}Ga_{0.47}As$. There is a small discrepancy between the measured value and the value calculated using Eqn (1).

D CONCLUSION

We have discussed the linear thermal expansion coefficient of $In_{1-x}Ga_xAs$ alloy. There are as yet no reliable experimental data on these materials.

REFERENCES

[1] K. Haruna, H. Maeta [in *Properties of Indium Phosphide*, EMIS Datareviews Series no.6 (INSPEC, IEE, 1991) ch.1 p.16-17]

[2] R. Bisaro, P. Merenda, T.P. Pearsall [*Appl. Phys. Lett. (USA)* vol.34 (1979) p.100-2]

[3] N.A. Bert, A.T. Gorelenok, S.G. Konnikov, V.E. Umanskii, A.S. Usikov [*Sov. Phys.-Tech. Phys. (USA)* vol.26 (1981) p.610-11]

[4] V.S. Krasil'nikov et al [*Sov. Phys.-Crystallogr. (USA)* vol.33 (1988) p.874-9]

[5] J. Bak-Misiuk, J. Wolf, U. Pietsch [*Phys. Status Solidi A (Germany)* vol.118 (1990) p.209-17]

[6] U. Pietsch, D. Marlow [*Phys. Status Solidi A (Germany)* vol.93 (1986) p.143-9]

[7] N.N. Sirota, Yu.I. Pashnitsev [*Dokl. Akad. Nauk SSSR (USSR)* vol.127 (1959) p.609-11]

[8] V.M. Glazov, S.N. Chizhevskaya, S.B. Evgen'ev [*Russ. J. Phys. Chem. (UK)* vol.43 (1969) p.201-5]

[9] H. Kressel, J.K. Butler [*Semiconductor Lasers and Heterojunction LEDs* (Academic Press, New York, 1977) ch.9 p.304-6]

[10] N.N. Sirota, A.M. Antyukhov, A.A. Sidorov [*Sov. Phys.-Dokl. (USA)* vol.29 (1984) p.662-5]

[11] J.C. Brice [in *Properties of Gallium Arsenide*, 2nd Edition, EMIS Datareviews Series no.2 (INSPEC, IEE, 1990) ch.1 p.18-19]

2.3 Bulk modulus of InGaAs

A.D. Prins and D.J. Dunstan

February 1993

A INTRODUCTION

The bulk modulus of a crystal is given by

$$B = -V\frac{dP}{dV} \tag{1}$$

Adiabatic bulk moduli may be obtained most accurately from the values of elastic constants determined from ultrasound velocity measurements, by the relation

$$B = \frac{c_{11} + 2c_{12}}{3} \tag{2}$$

The calculated difference between the adiabatic and isothermal bulk moduli is small. The adiabatic value is the larger of the two [1] but only by about 0.5% [2] which is well within experimental error.

B UNSTRAINED InGaAs

Neither the bulk moduli nor the elastic constants have been measured in the ternary InGaAs. Data exist for the two binary compounds GaAs and InAs and the bulk modulus of $In_xGa_{1-x}As$ can be obtained by linear interpolation [3]:

$$B(x) = xB_{InAs} + (1-x)B_{GaAs} \tag{3}$$

Values in the literature for the binary bulk moduli show about 6% scatter. TABLE 1 gives the average values, the standard deviations, σ, and the number of values found, n.

TABLE 1 Room temperature bulk moduli of GaAs and InAs (GPa).

Compound	B	σ	n
GaAs	75.5	0.8	15
InAs	59.1	1.6	3

Many authors underestimate, or indeed give no indication of, errors in the elastic constants. However, the work of Cottam and Saunders [4] for GaAs and Gerlich and Wolf [2] for InAs give values for B of 75.2 ± 1.2 GPa and 57.9 ± 2.2 GPa respectively. These errors agree well with the standard deviations of TABLE 1. From our earlier Datareview of the GaInAsP system [5] we recommend the average values of TABLE 1 with errors of ± 2 GPa, in good agreement with the values of [2] and [4].

Low-temperature data has been given for GaAs by [4] and [6] and for InAs by [2] and [7]. We recommend the interpolated and corrected data of [5] given in TABLE 2.

TABLE 2 Temperature dependence of GaAs and InAs bulk moduli (GPa).

Temperature	GaAs	InAs	Estimated error
300 K	75.5	59.1	± 2
77 K	76.9	62.1	± 2
4.2 K	77.1	63.4	± 2

Keyes [8] and Lam [9] proposed that the bulk moduli of the III-V semiconductors depend only on the lattice constant, not on the constituent elements. These rules provide an alternative interpolation for the ternaries and quaternaries but even for $In_{0.5}Ga_{0.5}As$ give values of B less than 2% below linear interpolation and therefore within experimental error.

Particularly for semiconductor heterostructures, relative values of the bulk moduli may be more important than absolute values. The present authors measured the bulk modulus of lattice-matched InGaAs and InGaAsP relative to that of InP. The results indicated little difference [10], showing that any differences should not be important to InGaAsP/InP heterostructure devices. However, neither Keyes' rule nor linear interpolation could account fully for the results, so that neither method predicts accurate values for the ternaries and quaternaries. Further work is required to measure them directly; until this is done linear interpolation will provide values as good as the binary values on which it is based.

C PRESSURE DERIVATIVE

Pressure derivatives of the bulk moduli, B, have been measured for GaAs [11] and InAs [12]. The recommended value for both binaries is given in [5] and is 4.5 ± 0.5. The same value may be recommended for $In_xGa_{1-x}As$.

D STRAINED InGaAs

The pressure dependence of B means that strained layers have bulk moduli different from those of unstrained material. For biaxial compression $In_xGa_{1-x}As/GaAs$:

$$\varepsilon_{xx} = \varepsilon_{yy} = \varepsilon_0 = 0.07x$$

and for a Poisson's ratio of $\nu = 1/3$

$$\varepsilon_{zz} = -\frac{2\nu}{1-\nu}\varepsilon_0 = -\varepsilon_0$$

The volume reduction is

$$\varepsilon_{vol} = \varepsilon_{xx} + \varepsilon_{yy} + \varepsilon_{zz} = \varepsilon_0$$

so that the grown-in hydrostatic pressure is

$$P_0 = \varepsilon_0 B$$

and the bulk modulus is

$$B = B_0 + B'P_0 = B_0 + B'\varepsilon_0 B$$

This gives the bulk modulus as a function of strain as

$$B(\varepsilon_0) = \frac{B_0}{1-\varepsilon_0 B'} \qquad (4)$$

For a strain of 1.5% this correction is twice the experimental error in B_0.

Eqn (4) comes from a consideration of the hydrostatic component of the strain of a pseudomorphic layer. The axial component of the strain should not affect the bulk modulus apart from small corrections due to higher order elastic constants [13]. However, a note of caution should be expressed for the elastic constants of strained InGaAs, and indeed for other strained materials. This is because the hydrostatic pressure coefficient of the bandgap is significantly lower than expected [13,14]. This behaviour is not understood [13] but one likely explanation requires anomalous elastic constants. If the volume reduces less than expected under applied hydrostatic pressure, in the presence of biaxial compression, the hydrostatic pressure coefficient will be reduced.

E CONCLUSION

Until the bulk moduli of the $In_xGa_{1-x}As$ system are measured directly the only way of estimating their values is to use interpolation methods between the measured binary values (Eqn (3) and TABLE 2). Effects of strain may be accounted for using Eqn (4) with the pressure derivative of the bulk modulus given in Section C. However these methods only give values as good as the determined binary values.

REFERENCES

[1] W.C. Overton [*J. Chem. Phys. (USA)* vol.37 (1962) p.116]

[2] D. Gerlich, M. Wolf [*High Pressure Science and Technology*, Eds B. Vodar, Ph. Marteau, vol.1 (Pergamon, Oxford, UK, 1980) p.506]

[3] S. Adachi [*J. Appl. Phys. (USA)* vol.53 no.12 (1982) p.8775-92]
[4] R.I. Cottam, G.A. Saunders [*J. Phys. C (UK)* vol.6 no.13 (9 July 1973) p.2105-18]
[5] A.D. Prins, D.J. Dunstan [in *Properties of InP*, EMIS Datareviews Series no.6 (INSPEC, IEE, 1991) p.373-5]
[6] C.W. Garland, K.C. Park [*J. Appl. Phys. (USA)* vol.33 (1962) p.759-60]
[7] R. Reifenberger, M.J. Keck, J. Trivisonno [*J. Appl. Phys. (USA)* vol.40 (1969) p.5403]
[8] R. Keyes [*J. Appl. Phys. (USA)* vol.33 (1962) p.3371]
[9] P.K. Lam, M.L. Cohen, G. Martinez [*Phys. Rev. B (USA)* vol.35 no.17 (15 June 1987) p.9190-4]
[10] A.D. Prins, D.J. Dunstan [*Semicond. Sci. Technol. (UK)* vol.4 no.4 (1989) p.239-40]
[11] H.J. Skimin, A. Jayaraman, P. Andreatch [*J. Appl. Phys. (USA)* vol.38 (1967) p.2362]
[12] [in *Landolt-Börnstein, Numerical Data and Functional Relationships in Science and Technology - Crystals and Solid-State Physics* (Springer, Berlin, Germany) vol.17 (1982)]
[13] V.A. Wilkinson et al [*Phys. Rev. B (USA)* vol.42 no.5 (15 August 1990) p.3113-19]
[14] V.A. Wilkinson, A.R. Adams [Datareview in this book: 3.2 Effect of pressure and temperature on band structure of InGaAs]

2.4 Optical and acoustic phonons in InGaAs

S. Adachi

December 1992

A INTRODUCTION

Phonons, representing the lattice vibrations occurring at a frequency ω_q, have an energy $\hbar\omega_q$ and have a momentum $\hbar q$, where $\hbar$ is the reduced Planck's constant. Two phonon branches result in a diatomic lattice system, a high energy branch for optical phonons (the two atoms in the unit cell move in opposite directions) and a lower energy branch for acoustic phonons (both the atoms in the unit cell move in phase with each other). We will discuss in Sections B and C some properties of the optical and acoustic phonons in $In_{1-x}Ga_xAs$ ternary system, respectively. In Section D the conclusions obtained from this contribution are briefly summarised.

B OPTICAL PHONONS

B1 Long-Wavelength Phonons: Mode Behaviour

Long-wavelength optical phonons in III-V ternary compounds exhibit either a 'one-mode' or a 'two-mode' behaviour (or more rigorously, three different types of mode behaviour, viz., 'one-', 'two-' and 'one-two-mode' (mixed-mode) behaviours). In a one-mode system, such as most I-VII mixed crystals, a single set of long-wavelength optical phonons appears. The one-mode phonon spectrum is approximately constant in strength but varies in energy as the alloy composition varies from one end member to the other. The two-mode systems, common for III-V ternaries, exhibit two distinct sets of optical modes with frequencies characteristic of each end member and strengths roughly proportional to the respective concentration. The one-two-mode system exhibits a single mode over only a part of the composition range, with two modes observed over the remaining range of composition.

The optical-phonon mode behaviour in $In_{1-x}Ga_xAs$ alloy has been studied experimentally by several groups [1-4]. TABLE 1 summarises these results. The first determination was obtained by Brodsky and Lucovsky [1] with infrared (IR) reflectivity at room temperature on thick (12 to 102 μm) epitaxial films. They reported two-mode behaviour in all of their samples. They emphasised the variation of the frequencies of the LO and TO phonons as a function of composition, in particular the decrease of the high-frequency LO and TO phonon frequencies with increasing InAs content. However, the highest value of x in their study was 0.78; in that sample the low-frequency mode was very weak, $\omega_{LO}-\omega_{TO} < 2\ cm^{-1}$.

Lucovsky and Chen [2] made a study of the IR reflectivity of not only lower-x (x = 0.16, 0.25, 0.47) but also higher-x (x = 0.84, 0.92) samples. For the samples with x = 0.84 and 0.92, only one IR band was observed, whereas in the other samples there were clearly two modes (i.e., one-two modes). A shoulder in the IR reflectivity near 260 cm^{-1} was also observed in the x = 0.84 and 0.92 samples that was attributed to zero boundary phonon modes made active by the breakdown of momentum conservation in these disordered alloy systems.

More recently, Yamazaki et al [3] have measured the IR spectra of $In_{1-x}Ga_xAs$ alloy ($0 \leq x \leq 1.0$) at near-normal incidence in the 180 - 400 cm^{-1} frequency range. The samples used were single-crystalline films grown on semi-insulating GaAs substrates using a Ga-In-$AsCl_2$-H_2 vapour-phase system. They observed only one IR band (GaAs-type) for $x \geq 0.8$ and two bands for the remaining range of composition. They also analysed the IR data on the basis of the modified cluster model including the effect of clusters and found that the model quantitatively explains the composition dependence of the phonon frequencies in both one- and two-mode regions. The value of clustering parameter β obtained by the analysis described above also made clear the reason why the one-mode region exists in the experimental results within the composition range $x \geq 0.8$.

Raman-scattering studies on $In_{1-x}Ga_xAs$ alloy have been made by Emura et al [4] over a wide range of the alloy composition. The samples were grown epitaxially by molecular-beam epitaxy on Fe-doped (100)InP substrates. Their results on the Raman spectra from the In-rich samples support the conclusion that this alloy system displays a one-two or a two-mode phonon nature. The composition variations of the Raman-active mode frequencies were found to be not linear. A least-squares calculation for the best fit to the experimental results to the second-order term of x gave the following expressions for each band (in cm^{-1}):

$$\omega_{LO} \text{ (InAs-type)} = 234.9 - 7.7x + 8.92x^2 \quad (1a)$$

$$\omega_{TO} \text{ (InAs-type)} = 233.7 - 16.5x \quad (1b)$$

$$\omega_{LO} \text{ (GaAs-type)} = 290.0 - 18.6x - 32.4x^2 \quad (1c)$$

$$\omega_{TO} \text{ (GaAs-type)} = 265.1 - 5.3x - 29.0x^2 \quad (1d)$$

TABLE 1 Behaviour of long-wavelength optical phonons in $In_{1-x}Ga_xAs$ alloy.
IR = Infrared reflectivity; RS = Raman scattering (in cm^{-1}).

Method	Mode	Sample	Ref
IR	Two mode	Epitaxial films (on GaAs)	Brodsky and Lucovsky [1]
IR	One-two mode	Melt-grown ingots (Polycrystalline)	Lucovsky and Chen [2]
IR	One-two mode	Epitaxial films (on GaAs)	Yamazaki et al [3]
RS	One-two or two mode	Epitaxial films (on InP)	Emura et al [4]

Long-wavelength phonon behaviours in $In_{1-x}Ga_xAs$ lattice-matched to InP (x = 0.47) have been studied by several authors using IR or Raman spectroscopy [4,5-13]. As shown in TABLE 2 and FIGURE 1, there is an overall agreement between the experimental results but some significant quantitative discrepancies still subsist. Pickering [7], using a two-mode fit to the reststrahlen bands, deduced InAs-type LO and TO frequencies of 230 and 224 cm^{-1} and GaAs-type LO and TO frequencies of 272 and 254 cm^{-1}, respectively. Pearsall et al [8] have identified the GaAs-type LO mode (270.5 cm^{-1}) and InAs-type TO mode (226 cm^{-1}) by Raman scattering in $In_{0.53}Ga_{0.47}As$ grown on (111)InP substrate. They identified a band at 254 cm^{-1} but did not attribute it to a TO mode. Also, they found a 244 cm^{-1} feature in this alloy and attributed it to alloy scattering. Raman studies [4,5,8] that identify the phonon modes in $In_{1-x}Ga_xAs$ have differences in identification of these bands as summarised by Nash et al [14].

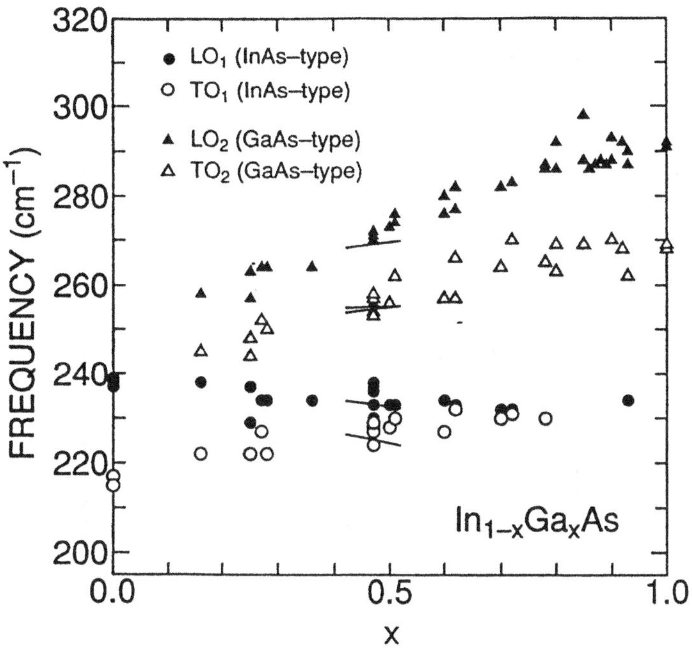

FIGURE 1 Plots of the InAs- and GaAs-type phonon frequencies as a function of alloy composition x for $In_{1-x}Ga_xAs$ alloy at room temperature. The solid lines represent the calculated results of Eqn (2).

There are two important limitations of the Raman studies which hinder the proper and consistent identification of phonon modes in $In_{1-x}Ga_xAs$ [13]. The first limitation is that the Raman studies backscatter off one crystal plane such as the growth plane. For a zinc-blende-type crystal such as $In_{1-x}Ga_xAs$, the LO phonon modes are only allowed in (100) backscattering, TO phonon modes are only allowed in (011) backscattering, and both LO and TO modes are allowed in (111) backscattering for specific polarisations. The use of at least two different scattering planes such as the (100) and (011) planes would identify all the allowable phonon modes. The second limitation is the neglect of using line-shape analysis on Raman spectra containing convoluted bands. Line-shape analysis is imperative in deconvolving complicated spectral responses and obtaining precise and accurate phonon frequencies. Discrepancies in identifying bands among the studies could be explained by visually hidden features.

Estrera et al [13] have recently identified the four allowable phonon modes in $In_{1-x}Ga_xAs$ near-lattice-matched to InP ($0.42 < x < 0.52$) using the selectivity of first-order Raman scattering off the (100) normal surface and the (011) cleaved plane and detailed line-shape analysis employing a sequential simplex optimisation procedure. The samples were 1.0 - 1.5 μm thick. They were grown by molecular-beam epitaxy or metalorganic molecular-beam epitaxy. The laser spot size for (011) scattering was 1 μm diameter. Backscattering on the (011) cleaved plane of 1.0 - 1.5 μm epitaxial films is possible because of the minimum spot size of 1 μm provided by a microscope. These authors found from a linear fitting procedure the relations between the Raman frequencies and alloy composition x determined from X-ray diffraction to be given by

$$\omega_{LO}\ (\text{InAs-type}) = 240.4 - 15.5x \qquad (2a)$$

$$\omega_{LO} \text{ (GaAs-type)} = 261.7 + 15.8x \tag{2b}$$

$$\omega_{TO} \text{ (GaAs-type)} = 252.5 + 5.6x \tag{2c}$$

TABLE 2 Frequencies of the InAs- (LO_1, TO_1) and GaAs-type (LO_2, TO_2) phonons in $In_{1-x}Ga_xAs$ measured at room temperature for different values of the Ga concentration. IR = Infrared reflectivity; RS = Raman scattering (in cm^{-1}).

x	Experiment	LO_1	TO_1	LO_2	TO_2	Ref
0.00	IR	238	215	-	-	[1]
	IR	238	215	-	-	[3]
	RS	237	-			[4]
	RS	239	217	-	-	[8]
0.16	IR	238	222	258	245	[2]
0.25	IR	237	222	263	248	[2]
	RS	229	-	257	244	[4]
0.27	IR	234	227	264	252	[3]
0.28	IR	234	222	264	250	[1]
0.36	RS	234	-	264	-	[4]
0.47	IR	233	228	271	257	[2]
	RS	238	-	271	253	[4]
	IR	230	224	272	254	[7]
	RS	237	227	270	254	[8]
	RS	236	229	272	258	[10]
	RS	222 ± 7	-	256	-	[11]
	RS	233	-	270	-	[12]
0.50	IR	233	228	273	256	[1]
0.51	IR	233	230	276	262	[3]
	RS	-	-	274	-	[10]
0.60	IR	234	227	276	257	[1]
	RS	-	-	280	-	[10]
0.62	IR	233	232	282	266	[3]
	RS	233	-	277	257	[4]
0.70	IR	232	230	282	264	[1]
0.72	IR	232	231	283	270	[3]
0.78	IR	-	230	287	265	[1]
	RS	-	-	286	-	[10]
0.80	IR	-	-	292	269	[3]
	RS	-	-	286	263	[8]
0.84	IR	-	-	298	269	[2]
0.85	RS	-	-	288	-	[10]
0.86	RS	-	-	286	-	[10]
0.87	RS	-	-	287	-	[10]
0.88	RS	-	-	288	-	[10]
0.89	RS	-	-	287	-	[10]
0.90	IR	-	-	293	270	[3]
	RS	-	-	288	-	[10]
0.92	IR	-	-	292	268	[2]
0.93	RS	234	-	287	262	[4]
	RS	-	-	290	-	[10]
1.00	IR	-	-	291	268	[1]
	IR	-	-	292	269	[3]
	RS	-	-	291	-	[4]
	RS	-	-	292	269	[8]
	RS	-	-	292	-	[10]

for (100) backscattering configuration, and

$$\omega_{TO}\ (\text{InAs-type}) = 236.5 - 24.0x \tag{2d}$$

$$\omega_{TO}\ (\text{GaAs-type}) = 247.2 + 15.7x \tag{2e}$$

for (011) backscattering configuration. They identified another phonon mode R* at 244 cm^{-1} which was attributed to an alloy disorder mode in these films. This mode was observed to be in all samples, geometries and polarisations studied which would indicate its direct relationship to a disorder mode verifying Pearsall's band identification [8].

B2 Zone-Boundary Phonons

The lattice vibrations in the crystals can be described in a framework that is analogous to the descriptions of electrons. The vibronic and and electronic states are best described in reciprocal space or **k** space. The phonon dispersion curves can be determined from neutron scattering or X-ray thermal diffuse scattering. First-order Raman scattering measurements give information only about phonon states with $\mathbf{q} = 0$ (Γ). This severe limitation can be overcome by using second-order Raman spectra; two phonons with $\mathbf{q}_1 + \mathbf{q}_2 \sim 0$ are then produced, and the complete phonon spectrum may become accessible. The phonon dispersion curves have been determined for GaAs by Waugh and Dolling [15] and by Strauch and Dorner [16] (neutron scattering) and for InAs by Carles et al [17] (second-order Raman scattering) and by Orlova [18] (X-ray thermal diffuse scattering). TABLE 3 summarises the zone-boundary (X, L) phonon frequencies for GaAs and InAs along with those for the zone-centre (Γ) phonons [15,18]. Not only the long-wavelength phonons but also the zone-boundary phonons will show a one-two- or a two-mode behaviour in $In_{1-x}Ga_xAs$ alloy. To our knowledge, however, detailed experimental data have not yet been reported to date on the zone-boundary phonon frequencies in this alloy.

TABLE 3 Zone-centre (Γ) and zone-boundary (X, L) phonons in GaAs and InAs at room temperature (in cm^{-1}).

Critical point	Phonon	GaAs	InAs
Γ	LO	285 [15]	241 [18]
	TO	267 [15]	220 [18]
X	LO	241 [15]	200 [18]
	TO	252 [15]	214 [18]
L	LO	238 [15]	203 [18]
	TO	261 [15]	220 [18]

B3 Effects of Temperature, Pressure and Stress

The effects of temperature on the optical phonon frequencies are due primarily to the thermal expansion of the crystal lattice. The dependence of phonon frequency $\omega(q,T)$ with temperature T can be written as [19]

$$\omega(q,T) = \omega_0(q)\left(1 - \frac{\alpha_L^2 a_0}{\kappa_L k_B} T\right) \tag{3}$$

where $\omega_0(\mathbf{q})$ is the phonon dispersion of the harmonic linear chain, a_0 is the lattice constant, α_L is the coefficient of linear expansion, κ_L is the compressibility of the linear chain and k_B is the Boltzmann's constant. This expression predicts a linear decrease in phonon frequency with an increase of T. It has been found experimentally that this is nearly the case for the Γ-point LO and TO phonons in GaAs [20,21], for the Γ-, X-, L-point LO and TO phonons in InAs [18], and for the GaAs- and InAs-type LO phonons in $In_{1-x}Ga_xAs$ alloy [10].

Anharmonic properties of solids are customarily described in terms of the mode Grüneisen parameter γ. As a measure of the volume (V) dependence of the phonon frequency $\omega_i(\mathbf{q})$ of the ith mode, the mode Grüneisen parameter $\gamma_i(\mathbf{q})$ is defined by

$$\gamma_i = -\frac{\partial \ln \omega_i}{\partial \ln V} \tag{4}$$

The mode Grüneisen parameter can thus be calculated from a knowledge of the phonon frequencies as a function of crystal volume. It is noted that the volume change (compression) can be connected with hydrostatic pressure through Murnaghan's equation [19]. We summarise in TABLE 4 the mode Grüneisen parameters obtained experimentally or theoretically for a number of critical points of GaAs and InAs. No detailed data on the mode Grüneisen parameters for $In_{1-x}Ga_xAs$ alloy has been reported to date.

TABLE 4 Mode Grüneisen parameters obtained for a number of critical points of GaAs and InAs.

Critical point	Phonon	GaAs	InAs
Γ	LO	1.23 ± 0.02^a [22]	1.06^a [23]
	TO	1.39 ± 0.02^a [22]	1.21^a [23]
X	LO	1.23^b [24]	1.04^b [24]
	TO	1.73 ± 0.07^a [22]	1.23^b [24]
L	LO	1.03^b [24]	1.56^b [24]
	TO	1.48 ± 0.15^a [22]	1.10^b [24]

a: Experimental
b: Calculated

Phonon frequency is a strong function of the hydrostatic pressure. This quantity is also strongly influenced by uniaxial (biaxial) stress. Such stress-induced frequency shifts of phonons can be essentially explained by the phonon deformation potentials (PDPs). The PDPs are important anharmonicity parameters useful in a number of physical and technological applications. Detailed expressions of PDPs and their consequences for the first-order Raman shifts in semiconductors were first represented by Cerdeira et al [25].

There have been many reports on the effects of uniaxial stress on the $\mathbf{q} = 0$ (Γ) optical phonon frequencies and their degeneracies in semiconductors [19]. The information obtained from such studies concerns primarily the PDPs. We list in TABLE 5 the dimensionless PDPs, $\underline{K}_{11}$ $(= p/\omega_i^2)$, $\underline{K}_{12}$ $(= q/\omega_i^2)$, $\underline{K}_{44}$ $(= r/\omega_i^2)$ and $\underline{K}_{11} - \underline{K}_{12}$ for GaAs and InAs determined experimentally [19]. It is recognised that the experimental data cover a large range of PDP values by different authors.

TABLE 5 PDPs for GaAs and InAs [19].

PDP	GaAs	InAs
$\underline{K}_{11}$ (LO)	-(1.0 - 2.0)	
$\underline{K}_{11}$ (TO)	-(1.4 - 2.4)	
$\underline{K}_{12}$ (LO)	-(1.5 - 2.7)	
$\underline{K}_{12}$ (TO)	-(1.6 - 3.0)	
$\underline{K}_{44}$ (LO)	-(0.5 - 0.6)	
$\underline{K}_{44}$ (TO)	-(0.2 - 0.9)	-0.76
$\underline{K}_{11} - \underline{K}_{12}$ (LO)	0.45 - 0.70	
$\underline{K}_{11} - \underline{K}_{12}$ (TO)	0.16 - 0.60	1.14

Anastassakis and Cardona [26] have investigated the effects of a uniaxial stress on the LO and TO phonon frequencies of In-hardened GaAs ($In_{0.004}Ga_{0.996}As$). They found that the singlet-doublet phonon frequency splittings are smaller in In-hardened GaAs especially for a stress along [111] than those in pure GaAs. They considered that a possible explanation of this effect may lie in a hardening of the internal stress parameter due, at least, to the increase in the lattice constant. The PDPs determined from this experiment are $\underline{K}_{44}$ (LO) = -0.16, $\underline{K}_{11} - \underline{K}_{12}$ (LO) = 0.83, $\underline{K}_{44}$ (TO) = -0.54 and $\underline{K}_{11} - \underline{K}_{12}$ (TO) = 0.30.

The PDP parameters have not yet been determined for In-rich $In_{1-x}Ga_xAs$ alloy. Experimentally, the relatively well-known quantity is the mode Grüneisen parameter $\gamma_i = -(\underline{K}_{11} + 2\underline{K}_{12})/6$. Since all known γ_i values of III-V and II-VI compounds are about 1 ~ 2 (see TABLE 4), we can take the same value for $In_{1-x}Ga_xAs$ ternary.

Once the PDP values are known (or properly assumed), a straightforward determination of strain in semiconductors is possible by measuring the strain-induced shifts of the $\mathbf{q} = 0$ phonon frequencies. Indeed, this kind of study has been extensively done on heteroepitaxial material systems, such as $In_{1-x}Ga_xAs$/InP [4], $In_{1-x}Ga_xAs$/GaAs [10,27-29], GaAs/sapphire [30], GaAs/Si [31,32], GaAs/InP [33,34], $In_{1-x}Ga_xAs$/GaAs and $In_{1-x}Al_xAs$/GaAs strained-layer superlattices [35-37].

C ACOUSTIC PHONONS

C1 Long-Wavelength Phonons: Sound Velocity

The characteristic of acoustic phonons is essentially the same as that of sound waves. If the density g and the elastic stiffness constant c_{ij} of a solid are known, one can calculate the bulk sound velocity v (long-wavelength, nondispersive) from the following general relation:

$$v = (c_{ij}/g)^{\frac{1}{2}} \tag{5}$$

If we neglect nonlinear terms in the equation of motion, pure longitudinal sound waves may propagate in the [100], [110] and [111] directions. Pure transverse waves may propagate not only in these directions, but also in the directions [hkl] (h=k; l arbitrary) and [hk0] (h and k arbitrary). We list in TABLE 6 expressions for the elastic constants c_{ij} which control the sound velocities (see Eqn (5)) propagating along the three major directions, [100], [110] and [111] in cubic, zinc-blende lattice.

TABLE 6 Propagation direction, direction or plane of polarisation, c_{ij} (Eqn (5)) and sound wave mode (longitudinal (LA) and transverse acoustic wave (TA1 and TA2)) in cubic, zinc-blende lattice.

Propagation direction	Direction or plane of polarisation	c_{ij}	Wave mode
[100]	<100>	c_{11}	LA
[100]	(100)	c_{44}	TA1, TA2
[110]	[110]	$(c_{11} + c_{12} + 2c_{44})/2$	LA
[110]	(110)	$(c_{11} - c_{12})/2$	TA1
[110]	(001)	c_{44}	TA2
[111]	[111]	$(c_{11} + 2c_{12} + 4c_{44})/3$	LA
[111]	(111)	$(c_{11} - c_{12} + c_{44})/3$	TA1, TA2

TABLE 7 also lists expressions in terms of the GaAs-composition fraction x for the calculated longitudinal (LA) and transverse (TA1, TA2) sound velocities along the [100], [110] and [111] directions in $In_{1-x}Ga_xAs$ alloy. In the calculation, we used the following composition dependence of crystal density and stiffness constants [38]: $g(x) = 5.67 - 0.35x$ (g/cm^3); $c_{11}(x) = 8.33 + 3.55x$ (10^{11} dyn/cm^2); $c_{12}(x) = 4.53 + 0.85x$ (10^{11} dyn/cm^2); $c_{44}(x) = 3.96 + 1.98x$ (10^{11} dyn/cm^2).

TABLE 7 Calculated sound velocities in $In_{1-x}Ga_xAs$ alloy (300 K).

Propagation direction	Wave mode	Sound velocity (10^5 cm/s)
[100]	LA	3.83 + 0.90x
[100]	TA1, TA2	2.64 + 0.70x
[110]	LA	4.28 + 0.95x
[110]	TA1	1.83 + 0.64x
[110]	TA2	2.64 + 0.70x
[111]	LA	4.42 + 0.97x
[111]	TA1, TA2	2.14 + 0.65x

C2 Zone-Boundary Phonons

As in the case of optical phonons, the zone-boundary acoustic phonon frequencies can be determined from neutron scattering, X-ray thermal diffuse scattering, higher-order Raman scattering or infrared optical spectroscopy. We list in TABLE 8 the zone-boundary (X, L) LA and TA phonon frequencies for GaAs and InAs determined from neutron [15] and X-ray thermal diffuse scattering [18], respectively.

TABLE 8 Zone-boundary (X, L) phonons in GaAs and InAs at room temperature (in cm^{-1}).

Critical point	Phonon	GaAs	InAs
X	LA	227 [15]	166 [18]
	TA	79 [15]	67 [18]
L	LA	209 [15]	154 [18]
	TA	62 [15]	37 [18]

Because of its definitive nature, acoustic phonons in mixed crystals may always show one-mode behaviour. However, no experimental data on the zone-boundary phonon frequencies of the acoustic branch in $In_{1-x}Ga_xAs$ alloy has been reported to date. The elastic properties can be attributed to the long-range Coulomb forces in solids. Thus, it seems that a linear interpolation can be successfully used for want of such quantities.

C3 Effects of Temperature and Pressure

Temperature dependence of the sound velocity arises mainly from its dependence on the elastic stiffness constants. The temperature-dependent stiffness data are reported in [39] (GaAs) and [40] (InAs).

We list in TABLE 9 the mode Grüneisen parameters for the long-wavelength acoustic phonons along [100], [110] and [111] directions in GaAs and InAs [41]. TABLE 10 also lists the mode Grüneisen parameters for the zone-boundary (X, L) acoustic phonons in GaAs and InAs [22,24]. To our knowledge, however, no numeric data on these quantities has been reported for $In_{1-x}Ga_xAs$ alloy to date.

TABLE 9 Mode Grüneisen parameters for the long-wavelength ($q \rightarrow 0$) acoustic phonons along [100], [110] and [111] directions in GaAs and InAs [41].

Direction	Phonon	GaAs	InAs
[100]	LA	1.30	1.40
	TA	0.53	0.13
[110]	LA	1.29	1.26
	TA2	-0.05	-0.47
[111]	LA	1.29	1.23
	TA	0.23	-0.16

TABLE 10 Mode Grüneisen parameters for the zone-boundary (X, L) acoustic phonons in GaAs and InAs.

Critical point	Phonon	GaAs	InAs
X	LA	1.18[a] [24]	1.31[a] [24]
	TA	-1.62 ± 0.05[b] [22]	-2.24[a] [24]
L	LA	1.24[a] [24]	0.70[a] [24]
	TA	-1.72 ± 0.15[b] [22]	-1.75[a] [24]

a: Calculated
b: Experimental

D CONCLUSION

Studies of the composition dependence of the long-wavelength (**q** = 0) optical phonon frequencies indicate that one-two- (mixed) or two-mode behaviour, most probably one-two-mode behaviour, occurs for the $In_{1-x}Ga_xAs$ alloy system. Numerical information about the sound wave (**q** = 0 acoustic phonon) velocities propagating along the three major directions [100], [110] and [111] in this alloy has been presented. These velocities are found to increase linearly with increasing GaAs mole fraction. The effects of temperature, pressure and stress (strain) on the lattice vibrational properties are also discussed for this alloy system.

REFERENCES

[1] M.H. Brodsky, G. Lucovsky [*Phys. Rev. Lett. (USA)* vol.21 (1968) p.990-3]

[2] G. Lucovsky, M.F. Chen [*Solid State Commun. (USA)* vol.8 (1970) p.1397-401]

[3] S. Yamazaki, A. Ushirokawa, T. Katoda [*J. Appl. Phys. (USA)* vol.51 (1980) p.3722-9]

[4] S. Emura, S. Gonda, Y. Matsui, H. Hayashi [*Phys. Rev. B (USA)* vol.38 (1988) p.3280-6]

[5] A. Pinczuk, J.M. Worlock, R.E. Nahory, M.A. Pollack [*Appl. Phys. Lett. (USA)* vol.33 (1978) p.461-3]

[6] P.M. Amirtharaj, G.D. Holah, S. Perkowitz [*Phys. Rev. B (USA)* vol.21 (1980) p.5656-60]

[7] C. Pickering [*J. Electron. Mater. (USA)* vol.10 (1981) p.901-18]

[8] T.P. Pearsall, R. Carles, J.C. Portal [*Appl. Phys. Lett. (USA)* vol.42 (1983) p.436-8]

[9] G.M. Zinger, I.P. Ipatova, A.I. Ryskin [*Sov. Phys.-Semicond. (USA)* vol.18 (1984) p.1324]

[10] K. Kakimoto, T. Katoda [*Jpn. J. Appl. Phys. (Japan)* vol.24 (1985) p.1022-9]

[11] R.K. Soni, S.C. Abi, K.P. Jain, M. Balkanski, S. Slempkes, J.L. Benchimol [*J. Appl. Phys. (USA)* vol.59 (1986) p.2184-8]

[12] R. Borroff, R. Merlin, C. Chin, P.K. Bhattacharya [*Appl. Phys. Lett. (USA)* vol.53 (1988) p.1652-3]

[13] J.P. Estrera et al [*Appl. Phys. Lett. (USA)* vol.61 (1992) p.1927-9]

[14] K.J. Nash, M.S. Skolnick, S.J. Bass [*Semicond. Sci. Technol. (UK)* vol.2 (1987) p.329-36]

[15] J.L.T. Waugh, G. Dolling [*Phys. Rev. (USA)* vol.132 (1963) p.2410-12]

[16] D. Strauch, B. Dorner [*J. Phys. C: Condens. Matter (UK)* vol.2 (1990) p.1457-74]

[17] R. Carles, N. Saint-Cricq, J.B. Renucci, M.A. Renucci, A. Zwick [*Phys. Rev. B (USA)* vol.22 (1980) p.4804-15]

[18] N.S. Orlova [*Phys. Status Solidi B (Germany)* vol.119 (1983) p.541-6]

[19] S. Adachi [*Physical Properties of III-V Semiconductor Compounds: InP, InAs, GaAs, GaP, InGaAs, and InGaAsP* (Wiley-Interscience, New York, 1992)]

[20] B. Jusserand, J. Sapriel [*Phys. Rev. B (USA)* vol.24 (1981) p.7194-207]

[21] J.R. Shealy, G.W. Wicks [*Appl. Phys. Lett. (USA)* vol.50 (1987) p.1173-5]

[22] See R. Trommer, H. Müller, M. Cardona, P. Vogl [*Phys. Rev. B (USA)* vol.21 (1980) p.4869-78]

[23] K. Aoki, E. Anastassakis, M. Cardona [*Phys. Rev. B (USA)* vol.30 (1984) p.681-7]

[24] D.N. Talwar, M. Vandevyver [*Phys. Rev. B (USA)* vol.41 (1990) p.12129-39]

[25] F. Cerdeira, C.J. Buchenauer, F.H. Pollak, M. Cardona [*Phys. Rev. B (USA)* vol.5 (1972) p.580-93]
[26] E. Anastassakis, M. Cardona [*Solid State Commun. (USA)* vol.64 (1987) p.543-7]
[27] G. Burns, C.R. Wie, F.H. Dacol, G.D. Pettit, J.M. Woodall [*Appl. Phys. Lett. (USA)* vol.51 (1987) p.1919-21]
[28] O. Brafman, D. Fekete, R. Sarfaty [*Appl. Phys. Lett. (USA)* vol.58 (1991) p.400-2]
[29] W.C. Tang, H.J. Rosen, S. Guha, A. Madhukar [*Appl. Phys. Lett. (USA)* vol.58 (1991) p.1644-6]
[30] T.P. Humphreys et al [*Jpn. J. Appl. Phys. (Japan)* vol.28 (1989) p.L1595-8]
[31] G. Landa, R. Carles, C. Fontaine, E. Bedel, A. Muñoz-Yagüe [*J. Appl. Phys. (USA)* vol.66 (1989) p.196-200]
[32] B. Roughani, M. Kallergi, J. Aubel, S. Sundaram [*J. Appl. Phys. (USA)* vol.66 (1989) p.4946-9]
[33] A. Freundlich, J.C. Grenet, G. Neu, G. Landa, R. Carles [*Appl. Phys. Lett. (USA)* vol.55 (1989) p.1558-60]
[34] M.-E. Pistol, M. Gerling, D. Hessman, L. Samuelson [*Phys. Rev. B (USA)* vol.45 (1992) p.3628-35]
[35] M. Nakayama, K. Kubota, T. Kanata, H. Kato, S. Chika, N. Sano [*J. Appl. Phys. (USA)* vol.58 (1985) p.4342-5]
[36] U.D. Venkateswaran et al [*Phys. Rev. B (USA)* vol.42 (1990) p.3100-8]
[37] C. Wu, P. Lao, S.C. Shen [*Appl. Phys. Lett. (USA)* vol.58 (1991) p.1491-3]
[38] S. Adachi [*J. Appl. Phys. (USA)* vol.53 (1982) p.8775-92]
[39] A.S. Jordan [*J. Cryst. Growth (Netherlands)* vol.49 (1980) p.631-42]
[40] R. Reifenberger, M.J. Keck, J. Trivisonno [*J. Appl. Phys. (USA)* vol.40 (1969) p.5403-4]
[41] See T. Soma, H.-M. Kagaya [*Phys. Status Solidi (Germany)* vol. 130 (1985) p.497-501]

CHAPTER 3

BAND STRUCTURE OF LATTICE-MATCHED AND STRAINED InGaAs ALLOYS

3.1 Energy gaps and band structure of lattice-matched and strained InGaAs

J. Singh

February 1993

A INTRODUCTION

In recent years with the advent of heteroepitaxy, the use of built-in strain to modify electronic properties of semiconductors has become widespread. Both electronic and optoelectronic devices have benefited from such band structure modifications [1-5]. The use of strain has been particularly exploited in the InGaAs system. High speed MODFETs and low threshold quantum well lasers have been demonstrated using this system. We will briefly review some generic and specific properties of strained layers produced by heteroepitaxy. We will focus in particular on how the band structure is modified in the InGaAs system.

B HETEROEPITAXY AND STRAIN

When a thin film of a semiconductor with lattice constant a_L is deposited on a semi-infinite substrate with lattice constant a_S, one can loosely characterise two regions. If the thickness of the overlayer is small, the layer is under biaxial strain where the in-plane lattice constant is the same as that of the substrate and the out of plane lattice constant is given by the Poisson effect. If the overlayer thickness is too large, the overlayer has a network of dislocations. The overlayer thickness separating these two regimes is called the critical thickness and varies inversely with the mismatch between the substrate and the overlayer. The below critical thickness regime is the regime of pseudomorphic epitaxy and will be the subject of this Datareview.

For systems of interest in the present work, the epitaxial semiconductor layer is biaxially strained in the plane of the substrate by an amount $\varepsilon_{||}$, and uniaxially strained in the perpendicular direction, by an amount $\varepsilon_{\perp}$. For a thick substrate, the in-plane strain of the layer is determined from the bulk lattice constants of the substrate material, a_S, and the layer material, a_L

$$\varepsilon_{||} = a_S/a_L - 1 = \varepsilon \qquad (1)$$

Since the layer is subjected to no stress in the perpendicular direction, the perpendicular strain $\varepsilon_{\perp}$ is simply proportional to $\varepsilon_{||}$:

$$\varepsilon_{\perp} = -\varepsilon_{||}/\sigma \qquad (2)$$

where the constant σ is known as Poisson's ratio.

Noting that there is no stress in the direction of growth it can be shown that for the strained layer grown on (001) substrate [6,7]

$$\sigma = c_{11}/2c_{12}$$

$$\varepsilon_{xx} = \varepsilon_{||},\ \varepsilon_{yy} = \varepsilon_{xx},\ \varepsilon_{zz} = (-2c_{12}/c_{11})\varepsilon_{||} \tag{3}$$

$$\varepsilon_{xy} = 0,\ \varepsilon_{yz} = 0,\ \varepsilon_{zx} = 0$$

where c_{11} and c_{12} are the force constants of the overlayer.

The effect of the strain is to break the usual cubic symmetry of the crystal lattice. Thus in the above case the x-z and y-z symmetries are broken while the x-y symmetry is still maintained. As a result of the strain and the symmetry loss there are two important consequences on band structure: i) the band edge positions and other high symmetry positions are altered; ii) important degeneracies are lifted. The first effect can be utilised to tailor the bandgap of the material, while the second effect can alter the density of states at band edges. As can be seen from Eqn (3), for (001) strained epitaxy, the strain tensor has no off-diagonal terms. Thus for this orientation there is no angular distortion of the unit cell. However, for epitaxy in other directions, there are off diagonal terms that can be exploited for many applications. For example for the (111) strained growth we have,

$$\sigma = \frac{c_{11} + 2c_{12} + 4c_{44}}{2c_{11} + 4c_{12} - 4c_{44}}$$

$$\varepsilon_{xx} = \left[\frac{2}{3} - \frac{1}{3}\left(\frac{2c_{11} + 4c_{12} - 4c_{44}}{c_{11} + 2c_{12} + 4c_{44}}\right)\right]\varepsilon_{||}$$

$$\varepsilon_{yy} = \varepsilon_{xx} \tag{4}$$

$$\varepsilon_{zz} = \varepsilon_{xx}$$

$$\varepsilon_{xy} = \left[-\frac{1}{3} - \frac{1}{3}\left(\frac{2c_{11} + 4c_{12} - 4c_{44}}{c_{11} + 2c_{12} + 4c_{44}}\right)\right]\varepsilon_{||}$$

$$\varepsilon_{yz} = \varepsilon_{xy}$$

$$\varepsilon_{zx} = \varepsilon_{yz}$$

We will now discuss the effect of strain on the band structure.

C CONSEQUENCES OF STRAIN ON BAND STRUCTURE

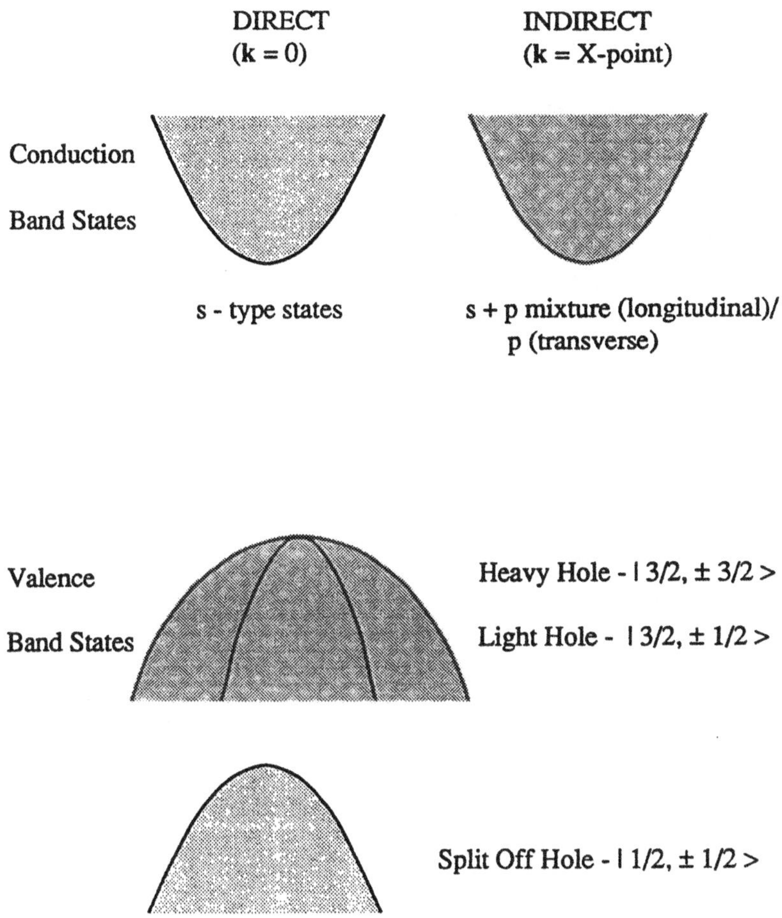

FIGURE 1 A schematic showing the nature of the central cell symmetry at the band edges of direct and indirect semiconductors.

Once the strain tensor is known, one can apply the deformation potential theory to calculate the effects of strain on various eigenstates in the Brillouin zone. The strain perturbation Hamiltonian is defined and its effects are calculated in the simple first order perturbation theory. In general we have [8-10]

$$H_{\varepsilon}^{\alpha\beta} = \sum_{ij} D_{ij}^{\alpha\beta} \varepsilon_{ij} \tag{5}$$

where D_{ij} is the deformation potential operator which transforms under symmetry operations as a second rank tensor. $D_{ij}^{\alpha\beta}$ are the matrix elements of D_{ij}.

We will summarise the effect of strain on band structure by examining various high symmetry points in the Brillouin zone. The symmetry of these states is shown in FIGURE 1.

C1 Conduction Band Edge for Direct Band Materials

The effect of the strain is to cause a shift in the energy given by

$$E^{(000)} = H^{\varepsilon} = D_{xx}\,(\varepsilon_{xx} + \varepsilon_{yy} + \varepsilon_{zz}) \tag{6}$$

where D_{xx} is conventionally denoted by $\Xi_d^{(000)}$.

C2 Conduction Band Edge for X-point-like States

$$\delta E^{(000)} = \delta E^{(\bar{1}00)} = D_{xx}\,\varepsilon_{xx} + D_{yy}\,(\varepsilon_{yy} + \varepsilon_{zz}) \tag{7}$$

Writing

$$D_{yy} = \Xi_d^{(100)} = D_{zz}$$

and

$$D_{xx} = \Xi_d^{(100)} + \Xi_u^{(100)} \tag{8}$$

we have,

$$\delta E^{(100)} = \delta E^{(\bar{1}00)} = \delta E' + \Xi_u^{(100)}\,\varepsilon_{xx} \tag{9}$$

$$\delta E^{(010)} = \delta E^{(0\bar{1}0)} = \delta E' + \Xi_u^{(100)}\,\varepsilon_{yy} \tag{10}$$

$$\delta E^{(001)} = \delta E^{(00\bar{1})} = \delta E' + \Xi_u^{(100)}\,\varepsilon_{zz} \tag{11}$$

where $\delta E' = \Xi_d^{(100)}\,(\varepsilon_{xx} + \varepsilon_{yy} + \varepsilon_{zz})$. For a strain tensor in which the diagonal elements are not equal, the 6-valley degeneracy can be lifted. This occurs for (001) pseudomorphic growth.

C3 Valence Band Edge

The valence band edges are made up of p-type states and the effect of the strain is to lift the degeneracy of the heavy hole and light hole states.

Restricting ourselves to the heavy hole (HH) and light hole (LH) states, the strain Hamiltonian can be written as (the state of ordering is | 3/2, 3/2>, | 3/2, -1/2>, | 3/2, 1/2>, and | 3/2, -3/2>)

$$H^{\varepsilon} = \begin{bmatrix} H^{\varepsilon}_{hh} & H^{\varepsilon}_{12} & H^{\varepsilon}_{13} & 0 \\ H^{\varepsilon *}_{12} & H^{\varepsilon}_{lh} & 0 & H^{\varepsilon}_{13} \\ H^{\varepsilon *}_{13} & 0 & H^{\varepsilon *}_{lh} & -H^{\varepsilon}_{12} \\ 0 & H^{\varepsilon *}_{13} & -H^{\varepsilon *}_{12} & H^{\varepsilon}_{hh}{}^{*} \end{bmatrix} \tag{12}$$

where the matrix elements are given by

$$H^{\varepsilon}_{hh} = a(\varepsilon_{xx} + \varepsilon_{yy} + \varepsilon_{zz}) - b[\varepsilon_{zz} - \frac{1}{2}(\varepsilon_{xx} + \varepsilon_{yy})]$$

$$H^{\varepsilon}_{lh} = a(\varepsilon_{xx} + \varepsilon_{yy} + \varepsilon_{zz}) + b[\varepsilon_{zz} - \frac{1}{2}(\varepsilon_{xx} + \varepsilon_{yy})]$$

$$H^{\varepsilon}_{12} = -d(\varepsilon_{xz} + i\varepsilon_{yz})$$

$$= < \frac{3}{2}, \frac{3}{2} |H_{\varepsilon}| \frac{3}{2}, \frac{1}{2} >$$

$$H^{\varepsilon}_{13} = \frac{\sqrt{3}}{2} b\,(\varepsilon_{yy} - \varepsilon_{xx}) + id\varepsilon_{xy}$$

$$= < \frac{3}{2}, \frac{3}{2} |H_{\varepsilon}| \frac{3}{2}, -\frac{1}{2} > \qquad (13)$$

Here, the quantities a, b, and d, are valence band deformation potentials. As discussed earlier, strains achieved by lattice mismatched epitaxial growth on (001) substrate can be characterised by $\varepsilon_{xx} = \varepsilon_{yy} = \varepsilon$, and $\varepsilon_{zz} = -(2c_{12}/c_{11})\varepsilon$. All of the diagonal strain terms are then zero. Using this information, we get

$$H^{\varepsilon}_{hh} = \left[2a\left(\frac{c_{11} - c_{12}}{c_{11}}\right) + b\left(\frac{c_{11} + 2c_{12}}{c_{11}}\right)\right]\varepsilon \qquad (14)$$

$$H^{\varepsilon}_{lh} = \left[2a\left(\frac{c_{11} - c_{12}}{c_{11}}\right) - b\left(\frac{c_{11} + 2c_{12}}{c_{11}}\right)\right]\varepsilon \qquad (15)$$

with all other terms zero.

An important consequence of the strain which is not hydrostatic in nature is that the HH, LH degeneracy is lifted. This has important consequences for the hole density of states.

From this discussion we can see that the band structure of a strained system is affected by strain as well as growth orientation. To see the importance of these changes we show a comparison of the bandgaps of strained and unstrained $In_xGa_{1-x}As$ grown on GaAs and strained and unstrained $In_{0.53+x}Ga_{0.47-x}As$ grown on InP based on the parameters for GaAs and InAs shown in TABLE 1.

i) $In_xGa_{1-x}As$ grown on (001) GaAs. Unstrained: The room temperature bandgap is [10],

$$E_g(x) = 1.43 - 1.53x + 0.45x^2 \text{ (eV)} \qquad (16)$$

$$\text{HH - LH splitting} = 0 \qquad (17)$$

TABLE 1 Some important properties of the GaAs and InAs semiconductors.
The values given are for 300 K.

Properties	GaAs	InAs
Lattice-Constant (Å)	5.6533	6.0583
Bandgap (ev)	1.423	0.356
Dielectric Constant ε_0	12.91	15.15
ε_∞	10.1	12.25
dEg/dP (10^{-12}eV cm^2/dyne)	11.7	10.0
Force Constants (dyne/cm^2)		
c_{11}	11.88×10^{11}	8.33×10^{11}
c_{12}	5.38×10^{11}	4.53×10^{11}
Piezoelectric Constant e_{14} (C/m^2)	−0.16	−0.05
Linear Expansion Coefficient (K^{-1})	6.86×10^{-6}	4.52×10^{-6}
Deformation Potentials		
b (eV)	−2.0	
d(eV)	−6.0	

Strained: For small x, the shifts are linear and are given by [11-14],

$$E_g(x) = 1.43 - 1.11x + 0.45x^2 \text{ (eV)} \tag{18}$$

$$\text{HH - LH splitting} = 0.42x \text{ (eV)} \tag{19}$$

ii) $In_{0.53+x}Ga_{0.47-x}As$ grown on (001) InP
Unstrained: For small x, the bandgap at 300°K is,

$$E_g(x) = 0.75 - 1.05x + 0.45x^2 \text{ (eV)} \tag{20}$$

$$\text{HH - LH splitting} = 0 \tag{21}$$

Strained:

$$E_g(x) \cong 0.75 - 0.63x + 0.45x^2 \text{ (eV)} \tag{22}$$

$$\text{HH - LH splitting} \equiv 0.42x \text{ (eV)} \tag{23}$$

There are some additional higher order corrections proportional to x^2 when the strain is large. However, for small strain used in real devices ($x \leq 0.2$), the above expressions are valid. In the case of InGaAs growth on an InP substrate, x can be positive (to produce a compressive strain) or negative (to produce a tensile strain). This can increase or decrease the bandgap over the unstrained bandgap. Also for compressive strain, the HH is above the LH while for tensile strain the situation is reversed as shown in FIGURE 2 [15]. It is important to note that the HH-LH splitting can be quite large compared to k_BT at room temperature, so that the strain splitting is quite significant for device applications.

It is also important to note that the bandgap of the strained layer depends upon growth orientation since the strain tensor changes as a function of orientation [16]. In FIGURE 3 we show the bandgap of $In_{0.2}Ga_{0.8}As$ grown on various orientations between (001) and (111) planes. The bandgap of the strained layer is 1.218 eV for (001) growth and 1.24 for (111) growth. The constrained gap is 1.128 eV.

C4 Built-in Electric Fields

We finally note that for strained epitaxy in which the off-diagonal terms of the strain tensor are not zero, one can have a strong built-in electric field due to the piezoelectric effect [17-19]. This field is given by:

$$F = \frac{\sqrt{3}\, e_{14}\varepsilon_{xy}}{\varepsilon_0\varepsilon_s} \text{ (V/m)} \tag{24}$$

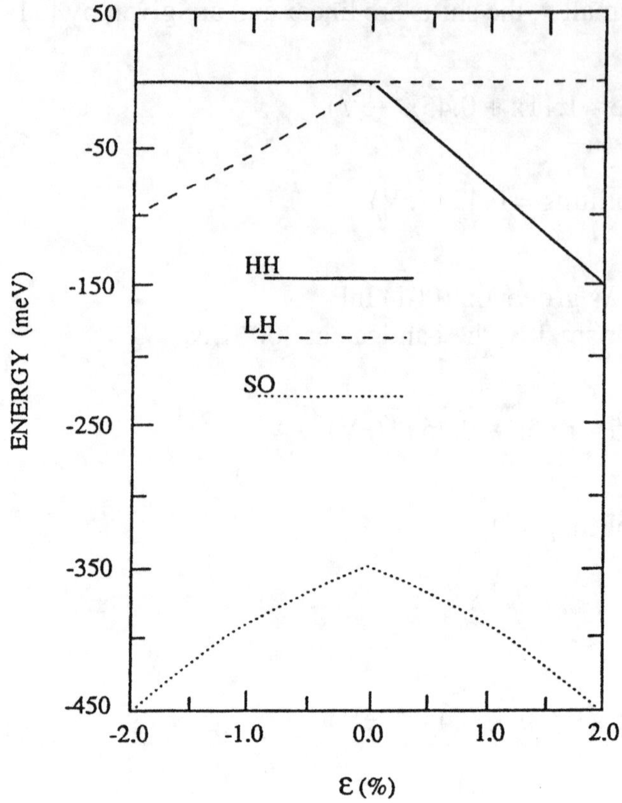

FIGURE 2 The variation of valence band edges with compressive (-ve ε) and tensile strain (+ve ε). The split off band measurement is also shown.

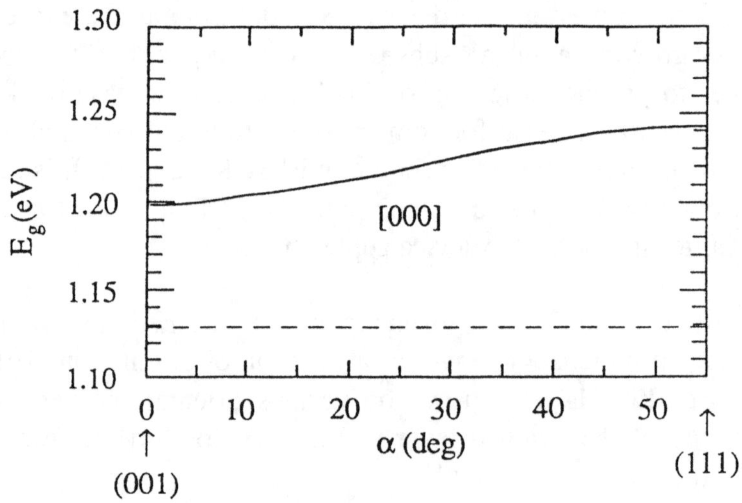

FIGURE 3 The variations of the bandgap of strained $In_{0.2}Ga_{0.8}As$ with substrate orientation. The substrate is GaAs. Also shown is the bandgap of unstrained $In_{0.2}Ga_{0.8}As$.

and for the InGaAs alloy the value turns out to be

$$F \sim 2.5 \times 10^7 \varepsilon_{xy} \text{ (V/cm)} \tag{25}$$

For InGaAs grown on (111) substrates with a mismatch of ~ 2%, this field is ~ 10^5 V/cm. This is a fairly large field and can be exploited to produce strong built-in fields without doping or an external bias.

D CONCLUSION

We have seen in this Datareview that lattice matched and strained alloys of InGaAs can provide a remarkable degree of bandgap tailoring. This is particularly true of the coherently strained InGaAs grown on GaAs or InP substrates. The strain can not only modify the bandgap, but more importantly can remove the HH, LH degeneracy and thus alter the density of states of the valence band. The effects of strain can be felt up to ~100 meV into the bands and are therefore quite significant for electronic and optoelectronic devices.

REFERENCES

[1] E. Yablanovitch, E.O. Kane [*J. Lightwave Technol. (USA)* vol.4 (1986) p.504]

[2] A.R. Adams [*Electron. Lett. (UK)* vol.22 (1986) p.249]

[3] J.M. Hinckley, J. Singh [*Appl. Phys. Lett. (USA)* vol.53 (1988) p.785]

[4] T.E. Zipperian, L.R. Dawson, T.J. Drummond, J.E. Schirber, J.J. Fritz [*Appl. Phys. Lett. (USA)* vol.52 (1988) p.975]

[5] C.P. Lee, H.T. Wang, G.T. Sullivan, N.H. Sheng, D.L. Miller [*IEEE Electron Device Lett. (USA)* vol.8 (1987) p.85]

[6] G.C. Osbourn [*J. Appl. Phys. (USA)* vol.53 (1982) p.1586]

[7] J.M. Hinckley, J. Singh [*Phys. Rev. B (USA)* vol.42 (1986) p.3546]

[8] G.L. Picus, G.E. Bir [*Induced Effects in Semiconductors* (John Wiley and Sons, New York, 1974)]

[9] G.L. Picus, G.E.Bir [*Sov. Phys.-Solid State (USA)* vol.1 (1959) p.1502]

[10] S. Adachi [*J. Appl. Phys. (USA)* vol.53 (1982) p.8775]

[11] H. Kato, N. Iguchi, S. Chika, M. Nakayama, N. Sato [*Jpn. J. Appl. Phys. (Japan)* vol.25 (1986) p.1327]

[12] E.P. O'Reilly [*Semicond. Sci. Technol. (UK)* vol.4 (1989) p.121]

[13] G. Li, D.Huang, U.K. Reddy, T.S. Henderson R. Houdre, H. Morkoc [J. Appl. Phys. *(USA)* vol.62 (1987) p.3366]

[14] J. Singh [*Physics of Semiconductors and Their Heterostructures* (McGraw Hill, New York, 1993)]

[15] J.P. Loehr [*Theoretical Studies of Pseudomorphic Quantum Well Optoelectronic Devices* (University of Michigan, Ph.D. Thesis, 1991)]

[16] J.M. Hinckley, J. Singh [*J. Appl. Phys. (USA)* vol.69 (1991) p.2694]

[17] C. Mailhiot, D.L. Smith [*J. Vac. Sci. Technol. A (USA)* vol.5 (1987) p.2060]

[18] C. Mailhiot, D.L. Smith [*Phys. Rev. B (USA)* vol.35 (1987) p.1242]

[19] P.K. Laurich, K. Elcess, C.G. Fonstad, J.G. Berry, C. Mailhiot, D.L. Smith [*Phys. Rev. Lett. (USA)* vol.62 (1989) p.649]

3.2 The effect of temperature and pressure on InGaAs band structure

V.A. Wilkinson and A.R. Adams

March 1993

A INTRODUCTION

Both temperature and pressure can cause considerable perturbations of the band structure of semiconductors. Here we review their effects on strained and unstrained InGaAs and try to suggest, where possible, which data form the most consistent picture.

Unless a phase transition occurs hydrostatic pressure maintains the symmetry of the crystal but reduces the lattice constant. In crystalline III-V materials this causes the direct bandgap to increase throughout the Brillouin zone and the allowed energy bands to broaden. This results in energy shifts with pressure which are characteristic of the conduction band minima. Of most interest and practical importance are their shifts with respect to the top of the valence band at the Brillouin zone centre. Approximately, these are: Γ at 1×10^{-1} eV/GPa, L at 4×10^{-2} eV/GPa and X at -1×10^{-2} eV/GPa.

Currently there is considerable interest in pseudomorphic, strained InGaAs layers. The built-in strain contains a hydrostatic component which modifies the band structure as described above and a shear component which removes degeneracies related to the cubic symmetry of the normal III-V lattice.

When the temperature is reduced the lattice constant also decreases as under hydrostatic pressure and the direct bandgap increases. The bandgap also increases due to the decrease of electron-phonon interactions and this is the dominant mechanism at temperatures above about 100 K.

The temperature variation of the band gap is commonly described by the empirical relation proposed by Varshni [1].

$$E(T) = E(O) - \alpha T^2/(T+\beta) \qquad (1)$$

where α and β are material dependent constants. More recently the semi-empirical Bose-Einstein expression has been proposed [2,3]:

$$E(T) = E_B - a_B\,(1+2/\{\exp(\Theta/T)-1\}) \qquad (2)$$

where a_B represents the strength of the electron-phonon interaction and Θ corresponds to the average phonon frequency.

B PRESSURE EFFECTS

The influence of pressure on the full alloy range has not been studied. However, some conclusions can be drawn from available data on GaAs, InAs and $In_{0.53}Ga_{0.47}As$. Interesting effects observed in strained InGaAs quantum wells are also discussed.

B1 Direct Gap E_o

B1.1 Unstrained InGaAs

Work on the pressure coefficient of the direct gap of GaAs has been reviewed by Adams [4] who suggested we may write:

$$dE_o/dP = 10.73 \pm 0.05 \times 10^{-2} \text{ eV/GPa [5]} \qquad (3)$$

The pressure coefficient of InAs has been measured by several authors [6-8]. Zallen and Paul [6] used reflectivity and concluded

$$dE_o/dP = 10 \times 10^{-2} \text{ eV/GPa} \qquad (4)$$

Accurate absolute values of pressure coefficients are difficult to obtain; however, these data suggest that InAs has approximately the same pressure coefficient as GaAs and therefore we might expect the intermediate alloy $In_xGa_{1-x}As$ to have the same coefficient.

The pressure coefficient of $In_{0.53}Ga_{0.47}As$ (lattice-matched to InP) has been investigated. Lambkin and Dunstan [9] studying the photoluminescence (PL) from an MOCVD-grown epilayer up to 7 GPa report a sub-linear behaviour described by 11×10^{-2} P - 27×10^{-4} P^2 at 300 K and essentially the same behaviour at 80 K. People et al [10] measuring the stimulated emission from an MBE grown InGaAs layer up to 3.5 GPa at room temperature report a value of 12.44×10^{-2} P - 6.8×10^{-4} P^2 (energy is in eV and pressure in GPa).

Given the available data we suggest that the pressure coefficient of the direct gap of bulk unstrained $In_xGa_{1-x}As$ should be taken as that of GaAs i.e. 10.7×10^{-2} eV/GPa but this may somewhat overestimate the pressure coefficient at very high pressures.

B1.2 Strained InGaAs

There is much current interest in strained $In_xGa_{1-x}As$ grown either on GaAs or InP. A dependence on x of $(10.7 - 6.0x) \times 10^{-2}$ eV/GPa for $x < 0.23$ is found for the PL energy from wide InGaAs/GaAs quantum wells (100 Å) grown on (100) orientated substrates. This gives a good approximation of the pressure coefficient of the direct gap of bulk strained $In_xGa_{1-x}As$ [11]. In this material system the wells are under compression and have an in-plane strain of -0.07x. The behaviour under pressure is not yet understood since the pressure coefficients of the binary components are very similar and strain is not predicted to have large effects on the hydrostatic pressure coefficients [11]. The pressure coefficients of bulk strained and unstrained InGaAs are plotted in FIGURE 1.

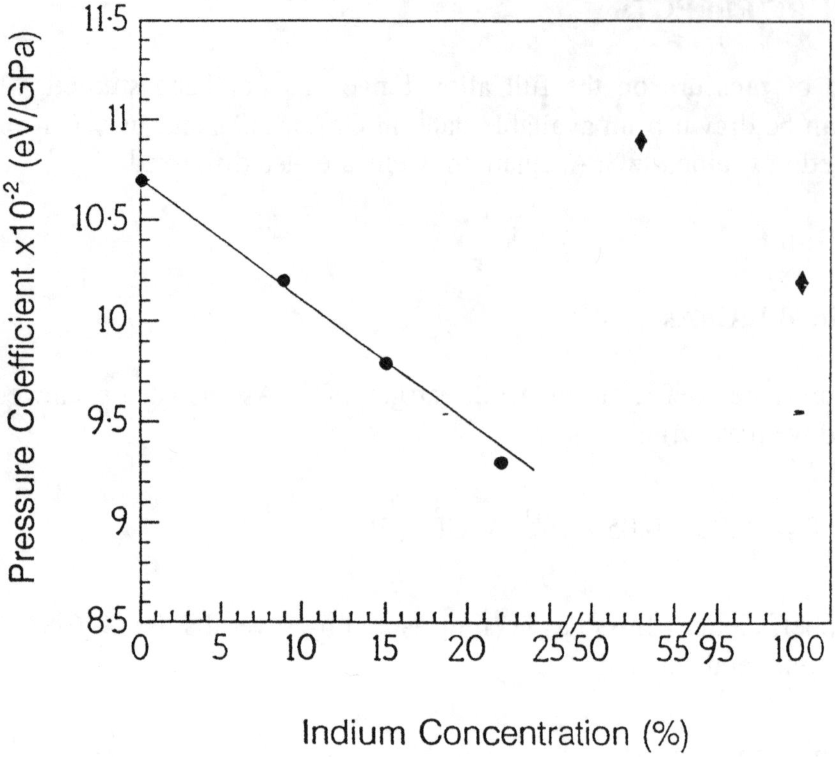

FIGURE 1 Pressure coefficients of 100 Å quantum wells of $In_xGa_{1-x}As$ of various compositions are plotted against the compositions x (solid circles). The solid line is a linear fit, (10.7 - 6.0x) x 10^{-2} eV/GPa. Also shown for comparison are the measured pressure coefficients of $In_{0.53}Ga_{0.47}As$ and InAs (solid diamonds). Reproduced from [11].

B2 Indirect Gaps

B2.1 Unstrained InGaAs

We are not aware of any data for the pressure dependence of the indirect gaps of InGaAs or InAs. Values for GaAs are given below, and should be a good approximation at least for low In content material:

$$dE_{\Gamma X}^{GaAs}/dP = -1.34 \times 10^{-2} \text{ eV/GPa [5]} \tag{5}$$

$$dE_{\Gamma L}^{GaAs}/dP = +2.8 \times 10^{-2} \text{ eV/GPa [12]} \tag{6}$$

B2.2 Strained InGaAs

The pressure coefficient of the Γ-$X_{x,y}$ gap has been reported for compressively strained $In_xGa_{1-x}As$ on (100) GaAs [13] as -2.6 x 10^{-2} eV/GPa from 4 K PL measurements for $0.09 < x < 0.23$. Strain along [001] splits the X-minima with X_x and X_y remaining degenerate and lying below X_z. This coefficient is twice that of GaAs and such a large change is not expected for small changes in composition. However, as shown in the case of the direct gap of pseudomorphic InGaAs, the influence of strain on the hydrostatic pressure coefficients is not yet fully understood.

B3 Heterostructures

B3.1 InGaAs/InP unstrained heterostructures

The pressure dependence of the Γ conduction-band offset in $In_{0.53}Ga_{0.47}As$/InP quantum wells has been determined from 4 K PL measurements and is found to decrease at $2.3 \pm 0.6 \times 10^{-2}$ eV/GPa [16].

B3.2 InGaAs/GaAs strained heterostructures

Several authors have reported the pressure coefficients of InGaAs/GaAs quantum wells of varying width [11,14,15]. The coefficient is found to decrease with increasing well width. This appears to be primarily due to the reduced pressure coefficient of the strained InGaAs well material as described in Section B1.

Hydrostatic pressure 4 K PL measurements have been conducted to determine spectroscopically the Γ band offsets in InGaAs/GaAs heterojunctions giving the fractional valence band offset, $Q_v = 0.4$ for $x < 0.2$ [13].

C TEMPERATURE EFFECTS

The available data on the temperature dependence of the direct gap E_o of $In_xGa_{1-x}As$ and typical data for GaAs and InAs are given in TABLE 1. The fitting parameters are defined in Section A.

Paul et al [25] have derived the following empirical expression for the temperature dependence of the bandgap of bulk InGaAs which provides a suitable interpolation between GaAs and InAs taking account of bandgap bowing:

$$E_g(x,T) = E_g^{GaAs}(0) - \alpha^{GaAs}T^2/(T+\beta^{GaAs})$$

$$+ [E_g^{InAs}(0) - \alpha^{InAs}T^2/(T+\beta^{InAs}) - E_g^{GaAs}(0)$$

$$+ \alpha^{GaAs}T^2/(T+\beta^{GaAs})]x - 0.475x(1-x) \qquad (7)$$

The effect of temperature on some other direct gaps in the Brillouin zone have been reported for GaAs and InAs. Typical values are given in TABLE 2. We are not aware of any data for InGaAs and can therefore only suggest that interpolated values are used. For small x the GaAs values will be reasonably accurate.

TABLE 1 Fitting parameters describing the temperature dependence of E_o for GaAs, $In_xGa_{1-x}As$ and InAs.

Material	Measurement Technique	E(O) (eV)	α 10^{-4} (eV/K)	β (K)	E_B (eV)	a_B (meV)	Θ (K)	Ref
GaAs	Spectral ellipsometry	1.517 ± 0.008	5.5 ± 1.3	225 ± 174	1.571 ± 0.023	57 ± 29	240 ± 102	[17]
GaAs	Photoreflectance (PR)	1.512 ± 0.005	5.1 ± 0.5	190 ± 82				[18]
GaAs	PL	1.52	10.6	671	1.5294 ± 0.0011	10.4 ± 1.1	102.4 ± 5.7	[19]
$In_{0.06}Ga_{0.94}As$ (strain relieved)	PR	1.420 ± 0.005	4.8 ± 0.4	200 ± 50	1.466 ± 0.014	44 ± 9	203 ± 45	[20]
$In_{0.15}Ga_{0.85}As$ (strain relieved)	"	1.285 ± 0.005	5.0 ± 0.4	231 ± 40	1.339 ± 0.015	53 ± 10	238 ± 50	[20]
$In_{0.21}Ga_{0.79}As$ strained QW (100 Å)	"	1.312 ± 0.005	4.8 ± 0.3	140 ± 40				[21]
$In_{0.53}Ga_{0.47}As$	PL	0.81	4.91	301				[22]
InAs	Transmission	0.43	3.16	93				[1]
InAs	PL	0.42	2.76	83				[23]
InAs	Transmission	0.42	2.5	75				[24]

TABLE 2 Fitting parameters describing the temperature dependence of various direct gaps in the Brillouin zone for GaAs and InAs.

Material and Gap	Measurement Technique	E(O) (eV)	α 10^{-4} (eV/K)	β (K)	E_B (eV)	a_B (meV)	Θ (K)	Ref
GaAs								
$E_o + W_o$	Spectral ellipsometry	1.851 ± 0.005	3.5 ± 0.4	225 (fixed)	1.907 ± 0.009	58 ± 7	240 (fixed)	[17]
E_1	"	3.041 ± 0.003	7.2 ± 0.2	205 ± 31	3.125 ± 0.009	91 ± 11	274 ± 30	[17]
E_o'	"	4.509 ± 0.008	4.0 ± 0.7	241 ± 177	4.563 ± 0.021	59 ± 26	323 ± 119	[17]
E_2	"	5.133 ± 0.021	6.6 ± 0.4	43 ± 66	5.161 ± 0.033	38 ± 33	114 ± 95	[17]
InAs								
E_1	Reflectance		5.4 ± 0.3	$\beta \equiv 0$ 300-500 K				[26]
E_1	Wavelength modulated reflectance		5	$\beta \equiv 0$ 80-300 K				[27]
E_2	"		5.6	$\beta \equiv 0$ 80-300 K				[27]

D CONCLUSION

The available data on the pressure coefficients and temperature dependence of the bandgaps of InGaAs have been reviewed. We hope that highlighting the scarcity of such data will encourage further work in the field.

REFERENCES

[1] Y.P. Varshni [*Physica (Netherlands)* vol.34 (1967) p.149]

[2] L. Viña, S. Logothetidis, M. Cardona [*Phys. Rev. B (USA)* vol.30 (1984) p.1979]

[3] S. Logothetidis, L. Viña, M. Cardona [*Phys. Rev. B (USA)* vol.31 (1985) p.947]

[4] A.R. Adams [in *Properties of Gallium Arsenide, 2nd Edition*, EMIS Datareviews Series no.2 (INSPEC, IEE, 1990) p.119]

[5] D.J. Wolford, J.A. Bradley [*Solid State Commun. (USA)* vol.53 (1985) p.1069]

[6] R. Zallen, W. Paul [*Phys. Rev. (USA)* vol.155 (1967) p.703]

[7] Y.F. Tsay, S.S. Mitra, B. Bendow [*Phys. Rev. B (USA)* vol.10 (1974) p.1476]

[8] Y.J. Jung, B.H. Kim, H.J. Lee, J.C. Woolley [*Phys. Rev. B (USA)* vol.26 (1982) p.3151]

[9] J.D. Lambkin, D.J. Dunstan [*Solid State Commun. (USA)* vol.67 (1988) p.827]

[10]] R. People, A. Jayaraman, K.W. Wecht, D.L. Sivco, A.Y. Cho [*Appl. Phys. Lett. (USA)* vol.52 (1988) p.2124]

[11] V.A. Wilkinson et al [*Phys. Rev. B (USA)* vol.42 (1990) p.3113]

[12] A. Kangarlu et al [*Phys. Rev. B (USA)* vol.38 (1988) p.9790]

[13] V.A. Wilkinson, A.D. Prins, D.J. Dunstan, L.K. Howard, M.T. Emeny [*J. Electron. Mater. (USA)* vol.20 (1991) p.509]

[14] A.D. Prins, J.D. Lambkin, K.P. Homewood, M.T. Emeny, C.R. Whitehouse [*High Press. Res. (USA)* vol.3 (1990) p.51]

[15] H.Q. Hou, L.J. Wang, R.M. Tang, J.M. Zhou [*Phys. Rev. B (USA)* vol.42 (1991) p.2926]

[16] J.D. Lambkin, D.J. Dunstan, E.P. O'Reilly, B.R. Butler [*J. Cryst. Growth (Netherlands)* vol.93 (1988) p.323]

[17] P. Lautenschlager, M. Garriga, S. Logothetidis, M. Cardona [*Phys. Rev. B (USA)* vol.35 (1987) p.9174]

[18] H. Shen et al [*Appl. Phys. Lett. (USA)* vol.53 (1988) p.1080]

[19] E. Grilli, M. Guzzi, R. Zamboni, L. Pavesi [*Phys. Rev. B (USA)* vol.45 (1992) p.1638]

[20] Z. Hang, D. Yan, F.H. Pollak, G.D. Pettit, J.M. Woodall [*Phys. Rev. B (USA)* vol.44 (1991) p.10546]

[21] Y.S. Huang et al [J. Appl. Phys. (USA) vol.70 (1991) p.7537]

[22] P.W. Yu, E. Kuphal [*Solid State Commun. (USA)* vol.49 (1984) p.907]

[23] Z.M. Fang, K.Y. Ma, D.H. Jaw, R.M. Cohen, G.B. Stringfellow [*J. Appl. Phys. (USA)* vol.67 (1990) p.7034]

[24] H.C. Casey, M.B. Panish [*Heterostructure Lasers, Pt. B* (Academic Press, New York, 1978) p.9]

[25] S. Paul, J.B. Roy, P.K. Basu [*J. Appl. Phys. (USA)* vol.69 (1991) p.827]

[26] F. Lukes, E. Schmidt [*Proc. 6th Int. Conf. Physics of Semicond.*, Exeter, UK, 1962 (The Institute of Physics and the Physical Society, London, 1966) p.339]

[27] R.R.L. Zucca, Y.R. Shen [*Phys. Rev. (USA)* vol.155 (1970) p.2668]

3.3 Impact ionisation coefficients in InGaAs

G.E. Stillman

August 1993

A INTRODUCTION

In many semiconductor device applications such as FETs or HBTs it is important to be able to design the device structure so as to avoid avalanche or Zener breakdown under normal operating conditions. In other devices such as IMPATT diodes or avalanche photodiodes, APDs, it is important to be able to predict the avalanche multiplication characteristics under specified electric field profiles so that optimum performance devices can be designed. The parameters that have to be known before this can be done are the electron and hole impact ionisation coefficients. In this Datareview we first describe the measurement techniques used to determine the impact ionisation coefficients for electrons and holes, α_n and β_p respectively [1]. Because of considerable confusion in the literature concerning the orientation dependence of α_n and β_p, the best current results for GaAs and InP are described. These results show that the rapid decrease in α_n with decreasing electric field at low electric fields described in earlier reports [2,3] does not exist. More recent detailed measurements of α_n and β_p for <100> and <111> InP are reviewed, which also show that α_n and β_p for these two spatial orientations are equal within the uncertainty in the electric field. There is no significant orientation dependence of α_n and β_p in either GaAs or InP. Finally, the measurements of α_n and β_p in $Ga_{0.47}In_{0.53}As$ that have been reported are summarised.

B MEASUREMENT TECHNIQUES

The most straightforward method for determining the impact ionisation coefficients in semiconductors is the measurement of the multiplication of photocurrent in p-n junctions as a function of bias voltage. Several experimental conditions must be met for this method to be reliable: 1) the multiplication due to pure electron injection, $M_n(V)$, and pure hole injection, $M_p(V)$, must be measured separately in the same device structure, but not in complementary devices (i.e., p^+n and n^+p); 2) the primary photocurrent (i.e., unmultiplied photocurrent) must be accurately determined as a function of bias voltage; 3) the spatial variation of the electric field should be slow enough so that the ionisation coefficients are only functions of the electric field; 4) the electric field must be accurately known as a function of position and bias voltage V; 5) the avalanche multiplication and electric field must be uniform across the active area of the device.

The ideal device structures that have the potential of meeting these requirements and the variation of the electric field F(x) in each of these structures are shown in FIGURE 1. The equations used for the analysis of photocurrent multiplication measurements on each of these structures to determine α_n and β_p have been given by Stillman and Wolfe [1]. Although these structures are conceptually simple, it has not been easy to meet the conditions listed above in many measurements, and refinements for the fabrication of these structures are discussed below.

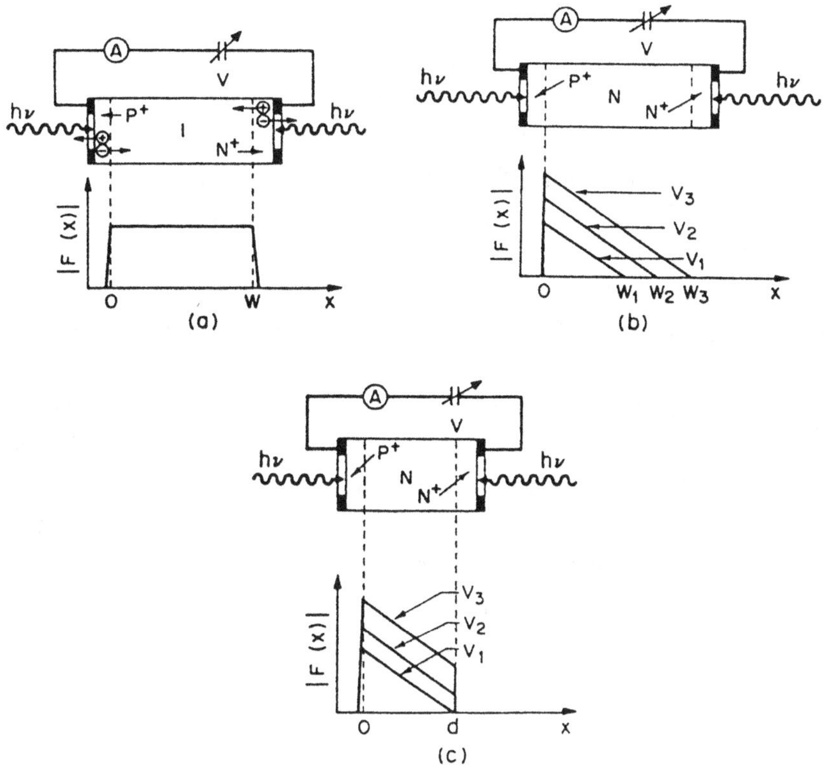

FIGURE 1 Different conceptual device structures used in ionisation rate measurements: (a) p-i-n diode; (b) abrupt p-n junction; (c) punchthrough diode (after Stillman and Wolfe [1]).

C IONISATION COEFFICIENTS IN GaAs

The first measurements in GaAs which did not assume that α_n and β_p were equal were made by Stillman et al [4], but because the above conditions were not met, these results were in error [1]. Many other measurements of α_n and β_p in GaAs, commonly based on measurements on a single diode only or several diodes from the same wafer with the same doping and structure, often gave conflicting results; this work has recently been reviewed by Capasso [5]. The results reported in 1977 by Pearsall et al [2,3] for α_n and β_p in <100>, <110>, and <111> oriented GaAs were made using the method shown in FIGURE 1(b). These measurements indicated that at low electric fields for both the <100> and <111> directions, α_n decreased rapidly with decreasing field. In contrast β_p showed no anisotropy, resulting in increasing values of $k = \beta_p/\alpha_n$. These results were interpreted in terms of the anisotropic nature of the conduction bands and the isotropic nature of the valence bands in GaAs. It was inferred that APDs with larger values of $k = \beta_p/\alpha_n$ could be obtained with <111> oriented devices. Subsequently, calculations of impact ionisation in GaAs and InP using the Monte Carlo technique by Shichijo and Hess [6] indicated that the anisotropy of α_n and β_p between the <100> and <111> directions in GaAs [6,7] and InP [8] was negligible, even at low electric fields.

In 1983, Bulman et al [9,10] reported detailed measurements on a large number of GaAs diodes with different doping levels. FIGURE 2 gives actual device structures used for these measurements. The device structure shown in FIGURE 2(a) can correspond to any of the configurations shown in FIGURE 1, depending on the thickness and doping of the various epitaxial layers and the substrate. The etched well in the substrate must be formed in order to achieve pure minority carrier injection from the substrate side, electrons for the case shown in FIGURE 2(a), since if the substrate is too thick the injected minority carriers will recombine before reaching the high field region of the device. This can result in mixed injection due to electroabsorption of radiative recombination emission in the high field region [9]. Because of the short diffusion lengths in III-V semiconductors, it is not possible to control the etched thickness of the substrate precisely enough to eliminate this effect reproducibly. The structure in FIGURE 2(b) was first used by Bulman et al [9] to enable reproducible control of all of the thicknesses by the epitaxial growth technique. For measurements on GaAs, the etch-stop layer was $Al_{0.35}Ga_{0.65}As$. The accurate control of the thickness of different regions of the device permits more precise control of the purity of the primary photocurrents. These measurements were made on 44 different devices over a wide electric field range, and the excellent reproducibility from device to device and wafer to wafer is evident in the data of FIGURE 3. There is no indication of the rapid decrease in α_n at low fields with decreasing field as previously reported by Pearsall et al [2,3]. These results are considered to be the most reliable available for GaAs; the best fit values which yield the curves for α_n and β_p shown in FIGURE 3 are given in TABLE 1.

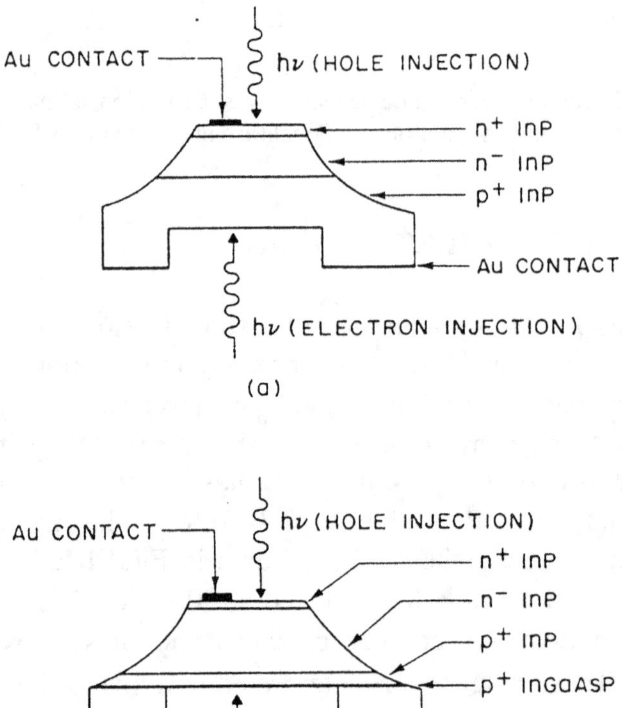

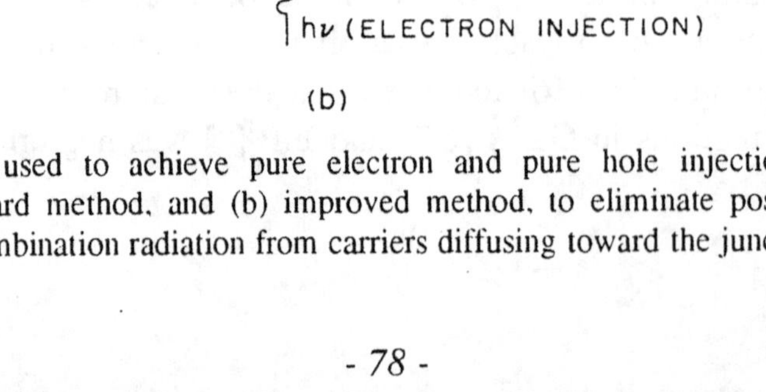

FIGURE 2 Structures used to achieve pure electron and pure hole injection in photomultiplication measurements: (a) standard method, and (b) improved method, to eliminate possible contamination from electroabsorption of recombination radiation from carriers diffusing toward the junction in the substrate (after Bulman et al [10]).

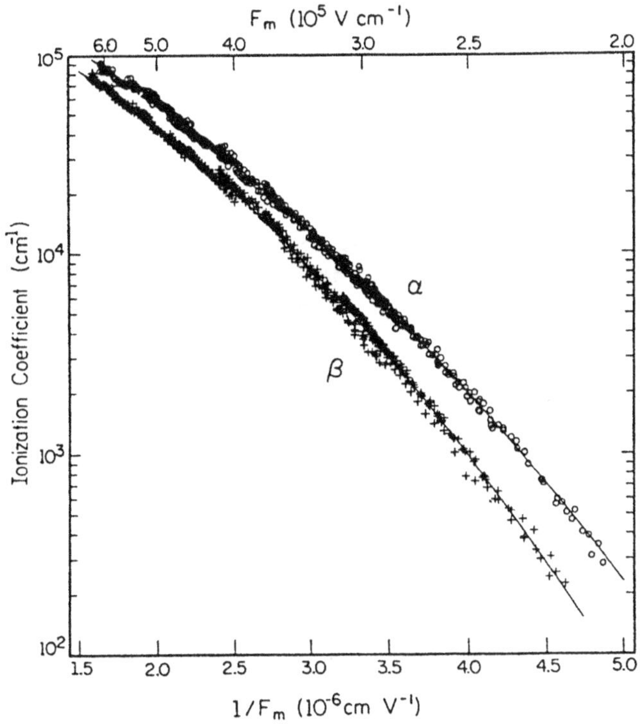

FIGURE 3 Ionisation rate versus reciprocal electric field in <100> GaAs at 300 K (from Bulman et al [9]).

TABLE 1 Least squares fit to impact ionisation coefficient data for <100> GaAs shown in FIGURE 3, with dead space corrections included, after Bulman et al [10].

$\alpha_{dsc}(F) = 1.899 \times 10^5 \exp(-(5.750 \times 10^5/F)^{1.82})$ (cm^{-1})
$\beta_{dsc}(F) = 2.215 \times 10^5 \exp(-(6.570 \times 10^5/F)^{1.75})$ (cm^{-1})

D IONISATION COEFFICIENTS IN InP

The same etch-stop structure used by Bulman et al for GaAs was also used by Cook et al [11] to measure α_n and β_p in <100> InP. Their results are shown in FIGURE 4. These results show remarkable consistency among the different devices examined and provide information over a much wider electric field range than previously reported. A set of best fit curves to the data for a typical device from each of the three wafers is given by the equations in TABLE 2 along with the applicable electric field ranges. The values of α_n and β_p vary exponentially with 1/F at low fields and with $1/F^2$ at high fields. The ratio $k = \beta_p/\alpha_n$ decreases from 4.0 to 1.3 as the electric field increases from 2.4 to 7.7 x 10^5 V/cm. These results are in fair agreement with those of Kao and Crowell [12] for low fields and with those of Umebu et al [13] for high fields, but are significantly higher than those reported by Armiento et al [14]. To check the self-consistency of the analysis, the breakdown voltage for each device was calculated by applying these results to the same structure. The calculated breakdown voltages were found to agree within a few percent for those observed experimentally [9]. These data are considered to be the most reliable results for InP. The parameters for the best fit curves to this ionisation coefficient data for <100> InP, shown by the curves in FIGURE 4, are given in TABLE 2.

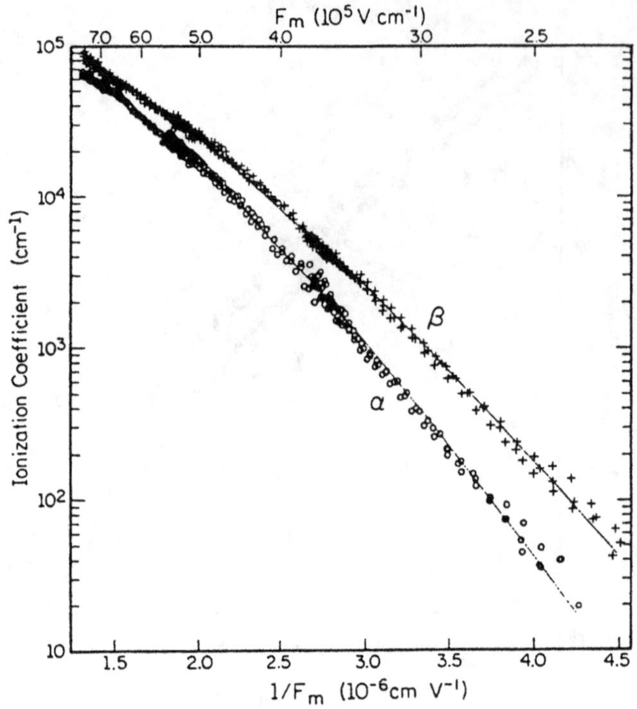

FIGURE 4 Ionisation rates versus reciprocal electric field in <100> InP at 300 K (from Cook et al [11]).

TABLE 2 Best fit curves to the ionisation coefficient data for three <100> InP devices with depletion layer doping levels of 1.2 x 10^{15}, 3.0 x 10^{16}, and 1.2 x 10^{17} cm^{-3}, shown in FIGURE 4, from Cook et al [11].

Doping level (cm^{-3})	Field range 10^5 (V/cm)	α (cm^{-1})	β (cm^{-1})
1.2 x 10^{15}	2.4 - 3.8	1.12 x 10^7 exp (-3.11 x 10^6/F)	4.79 x 10^6 exp (-2.55 x 10^6/F)
3.0 x 10^{16}	3.6 - 5.6	2.93 x 10^6 exp (-2.64 x 10^6/F)	1.62 x 10^6 exp (-2.11 x 10^6/F)
1.2 x 10^{17}	5.3 - 7.7	2.32 x 10^5 exp (-7.16 x 10^{11}/F^2)	2.48 x 10^5 exp (-6.23 x 10^{11}/F^2)

Tabatabaie et al [19] extended these measurements for <111> InP. Devices from three different wafers with active layer doping concentrations of 2 x 10^{14}, 2 x 10^{15}, and 4 x 10^{16} cm^{-3} were measured, and the results are summarised in FIGURE 5. This figure shows the dependence of α and β on the inverse electric field using a semi-logarithmic scale. Impact ionisation coefficients for (100) InP are also plotted on the same graph for comparison. The ionisation rates along the two orientations differ by only a five percent uncertainty in the measured value of the peak electric field which causes a slight shift along the horizontal coordinate. Isotropic ionisation coefficients were also observed in InP by Armiento and Groves in their comparisons of measurements on the <100> and the <110> orientations [15]. The rapid falloff of α_n at low electric fields observed in similar measurements in <111> GaAs by Pearsall et al is not evident in these data for InP; the ionisation coefficients in InP are isotropic for the range of the electric field values which are used for practical APDs. The lack of orientation dependence of β/α indicates that ballistic electrons do not account for a significant percentage of the ionising carriers reaching threshold, in agreement with Monte Carlo simulations of Brennan and Hess [7,8]. Furthermore, as far as the β/α ratio is concerned, there is no preferred orientation for the fabrication of low noise APDs. However, depending on the structure and fabrication techniques, technological considerations may strongly favour the (100) orientation.

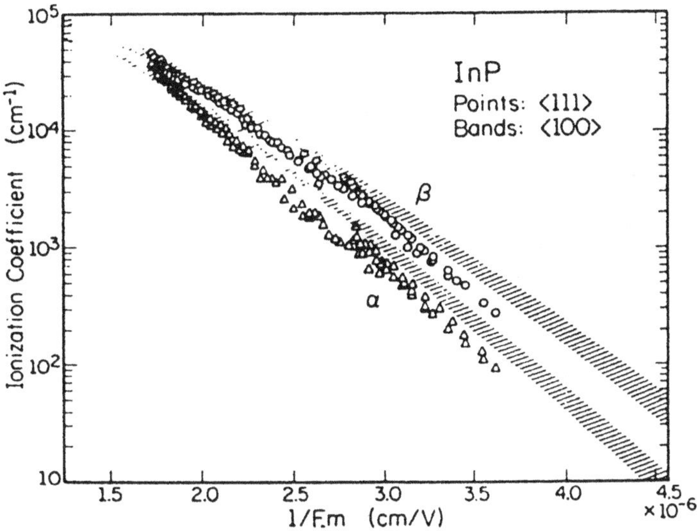

FIGURE 5 Comparison of the (111) and (100) oriented impact ionisation coefficients as functions of the inverse electric field (from Tabatabaie et al [19]).

E IONISATION COEFFICIENTS IN $Ga_{0.47}In_{0.53}As$

The impact ionisation coefficients have been measured for GaInAs by Pearsall [16] and Osaka et al [17] using a structure similar to that in FIGURE 2(a), except that, since an InP substrate was used, pure injection of minority carriers from the back side of the device could be measured by using incident photons that were not absorbed by the substrate but were strongly absorbed by the $Ga_{0.47}In_{0.53}As$. As a result, it was not necessary to etch a well in the substrate to obtain pure injection. The results of Pearsall are shown in FIGURE 6, and the results of Osaka et al are shown in FIGURE 7. Both measurements show that α_n is greater than β_p, but Pearsall's data are about one order of magnitude larger than the results of Osaka et al. Both sets of data give nearly the same ratio α_n/β_p. Although Osaka et al comment that 'at this stage it is not known which data are more realistic', Osaka and Mikawa used the values from [17] in a study of the noise of InP/GaInAsP/GaInAs APDs [18]; they obtained very good agreement between their calculated excess noise factors and experimentally measured excess noise factors for planar In/GaInAsP/GaInAs APDs. This good agreement gives strong support for the validity of the ionisation coefficients determined by Osaka et al [17]. Pearsall [16] also noted that measurements of the ionisation rate ratio with the electric field oriented along the <111> directions have a larger ratio of $\alpha_n/\beta_p = 5$. However, based on the discussion of the lack of orientation dependence of α_n and β_p in GaAs and InP discussed above, it seems unlikely that there is actually any anisotropy in the ionisation coefficients for electric fields in the <100> and <111> directions in $Ga_{0.47}In_{0.53}As$.

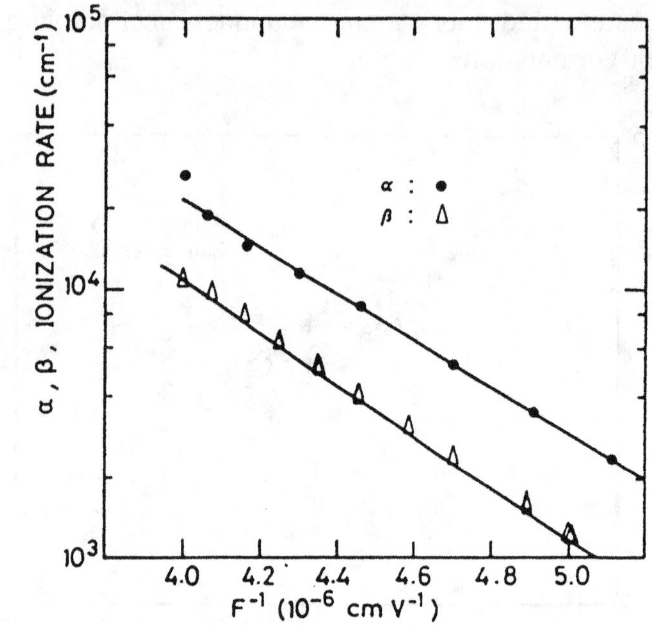

FIGURE 6 Measured impact ionisation rates for electrons (•) and holes (Δ) at 295 K in $Ga_{0.47}In_{0.53}As$ as a function of the inverse maximum electric field which occurs at the junction (from Pearsall [16]).

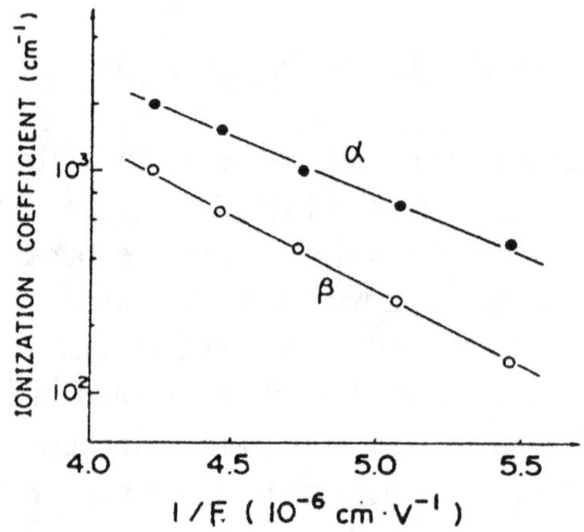

FIGURE 7 Derived electron and hole ionisation coefficients in $Ga_{0.47}In_{0.53}As$ by Osaka et al [17].

REFERENCES

[1] G.E. Stillman, C.M. Wolfe [in *Semiconductors and Semimetals*, vol.12, Eds R.K. Willardson, A.C. Beer (Academic Press Inc., New York, 1977) p.291-393]

[2] T.P. Pearsall, R.E. Nahory, J.R. Chelikowsky [*Phys. Rev. Lett. (USA)* vol.39 (1977) p.295]

[3] T.P. Pearsall, F. Capasso, R.E. Nahory, M.A. Pollack, J.R. Chelikowsky [*Solid-State Electron. (UK)* vol.21 (1978) p.297]

[4] G.E. Stillman, C.M. Wolfe, A.G. Foyt, W.T. Lendley [*Appl. Phys. Lett. (USA)* vol.24 (1974) p.8]

[5] T. Capasso [*Semicond. Semimet. (USA)* vol.22 pt.D (Academic Press Inc., New York, 1985) p.1-172]

[6] H. Shichijo, K. Hess [*Phys. Rev. B (USA)* vol.23 (1981) p.4197]

[7] K.F. Brennan, K. Hess [*Solid-State Electron. (UK)* vol.27 (1984) p.347]

[8] K.F. Brennan, K. Hess [*Phys. Rev. B (USA)* vol.39 (1984) p.5581]

[9] G.E. Bulman, V.M. Robbins, K.F. Brennan, K. Hess, G.E. Stillman [*IEEE Electron Device Lett. (USA)* vol.4 (1983) p.181]

[10] G.E. Bulman, V.M. Robbins, G.E. Stillman [*IEEE Trans. Electron Devices (USA)* vol.32 (1985) p.2454]

[11] L.W. Cook, G.E. Bulman, G.E. Stillman [*Appl. Phys. Lett. (USA)* vol.40 (1982) p.589]

[12] C.W. Kao, C.R. Crowell [*Solid-State Electron. (UK)* vol.23 (1980) p.881]

[13] I. Umebu, A.N.M.M. Choudhury, P.N. Robson [*Appl. Phys. Lett. (USA)* vol.36 (1980) p.302]

[14] C.A. Armiento, S.H. Groves, C.E. Hurwitz [*Appl. Phys. Lett. (USA)* vol.35 (1979) p.333]

[15] C.A. Armiento, S.H. Groves [*Appl. Phys. Lett. (USA)* vol.43 (1983) p.198]

[16] T.P. Pearsall [*Appl. Phys. Lett. (USA)* vol.36 (1980) p.218]

[17] F. Osaka, T. Mikawa, T. Kaneda [*IEEE J. Quantum Electron. (USA)* vol.21 (1985) p.1326]

[18] F. Osaka, T. Mikawa [*IEEE J. Quantum Electron. (USA)* vol.22 (1986) p.471]

[19] N. Tabatabaie, V.M. Robbins, N. Pan, G.E. Stillman [*Appl. Phys. Lett. (USA)* vol.46 (1985) p.182]

3.4 Heterojunctions of InGaAs and band offsets

S. Adachi

January 1993

A INTRODUCTION

One of the most important parameters for the design and analysis of heterojunction and quantum-well electronic and optoelectronic devices is the heterojunction band offset. The band offset is a consequence of the difference between the energy gaps of two semiconductors. The energy-gap difference is distributed between a conduction-band offset ΔE_c and a valence-band offset ΔE_v. In a type I (straddling lineup), we obtain

$$\Delta E_g = \Delta E_c + \Delta E_v \tag{1}$$

while for type II (broken-gap and staggered lineups) the relation is given by

$$\Delta E_g = |\Delta E_c + \Delta E_v| \tag{2}$$

where ΔE_g is the bandgap difference.

A large amount of experimental and theoretical work has been accumulated to address the heterojunction band offsets and related phenomena in InGaAs-based heterojunction systems. We will see in Section B some results of the lattice-matched heterojunction system. Results of the lattice-mismatched system are presented in Section C. Finally, the conclusions obtained from this contribution are briefly summarised in Section D.

B LATTICE-MATCHED HETEROJUNCTION SYSTEM

B1 $In_{1-x}Ga_xAs_yP_{1-y}$/InP($In_{0.53}Ga_{0.47}As$)

The lattice constant a_0 is known to obey Vegard's law well, that is, to vary linearly with alloy composition, and such is the case for $In_{1-x}Ga_xAs$ and $In_{1-x}Ga_xAs_yP_{1-y}$ alloys [1]. Using this law, the lattice-matching relation between the composition fractions x and y for $In_{1-x}Ga_xAs_yP_{1-y}$ lattice-matched to InP can be written as [1]

$$x = \frac{0.1896y}{0.4176 - 0.0125y} \quad (0 \le y \le 1.0) \tag{3}$$

Eqn (3) gives the end-point compositions x = 0, y = 0 for InP and x = 0.468, y = 1.0 for $In_{0.532}Ga_{0.468}As$ ($In_{0.53}Ga_{0.47}As$). The experimental techniques used for determining conduction-band offset ΔE_c and valence-band offset ΔE_v can be broadly divided into two distinct categories: optical and electrical. There have been several reports in the $In_{1-x}Ga_xAs_yP_{1-y}$($In_{0.53}Ga_{0.47}As$)/InP heterojunction system [2-11]. In TABLE 1 we summarise the results obtained from these works.

TABLE 1 Band-offset ratio ΔE_c:ΔE_v in the $In_{1-x}Ga_xAs_yP_{1-y}$/InP heterojunction.
PL = Photoluminescence, C-V = Capacitance-voltage technique,
PLE = Photoluminescence excitation spectroscopy,
PC = Photoconductive response.

ΔE_c:ΔE_v	Technique	Ref
67:33	PL	[2]
39:61	C-V	[5,7]
65:35	PL	[6]
37:63	PLE	[8]
36:64	PC	[9]
43:57	PLE	[10,11]

It is clear from TABLE 1 that there is a considerable disagreement in the band offsets determined by these authors. Langer et al [12] have used a fact that transition-metal impurity levels in semiconductors may serve as a reference in bandgap alignment in various semiconductor heterojunctions, and obtained the valence-band offset ΔE_v at 77 K in the $In_{1-x}Ga_xAs_yP_{1-y}$/InP heterojunction that can be written as [1]

$$\Delta E_v(y) = 502y - 152y^2 \text{ meV} \tag{4a}$$

The corresponding ΔE_c can be expressed, using the 77 K E_g (bandgap energy) versus y data in [13], as [1]

$$\Delta E_c(y) = 268y + 3y^2 \text{ meV} \tag{4b}$$

The band offsets in the $In_{1-x}Ga_xAs_yP_{1-y}/In_{0.53}Ga_{0.47}As$ heterojunction system can then be written as

$$\Delta E_v(y) = 350 - 502y + 152y^2 \text{ meV} \tag{5a}$$

$$\Delta E_c(y) = 271 - 268y + 3y^2 \text{ meV} \tag{5b}$$

Eqns (4) and (5) provide the band-offset ratio of ΔE_c:ΔE_v = 44:56 in the $In_{0.53}Ga_{0.47}As$/InP heterojunction (y = 1.0). Since the temperature variations of the bandgap energies are very similar between $In_{1-x}Ga_xAs_yP_{1-y}$ and $In_{1-x}Ga_xAs$ (InP) [1], the band-offset values and offset ratio in these heterojunction systems can be successfully assumed to be independent of temperature.

B2 $In_{0.53}Ga_{0.47}As/InP$

The $In_{0.53}Ga_{0.47}As/InP$ heterojunction system has broader interest since it is used for various optoelectronic and transport device applications. The band offsets in this heterojunction system have been studied both experimentally [9,14-33] and theoretically [34-38]. In TABLE 2 we show the band offsets (ΔE_c, ΔE_v) and offset ratio (ΔE_c:ΔE_v) in this heterojunction system obtained experimentally by different authors.

TABLE 2 Experimentally determined band offsets (ΔE_c, ΔE_v) and offset ratio (ΔE_c:ΔE_v) in the $In_{0.53}Ga_{0.47}As/InP$ heterojunction. PC = Photoconductive response, I-V = Current-voltage technique, C-V = Capacitance-voltage technique, AS = Admittance spectroscopy, OA = Optical absorption, PL = Photoluminescence, PLE = Photoluminescence excitation spectroscopy, ER = Electroreflectance, IP = Internal photoemission, XPS = X-ray photoemission spectroscopy, UPS = Ultraviolet photoemission spectroscopy.

ΔE_c (meV)	ΔE_v (meV)	ΔE_c:ΔE_v	Technique	Ref
219	390	36:64	C-V	[9]
220 ± 20	400		C-V, I-V, PC	[14]
30 - 190			C-V, AS	[15]
		50:50	OA, PL	[17]
230	380	38:62	PL, PLE, PC, ER	[18]
	0	100:0	C-V	[19]
250 ± 10	346 ± 10	42:58	AS	[21]
265	325	45:55	OA, PL	[22]
235 ± 20		38:62	OA	[23]
230 ± 40			C-V	[24]
220			C-V	[25]
210 ± 20			AS	[26]
203 ± 15[a]			IP	[27]
	250	58:42	XPS	[28]
217 ± 5	388	36:64	C-V	[29]
185 ± 5			I-V	[31]
	340	43:57	XPS	[32]
	440 ± 50[b], 260 ± 50[c]		UPS	[33]

a: At 300 K
b: InGaAs grown on InP
c: InP grown on InGaAs

As can be seen in TABLE 3, the offset ratio ΔE_c:ΔE_v determined widely spreads from 100:0 to 36:64. Using the capacitance-voltage (C-V) method proposed by Kroemer et al [39], information regarding the heterojunction position, fixed charge (σ_i) and band offset ΔE_c can all be easily determined. However, there are several phenomena which lead to inaccuracies in determining ΔE_c and related parameters by C-V methods [25]. Values derived from C-V methods, therefore, do not agree with each other [9,14-16,19,20,24,25,29]. Since the energy-gap difference ΔE_g between $In_{0.53}Ga_{0.47}As$ and InP shows no strong temperature dependence [1], the offset values can be considered to be independent of temperature. It has, however, been observed [15,16,20,24] that both σ_i and ΔE_c measured by C-V techniques apparently depend strongly on both temperature and measurement modulation frequency, contradicting the fact that ΔE_c is an inherent property of the heterojunction. In order to study this problem, the effects of shallow as well as deep traps have been considered by Leu and Forrest [25] by analytical as well as numerical simulation methods. They showed that the original equations of Kroemer et al [39] for the determination of σ_i and heterojunction diffusion potential appear to be valid only for heterojunctions with a low density of shallow interface traps. They also showed that at low measurement frequency, the measured value of ΔE_c is nearly independent of σ_i, and this value is near to the exact value of ΔE_c measured in the absence of interface traps. However, at high measurement frequency or low temperature, the measured ΔE_c decreased with increasing σ_i or increasing trap energy depth from the conduction-band minimum.

Studies of the valence-band offset ΔE_v at an $In_{0.53}Ga_{0.47}As$/InP heterostructure by X-ray have been published [32]. The value reported for InP overlayers on top of thick $In_{0.53}Ga_{0.47}As$ layers is 340 ± 50 meV. The results obtained by ultraviolet photoemission spectroscopy on both direct and inverse $In_{0.53}Ga_{0.47}As$/InP interfaces have also been reported [33]. The measured values were 440 ± 50 meV for the direct ($In_{0.53}Ga_{0.47}As$ grown on InP) interface and 260 ± 50 meV for the inverse interface. This band-offset asymmetry is attributed to a difference in the chemical nature of the two interfaces, deduced from the detailed analysis of the core-level spectra.

The band offsets in the $In_{0.53}Ga_{0.47}As$/InP heterojunction have been calculated theoretically by several authors [34-38]. We summarise in TABLE 3 these results. The calculated values are found to be in reasonable agreement with most of the available experiments (see TABLE 2).

TABLE 3 Theoretically determined band offsets (ΔE_c, ΔE_v) and offset ratio (ΔE_c:ΔE_v) in the $In_{0.53}Ga_{0.47}As$/InP heterojunction. PP = Pseudopotential method, LDA = Local-density functional approach, QW = Quantum-well energy level, NL = Neutrality level, DL = Dielectric level, MS = Model-solid approach, SE = Semiempirical estimation.

ΔE_c (meV)	ΔE_v (meV)	ΔE_c:ΔE_v	Method	Ref
	310 ± 0		PP	[34]
	410 ± 50		LDA	[35,37]
		40:60	QW	[36]
	260		NL	[37]
	250		DL	[37]
	350		MS	[37]
248			SE	[38]

B3 $In_{0.53}Ga_{0.47}As/In_{0.52}Al_{0.48}As$

Like $In_{0.53}Ga_{0.47}As$, $In_{0.52}Al_{0.48}As$ can be grown lattice-matched on InP. The bandgap discontinuity or difference ΔE_g between $In_{0.53}Ga_{0.47}As$ ($E_g = 0.75$ eV) and $In_{0.52}Al_{0.48}As$ ($E_g = 1.47$ eV) is about 0.72 eV [1]. This large bandgap discontinuity makes the heterojunction system $In_{0.53}Ga_{0.47}As/In_{0.52}Al_{0.48}As$/InP attractive for possible fabrication of optoelectronic and hot-electron injection devices. The ΔE_c values between $In_{0.53}Ga_{0.47}As$ and

$In_{0.52}Al_{0.48}As$ have been reported by several authors [40-49]. These results are summarised in TABLE 4. The experimental data suggest that $\Delta E_c \sim 500$ meV, $\Delta E_v \sim 200$ meV, and ΔE_c:ΔE_v ~ 70:30. The ΔE_v value calculated by Hybertsen [37] is about 170 meV, in reasonable agreement with the experimental data.

TABLE 4 Experimentally determined band offset (ΔE_c) and offset ratio (ΔE_c:ΔE_v) in the $In_{0.53}Ga_{0.47}As/In_{0.52}Al_{0.48}As$ heterojunction. I-V = Current-voltage technique, C-V = Capacitance-voltage technique, PL = Photoluminescence, PLE = Photoluminescence excitation spectroscopy, EBER = Electron-beam electroreflectance.

ΔE_c (meV)	ΔE_c:ΔE_v	Technique	Ref
520		I-V	[40]
500 ± 50	71(±7):29	C-V	[41]
520	70:30	PL	[42]
500	71:29	PLE	[44]
510 ± 40	70:30	I-V	[45]
553 ± 20	72:28	I-V	[46]
510 ± 20		PLE	[47]
519		I-V, C-V	[48]
	65(±0.1):35	PL, PLE, EBER	[49]

B4 $In_{0.53}Ga_{0.47}As/In_{0.52}(Al_xGa_{1-x})_{0.48}As$

Sugiyama et al [46] have studied the conduction-band discontinuities ΔE_c of $In_{0.53}Ga_{0.47}As/In_{0.52}(Al_xGa_{1-x})_{0.48}As$ heterostructures by measuring the current-voltage characteristics through the barrier structure as a function of temperature in the range 77 - 300 K. They found that the ΔE_c varies almost linearly with Al fraction x ($0 \leq x \leq 1.0$)

$$\Delta E_c(x) = 0.53x \text{ eV} \tag{6}$$

They also found that the band-offset ratio in this heterojunction system does not depend on Al fraction x:

$$\Delta E_c:\Delta E_v = 72(\pm 7):28 \tag{7}$$

The conduction-band discontinuities ΔE_c of $In_{0.53}Ga_{0.47}As/In_{0.52}(Al_xGa_{1-x})_{0.48}As$ heterostructures have also been determined by Sandhu et al [47] using photoluminescence and photoluminescence excitation spectroscopies. These authors also conclude that the offset ratio is independent of x and is given by

$$\Delta E_c:\Delta E_V = 73(+3):27 \tag{8}$$

B5 $In_{0.52}Al_{0.48}As/InP$

The $In_{0.52}Al_{0.48}As/InP$ interface is the least well characterised [37,50-52]. The staggered (type II) alignment is generally agreed on. Two experiments provide values for the band offsets: low-temperature luminescence from a series of superlattices with varying period implied the values $\Delta E_c \sim 390$ meV and $\Delta E_v \sim 290$ meV with an effective bandgap of 1.06 eV [51]; X-ray photoemission gave $\Delta E_c \sim 250$ meV and $\Delta E_v \sim 160$ meV [52]. First-principles calculation within the local-density functional approach [37] gave $\Delta E_v = 250$ meV, in reasonable agreement with the experimental data [51].

C LATTICE-MISMATCHED HETEROJUNCTION SYSTEM

Semiconductor layers may be grown pseudomorphically on a substrate of different lattice constant provided the layer is below a critical thickness, although the value and indeed the nature of the critical thickness is still the subject of debate. Strained-layer structures offer two broad classes of advantage over lattice-matched systems: firstly, new bandgaps are accessible through the greater freedom in choice of alloy compositions; and secondly, the strain splitting of the valence band at the Γ point allows for valence-band engineering with the potential for the highest valence-band state to have light-hole character in the plane of the well [1]. The band offset in strained-layer systems depends not only on the heterojunction semiconductors involved but also on the amount of mismatch strain at the interface. The strain field in the strained-heterojunction wells and barriers can, therefore, lead to several plausible configurations of the quantising superpotentials. The large tetragonal elastic strains in the strained layers can produce marked effects on the electronic properties through the deformation-potential interactions [1]. The hydrostatic component of the strain causes shifts in the bulk energy levels of the layers, and the shear component causes splittings of certain degenerate valence-band levels. The energy levels of the layers are, therefore, determined including the strain modification of the layer structure.

C1 $In_{1-x}Ga_xAs/GaAs$

The lattice-mismatched $In_{1-x}Ga_xAs/GaAs$ heterojunction system is an important material system for high-speed and optoelectronic device applications. There are many reports on the experimental determination of the band offsets in this heterojunction system [53-83]. We summarise in TABLE 5 the ratios of the ΔE_c to ΔE_v (heavy holes) determined experimentally.

There appears to be a controversy on the band alignments in the strained-layer $In_{1-x}Ga_xAs/GaAs$ heterojunction system. No composition dependence of the offset ratio has been reported in [62,67,70,74,75,78,82,83], while the composition-dependent offset values have been reported in [65,68,72,77,80].

TABLE 5 Experimentally determined offset ratio, $\Delta E_c:\Delta E_v$ (heavy holes), in the $In_{1-x}Ga_xAs/GaAs$ heterojunction. I-V = Current-voltage technique, C-V = Capacitance-voltage technique, PC = Photoconductive response, OA = Optical absorption, BT = Bipolar transistor characteristic, LS = Light scattering, PL = Photoluminescence, PR = Photoreflectance, PLE = Photoluminescence excitation spectroscopy.

$\Delta E_c:\Delta E_v$	x	Technique	Ref
83:17	0	I-V, C-V, PC	[53]
68:32	0.83, 0.85	OA	[55]
35:65	0.92, 0.95, 0.98	BT	[57]
70:30	0.807, 0.85, 0.87	OA	[58]
60:40(±4)	0.95	LS	[60,63]
83(±6):17	x>0.6	PC, PL	[62]
90:10	0	PC, PL	[62]
70:30(±5)	0.88	PR	[64]
62(±8):38	0.88	PL	[65]
36(±12):64	0.96	PL	[65]
50:50	0.85, 0.86	PL	[66]
70:30	0.80 - 0.91	OA, PL	[67]
70:30	0.86	PC	[68]
40:60	0.91	PC	[68]
58:42(±3)	0.75 - 0.95	PL, PR	[69]
65:35	0.7 - 1.0	OA	[70]
70:30	0.81, 0.85	PR	[71]
77:23	x>0.85	OA, PLE	[72]
83:17	0.85	PL	[73]
60:40	0.75, 0.79, 0.82	OA, LS, PL	[74]
61(±1.5):39	0.747 - 0.927	OA, PLE	[75]
52:48(±5)	0.87	OA, PL	[76]
60:40	0.72, 0.82, 0.91	PL	[78]
65(±7):35	0.79	PR	[81]
60:40	0.5, 0.86, 0.90	PL	[82]
60:40(±2)	0.77, 0.85, 0.91	PL	[83]

At a type I interface, both electrons and holes near the junction tend to migrate to the smaller-gap semiconductor. Potential barriers in this case exist for both electrons and holes in going from the smaller-gap semiconductor to the larger-gap semiconductor. At a type II interface, on the other hand, electrons and holes tend to separate, each going into a different material. All reports suggest that the heavy holes are confined in the $In_{1-x}Ga_xAs$ well region (i.e., type I). However, there are different opinions on whether the light-hole valence-band state is localised in the $In_{1-x}Ga_xAs$ well [60,63,74,75] or in the GaAs barriers as in a type II heterostructure [55,58,71,72]. This controversy may come from the fact that the light-hole state is weakly bound in the $In_{1-x}Ga_xAs/GaAs$ well or barrier layer at least for the lower In concentrations [75].

There are no reported values for the band-offset values in strained $In_{1-x}Ga_xAs$/GaAs for lower x values, except for x = 0, that is, InAs on GaAs. The InAs/GaAs heterojunction system has about 7% lattice mismatch and the corresponding critical thickness for a generation of misfit dislocations is as thin as 2 monolayers. Using X-ray photoelectron spectroscopy, Kowalczyk et al [84] have determined the valence- and conduction-band offsets in the InAs/GaAs(100) heterojunction to be 0.90 ± 0.07 and 0.17 ± 0.07 eV, respectively. These values correspond to the offset ratio of $\Delta E_c:\Delta E_v$ = 84(±6):16. Kowalczyk et al used in their study an InAs/GaAs single heterojunction with a 20 Å thick InAs layer, which is much thicker than the critical thickness for dislocation generation. Hence, the lattice strain is considered to be relaxed in their sample. Hirakawa et al [85] have recently studied heterojunction band offsets at highly strained InAs/GaAs(100) interfaces, using in situ X-ray photoemission spectroscopy with an emphasis on the effects of lattice strain. They examined two extreme cases: an InAs layer pseudomorphically grown on a GaAs substrate and a GaAs layer grown on an InAs substrate. The valence-band offsets were found to be very different: 0.53 eV for an InAs/GaAs interface (type I) and -0.16 eV for a GaAs/InAs interface (type II). These results clearly indicate a large effect of internal strain on the valence-band offset in this heterojunction system.

Theoretical studies on the band offsets in the $In_{1-x}Ga_xAs$(InAs)/GaAs heterojunction system have also been carried out by several authors [38,86-90].

C2 $In_{1-x}Ga_xAs/Al_yGa_{1-y}As$

The band-offset problems in the pseudomorphic $In_{1-x}Ga_xAs/Al_yGa_{1-y}As$ heterostructure grown on GaAs have been studied experimentally by several groups [61,75,83,91-96]. We summarise in TABLE 6 the results of these studies.

TABLE 6 Experimentally determined offset ratio, $\Delta E_c:\Delta E_v$ (heavy holes), in the $In_{1-x}Ga_xAs/Al_yGa_{1-y}As$ heterojunction. PL = Photoluminescence, PC = Photoconductive response, OA = Optical absorption, PLE = Photoluminescence excitation spectroscopy, RT = Resonant tunnelling, TP = Transport property.

$\Delta E_c:\Delta E_v$	x	y	Technique	Ref
67(±8):33	0.65 - 0.88	0.20 - 0.35	PL, PC	[61,92]
62:38	0.86	0.02	OA, PLE	[75]
100:0	0.8	0.25	RT	[91]
76:24	0.85	0.15	TP	[94]
(ΔE_c = 300±10 meV)	0.85	0.18	TP	[95]
(ΔE_v = 157±25 meV)	0.83	0.32	PL	[96]

Wilkinson et al [83] have found from photoluminescence under high hydrostatic pressure that the conduction-band offset linearly increases with increasing In concentration from $\Delta E_c/\Delta E_g$ = 30% (0% In) to 33.5% (20% In) for y = 0.10 and 0.24. Linear increase of $\Delta E_c/\Delta E_g$ with increasing In concentration has also been found by Debbar et al [93] from thermionic-emission-current measurements in $In_{11-x}Ga_xAs/Al_{0.2}Ga_{0.8}As$ quantum wells. These data show that the percentage band offset $\Delta E_c/\Delta E_g$ increases from 62% for 1 - x = 0.07 to 70% at 1 - x = 0.18. A linear interpolation of the data also leads to $\Delta E_c/\Delta E_g$ = 58% at x = 0, which is close to the generally accepted value (Section C1).

Arent [97] has calculated the valence-band offsets for pseudomorphically strained $In_{1-x}Ga_xAs/Al_yGa_{1-y}As$ heterointerfaces within the virtual-crystal approximation as a function of both In and Al contents. The strained band-offset ratios were found to be variable and nonlinear over the entire composition range and vary dramatically with Al content. His calculation also showed that type-II heterostructures formed with the strained light-hole valence band are possible, but only in the case of $In_{1-x}Ga_xAs/GaAs$ (y = 0) heterostructures with In content x > ~ 0.40.

C3 $In_{1-x}Ga_xAs/In_{1-y}Ga_yAs$

The strained-layer $In_{1-x}Ga_xAs/In_{1-y}Ga_yAs$ superlattices (x = 0.59/y = 0.37 and x = 0.54/y = 0.40) grown on (100)InP have been studied by Quillec et al [98]. The photoluminescence experiments performed by them, however, do not allow us to deduce the band offset precisely. Priester et al [86] have considered theoretically the bandgap alignments of such superlattices and obtained reasonable agreement with Quillec's experimental results.

C4 $In_{1-x}Ga_xAs/InP$

Superlattice structures of $In_{1-x}Ga_xAs/InP$ with x ranging from 0 to 1.0 are an excellent system for the study of the strain effects on the energy-band lineups because the magnitude, as well as sign, of strain can be systematically varied between +3.5% (x = 0) to -3.2% (x = 1.0). Previous work on $In_{1-x}Ga_xAs$ grown on GaAs (Section C1) does not permit the same degree of flexibility. Results from optical and electrical investigations show [99] that most of the change in bandgap discontinuity between the InP and $In_{1-x}Ga_xAs$ components of the strained-layer superlattice takes place in the conduction band. The conduction-band offsets determined from admittance spectroscopy are 315 ± 25 meV for x = 0.31 and 175 ± 25 meV for x = 0.63 [26]. Also, a change in the conduction-band offset sign has been observed near x = 0.8 where the strained-layer superlattice undergoes a type transition [99].

Theoretical studies on the bandgap alignments for pseudomorphic $In_{1-x}Ga_xAs/InP$ heterostructures have also been carried out by Hrivnák [38] and People [100]. It has been confirmed [38] that the transition from type-I to type-II superlattice occurs for strained layers of this heterojunction system (x > 0.8).

C5 $In_{1-x}Ga_xAs/In_{0.52}Al_{0.48}As$

Composition dependence of the conduction-band offset ΔE_c in an $In_{1-x}Ga_xAs/In_{0.52}Al_{0.48}As$ pseudomorphic heterostructure has been measured as a function of GaAs mole fraction over the range $0.36 \leq x \leq 0.56$ using current-voltage and capacitance-voltage measurements [48]. The results show a monotonic decrease of effective ΔE_c with increasing x according to $\Delta E_c \sim 0.638 - 0.254x$ for $x \geq 0.46$ and an abrupt shift to $\Delta E_c \sim 0.831 - 0.487x$ for $x \leq 0.42$.

D CONCLUSION

The $In_{1-x}Ga_xAs$-based heterojunction system can be divided into two distinct categories: lattice-matched (unstrained-layer) and lattice-mismatched (strained-layer) systems. We reviewed here the band offsets in both lattice-matched ($In_{1-x}Ga_xAs_yP_{1-y}/InP(In_{0.53}Ga_{0.47}As)$,

$In_{0.53}Ga_{0.47}As/InP$, $In_{0.53}Ga_{0.47}As/In_{0.52}Al_{0.48}As$, $In_{0.53}Ga_{0.47}As/In_{0.52}(Al_xGa_{1-x})_{0.48}As$ and $In_{0.52}Al_{0.48}As/InP$) and lattice-mismatched systems ($In_{1-x}Ga_xAs/GaAs$, $In_{1-x}Ga_xAs/Al_yGa_{1-y}As$, $In_{1-x}Ga_xAs/In_{1-y}Ga_yAs$, $In_{1-x}Ga_xAs/InP$ and $In_{1-x}Ga_xAs/In_{0.52}Al_{0.48}As$).

The usual method for band-offset determination implies fitting of the experimental and calculated energies of optical transitions between the valence- and conduction-band sublevels in the quantum wells. However, a low sensitivity of the transition energies to the band offset makes its determination rather ambiguous. A choice of model parameters also largely influences the offset value. The resultant offset values varied widely according to different authors.

The band offset is principally not dependent on doping concentration. Since the temperature variations of the bandgap energies are very similar among the III-V materials, the band offsets and offset ratios in the III-V heterojunction systems can be successfully assumed to be independent of temperature. Orientation dependence of the band offsets in $Al_xGa_{1-x}As/GaAs$ heterojunctions were studied both theoretically and experimentally. All these studies indicated that the band offset for this heterojunction system is independent of crystal orientation. Also, there are different opinions about $Al_xGa_{1-x}As/GaAs$ heterojunctions as to whether the band offset is commutative (i.e., independent of growth sequence) or not. No detailed study has, however, been carried out on $In_{1-x}Ga_xAs$-based heterojunctions, except $In_{0.53}Ga_{0.47}As/InP$.

A potentially important heterojunction formation effect is lattice strain, and such is especially the case for the strained-layer heterojunction system. Although strain-induced modifications of bulk energy-band structures are relatively well characterised for strained semiconductor heterolayers, the effects of strain on band offsets are not yet well understood. The strain dependence of the band offsets is directly relevant to the interpretation of experimental data, thus the study of basic heterojunction physics, and also the modelling of heterojunction devices. Despite significant progress, there are still many unresolved or poorly understood issues concerning the bandgap alignments in strained-layer heterojunctions. Further experimental and theoretical studies will be necessary to better understand the energy-band offsets not only in strained-layer heterojunctions but also in unstrained-layer heterojunctions.

REFERENCES

[1] S. Adachi [*Physical Properties of III-V Semiconductor Compounds: InP, InAs, GaAs, GaP, InGaAs, and InGaAsP* (Wiley-Interscience, New York, 1992)]

[2] R. Chin, N. Holonyak Jr., S.W. Kirchoefer, R.M. Kolbas, E.A. Rezek [*Appl. Phys. Lett. (USA)* vol.34 (1979) p.862-4]

[3] Y. Takanashi, Y. Horikoshi [*Inst. Phys. Conf. Ser. (UK)* no.63 (1982) ch.6 p.263-8]

[4] M. Nakao, S. Yoshida, S. Gonda [*Solid State Commun. (USA)* vol.49 (1984) p.663-6]

[5] S.R. Forrest, P.H. Schmidt, R.B. Wilson, M.L. Kaplan [*Appl. Phys. Lett. (USA)* vol.45 (1984) p.1199-201]

[6] P.E. Brunemeier, D.G. Deppe, N. Holonyak Jr. [*Appl. Phys. Lett. (USA)* vol.46 (1985) p.755-7]

[7] S.R. Forrest, P.H. Schmidt, R.B. Wilson, M.L. Kaplan [*J. Vac. Sci. Technol. B (USA)* vol.4 (1986) p.37-44]

[8] E.A. Montie, P.J.A. Thijs, G.W. `tHooft [*Appl. Phys. Lett. (USA)* vol.53 (1988) p.1611-13]

[9] M. Zachau et al [*Superlattices Microstruct. (UK)* vol.5 (1989) p.19-21]

[10] B. Soucail, P. Voisin, M. Voos, D. Rondi, J. Nagle, B. de Crémoux [*Superlattices Microstruct. (UK)* vol. 8 (1990) p.279-82]

[11] B. Soucail, P. Voisin, M. Voos, D. Rondi, J. Nagle, B. de Crémoux [*Semicond. Sci. Technol. (UK)* vol.5 (1990) p.918-20]

[12] J.M. Langer, C. Delerue, M. Lannoo, H. Heinrich [*Phys. Rev. B (USA)* vol.38 (1988) p.7723-39]

[13] T.P. Pearsall, L. Eaves, J.C. Portal [*J. Appl. Phys. (USA)* vol.54 (1983) p.1037-47]

[14] S.R. Forrest, O.K. Kim [*J. Appl. Phys. (USA)* vol.52 (1981) p.5838-42]

[15] S.R. Forrest, O.K. Kim [*J. Appl. Phys. (USA)* vol.53 (1982) p.5738-45]

[16] M. Ogura, M. Mizuta, K. Onaka, H. Kukimoto [*Jpn. J. Appl. Phys. (Japan)* vol.22 (1983) p.1502-9]

[17] H. Temkin, M.B. Panish, P.M. Petroff, R.A. Hamm, J.M. Vandenberg, S. Sumski [*Appl. Phys. Lett. (USA)* vol.47 (1985) p.394-6]

[18] M.S. Skolnick et al [*Semicond. Sci. Technol. (UK)* vol.1 (1986) p.29-40]

[19] K. Steiner et al [*Surf. Sci. (Netherlands)* vol.174 (1986) p.331-6]

[20] K. Kazmierski, P. Philippe, P. Poulain, B. de Cremoux [*J. Appl. Phys. (USA)* vol.61 (1987) p.1941-6]

[21] D.V. Lang et al [*Appl. Phys. Lett. (USA)* vol.50 (1987) p.736-8]

[22] D.J. Westland et al [*Appl. Phys. Lett. (USA)* vol.50 (1987) p.839-41]

[23] M.S. Skolnick et al [*Appl. Phys. Lett. (USA)* vol.51 (1987) p.24-6]

[24] P.S. Whitney, C.G. Fonstad [*J. Cryst. Growth (Netherlands)* vol.83 (1987) p.219-32]

[25] L.Y. Leu, S.R. Forrest [*J. Appl. Phys. (USA)* vol.64 (1988) p.5030-40]

[26] R.E. Cavicchi et al [*Appl. Phys. Lett. (USA)* vol.54 (1989) p.739-41]

[27] M.A. Haase, N. Pan, G.E. Stillman [*Appl. Phys. Lett. (USA)* vol.54 (1989) p.1457-9]

[28] J.R. Waldrop, R.W. Grant, E.A. Kraut [*J. Vac. Sci. Technol. B (USA)* vol.7 (1989) p.815-9]

[29] M.T. Furtado, M.S.S. Loural, A.C. Sachs, P.J. Shief [*Superlattices Microstruct. (UK)* vol.5 (1989) p.507-10]

[30] D. Gauthier, J.C. Portal, M. Razeghi [*Semicond. Sci. Technol. (UK)* vol.4 (1989) p.218-19]

[31] A.W. Higgs, H.J. Hutchinson, L.L. Taylor, N. Apsley, S.J. Bass [*Semicond. Sci. Technol. (UK)* vol.5 (1990) p.581-5]

[32] J.R. Waldrop, E.A. Kraut, C.W. Farley, R.W. Grant [*J. Appl. Phys. (USA)* vol.69 (1991) p.372-8]

[33] J.P. Landesman et al [*Appl. Phys. Lett. (USA)* vol.60 (1992) p.1241-3]

[34] M. Peressi, S. Baroni, A. Baldereschi, R. Resta [*Phys. Rev. B (USA)* vol.41 (1990) p.12106-110]

[35] M.S. Hybertsen [*J. Vac. Sci. Technol. B (USA)* vol.8 (1990) p.773-8]

[36] B.R. Nag, S. Mukhopadhyay [*Appl. Phys. Lett. (USA)* vol.58 (1991) p.1056-8]

[37] M.S. Hybertsen [*Appl. Phys. Lett. (USA)* vol.58 (1991) p.1759-61]

[38] L. Hrivnák [*Phys. Status Solidi A (Germany)* vol.124 (1991) p.K111-16]

[39] H. Kroemer, Wu-Yi Chin, J.S. Harris Jr., D.D. Edwall [*Appl. Phys. Lett. (USA)* vol.36 (1980) p.295-7]

[40] D.V. Morgan, K. Board, C.E.C. Wood, L.F. Eastman [*Phys. Status Solidi A (Germany)* vol.72 (1982) p.251-60]

[41] R. People, K.W. Wecht, K. Alavi, A.Y. Cho [*Appl. Phys. Lett. (USA)* vol.43 (1983) p.118-21]
[42] D.F. Welch, G.W. Wicks, L.F. Eastman [*J. Appl. Phys. (USA)* vol.55 (1984) p.3176-9]
[43] F. Capasso et al [*Appl. Phys. Lett. (USA)* vol.48 (1986) p.1294-6]
[44] J. Wagner, W. Stolz, K. Ploog [*Phys. Rev. B (USA)* vol.32 (1985) p.4214-16]
[45] C.K. Peng, A. Ketterson, H. Morkoç, P.M. Solomon [*J. Appl. Phys. (USA)* vol.60 (1986) p.1709-12]
[46] Y. Sugiyama, T. Inata, T. Fujii, Y. Nakata, S. Muto, S. Hiyamizu [*Jpn. J. Appl. Phys. (Japan)* vol.25 (1986) p.L648-50]
[47] A. Sandhu, Y. Nakata, S. Sasa, K. Kodama, S. Hiyamizu [*Jpn. J. Appl. Phys. (Japan)* vol.26 (1987) p.1709-12]
[48] J.-H. Huang, T.Y. Chang, B. Lalevic [*Appl. Phys. Lett. (USA)* vol.60 (1992) p.733-5]
[49] R.F. Kopf, M.H. Herman, M.L. Schnoes, A.P. Perley, G. Livescu, M. Ohring [*J. Appl. Phys. (USA)* vol.71 (1992) p.5004-11]
[50] H. Kroemer, G. Griffiths [*IEEE Electron Device Lett. (USA)* vol.4 (1983) p.20-2]
[51] L. Aina, M. Mattingly, L. Stecker [*Appl. Phys. Lett. (USA)* vol.53 (1988) p.1620-2]
[52] J.R. Waldrop, E.A. Kraut, C.W. Farley, R.W. Grant [*J. Vac. Sci. Technol. B (USA)* vol.8 (1990) p.768-72]
[53] P.K. Bhattacharya, H.-J. Bühlmann, M. Ilegems, P. Schmid, H. Melchior [*Appl. Phys. Lett. (USA)* vol.41 (1982) p.449-51]
[54] J.Y. Marzin, E.V.K. Rao [*Appl. Phys. Lett. (USA)* vol.43 (1983) p.560-2]
[55] J.Y. Marzin, M.N. Charasse, B. Sermage [*Phys. Rev. B (USA)* vol.31 (1985) p.8298-301]
[56] N.G. Anderson, W.D. Laidig, R.M. Kolbas, Y.C. Lo [*J. Appl. Phys. (USA)* vol.60 (1986) p.2361-7]
[57] L.P. Ramberg et al [*J. Appl. Phys. (USA)* vol.61 (1997) p.1234-6]
[58] G. Ji, D. Huang, U.K. Reddy, T.S. Henderson, R. Houdré, H. Morkoç [*J. Appl. Phys. (USA)* vol.62 (1987) p.3366-73]
[59] J. Hwang, P. Pianetta, C.K. Shih, W.E. Spicer, Y.-C. Pao, J.S. Harris Jr. [*Appl. Phys. Lett. (USA)* vol.51 (1987) p.1632-3]
[60] J. Menéndez et al [*Phys. Rev. B (USA)* vol.36 (1987) p.8165-8]
[61] T.G. Andersson, V. Kulakovski, Z.-G. Chen, A. Uddin, J. Vallin, J. Westin [*J. Phys. Colloq. (France)* vol.48 (1987) p.C5-163-7]
[62] T.G. Andersson, Z.G. Chen, V.D. Kulakovskii, A. Uddin, J.T. Vallin [*Phys. Rev. B (USA)* vol.37 (1988) p.4032-8]
[63] J. Menéndez, A. Pinczuk [*IEEE J. Quantum Electron. (USA)* vol.24 (1988) p.1698-711]
[64] S.H. Pan et al [*Phys. Rev. B (USA)* vol.38 (1988) p.3375-82]
[65] M.J. Joyce, M.J. Johnson, M. Gal, B.F. Usher [*Phys. Rev. B (USA)* vol.38 (1988) p.10978-80]
[66] R.L.S. Devine, W.T. Moore [*Solid State Commun. (USA)* vol.65 (1988) p.177-9]
[67] K.F. Huang, K. Tai, S.N.G. Chu, A.Y. Cho [*Appl. Phys. Lett. (USA)* vol.54 (1989) p.2026-8]
[68] P.W. Yu et al [*Appl. Phys. Lett. (USA)* vol.54 (1989) p.2230-2]
[69] D.J. Arent, K. Deneffe, C. Van Hoof, J. De Boeck, G. Borghs [*J. Appl. Phys. (USA)* vol.66 (1989) p.1739-47]
[70] S. Niki, C.L. Lin, W.S.C. Chang, H.H. Wieder [*Appl. Phys. Lett. (USA)* vol.55 (1989) p.1339-41]

[71] U.K. Reddy, G. Ji, T. Henderson, D. Huang, R. Houdré, H. Morkoç [*J. Vac. Sci. Technol. B (USA)* vol.7 (1989) p.1106-10]

[72] D. Gershoni, J.M. Vandenberg, S.N.G. Chu, H. Temkin, T. Tanbun-Ek, R.A. Logan [*Phys. Rev. B (USA)* vol.40 (1989) p.10017-20]

[73] W. Xinghua, R. Laiho [*Superlattices Microstruct. (UK)* vol.5 (1989) p.79-81]

[74] J.-P. Reithmaier, R. Höger, H. Riechert, A. Heberle, G. Abstreiter, G. Weimann [*Appl. Phys. Lett. (USA)* vol.56 (1990) p.536-8]

[75] J.-P. Reithmaier, R. Höger, H. Riechert, P. Hiergeist, G. Abstreiter [*Appl. Phys. Lett. (USA)* vol.57 (1990) p.957-9]

[76] H.Q. Hou, Y. Segawa, Y. Aoyagi, S. Namba, J.M. Zhou [*Solid State Commun. (USA)* vol.74 (1990) p.159-63]

[77] X. Marie et al [*J. Appl. Phys. (USA)* vol.69 (1991) p.812-15]

[78] Y. Zou et al [*Appl. Phys. Lett. (USA)* vol.58 (1991) p.601-3]

[79] X. Letartre, D. Stievenard, E. Barbier [*Appl. Phys. Lett. (USA)* vol.58 (1991) p.1047-9]

[80] B. Jogai [*Appl. Phys. Lett. (USA)* vol.59 (1991) p.1329-31]

[81] Y.S. Huang et al [*J. Appl. Phys. (USA)* vol.70 (1991) p.7537-42]

[82] S. Marcinkevicius, G. Ambrazevicius, T. Lideikis, K. Naudzius [*Solid State Commun. (USA)* vol.79 (1991) p.889-92]

[83] V.A. Wilkinson, A.D. Prins, D.J. Dunstan, L.K. Howard [*J. Electron. Mater. (USA)* vol.20 (1991) p.509-16]

[84] S.P. Kowalczyk, W.J. Schaffer, E.A. Kraut, R.W. Grant [*J. Vac. Sci. Technol. (USA)* vol.20 (1982) p.705-8]

[85] K. Hirakawa, Y. Hashimoto, K. Harada, T. Ikoma [*Phys. Rev. B (USA)* vol.44 (1991) p.1734-40]

[86] C. Priester, G. Allan, M. Lannoo [*Phys. Rev. B (USA)* vol.38 (1988) p.9870-3]

[87] M. Oloumi, C.C. Matthai [*J. Phys., Condens. Matter (UK)* vol.1 (1989) p.SB211-12]

[88] B. Jogai, P.W. Yu [*Phys. Rev. B (USA)* vol.41 (1990) p.12650-8]

[89] N.G. Anderson, S.D. Jones [*J. Appl. Phys. (USA)* vol.70 (1991) p.4342-56]

[90] M. Oloumi, C.C. Matthai [*J. Phys., Condens. Matter (UK)* vol.3 (1991) p.9981-7]

[91] M.A. Reed, J.W. Lee [*Appl. Phys. Lett. (USA)* vol.50 (1987) p.845-7]

[92] T.G. Andersson, Z.G. Chen, V.D. Kulakovskii, A. Uddin, J.T. Vallin [*Solid State Commun. (USA)* vol.64 (1987) p.379-82]

[93] N. Debbar, D. Biswas, P. Bhattacharya [*Phys. Rev. B (USA)* vol.40 (1989) p.1058-63]

[94] S.Y. Lin, D.C. Tsui, H. Lee, D. Ackley [*Appl. Phys. Lett. (USA)* vol.55 (1989) p.2211-13]

[95] H. Tomozawa, K. Matsuzaki, H. Hasegawa, H. Ohno [*Inst. Phys. Conf. Ser. (UK)* no.106 (1989) ch.6 p.369-74]

[96] R. People, A. Jayaraman, S.K. Sputz, J.M. Vandenberg, D.L. Sivco, A.Y. Cho [*Phys. Rev. B (USA)* vol.45 (1992) p.6031-6]

[97] D.J. Arent [*Phys. Rev. B (USA)* vol.41 (1990) p.9843-9]

[98] M. Quillec, J.Y. Marzin, J. Primot, G. Le Roux, J.L. Benchimol, J. Burgeat [*J. Appl. Phys. (USA)* vol.59 (1986) p.2447-50]

[99] D. Gershoni, H. Temkin, J.M. Vandenberg, S.N.G. Chu, R.A. Hamm, M.B. Panish [*Phys. Rev. Lett. (USA)* vol.60 (1988) p.448-51]

[100] R. People [*J. Appl. Phys. (USA)* vol.62 (1987) p.2551-3]

3.5 Electron and hole effective masses in lattice-matched and strained InGaAs

J. Singh

February 1993

A INTRODUCTION

The response of carriers in semiconductors to external applied stimuli depends directly on the electron or hole masses. The ability to modify the carrier masses, therefore, has profound implications for electronic and optoelectronic devices. In recent years, strained epitaxy has emerged as a powerful technique to modify the carrier effective mass. In this Datareview we will examine the role that strain plays in the InGaAs electron and hole mass modification.

B THEORETICAL FORMALISM

The theoretical basis for the results presented in this Datareview are described in [1]. We will therefore briefly discuss the physics only as it relates to the carrier masses. Focusing on the conduction band effective mass of a direct gap semiconductor, we note that in the k•p approximation the carrier mass can be described by an expression of the form [2,3]

$$\frac{1}{m_c^*} = \frac{1}{m_o} + \frac{2P_{cv}^2}{m_o^2}\frac{1}{3}\left[\frac{2}{E_{g\Gamma}} + \frac{1}{E_{g\Gamma} + \Delta}\right] \tag{1}$$

where m_o is the free electronic mass, P_{cv} is the momentum matrix element between the conduction band (s-like) state and valence band (p-like) state, $E_{g\Gamma}$ is the direct bandgap and Δ is the spin-orbit splitting. The element P_{cv} is almost unchanged as one goes from GaAs to InAs [4], so that the carrier mass is only affected by the bandgap values. In [1] we had discussed how the bandgaps change as a function of alloy composition and strain. In FIGURE 1 we show the calculated results for the electron masses in unstrained InGaAs and InGaAs grown along (001) on GaAs and InP substrates [5].

Several observations can be made from the results of FIGURE 1. The unstrained alloy mass decreases with increased In composition as a result of the changed bandgap. In the strained pseudomorphic form the bandgap of InGaAs increases as discussed in [1] and the carrier masses increase. It is important to note that the carrier masses are altered by ~ 10 - 15% in the range of technological interest i.e. excess In of ~ 20%. Experimental points are available for the unstrained alloy [6], but there is relatively little available for the strained InGaAs. Shubnikov-de-Haas work on the strained system suggests that the theoretical results shown in FIGURE 1 are quite reasonable [7].

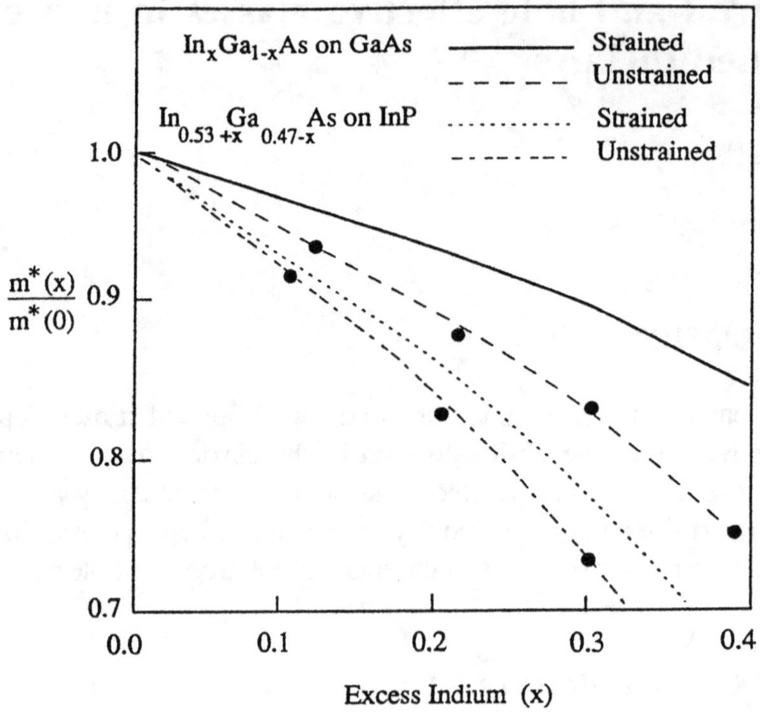

FIGURE 1 Change in the in-plane electron mass in the $In_xGa_{1-x}As$ system lattice matched to GaAs, and the $In_{0.53+x}Ga_{0.47-x}As$ on InP. The results are shown for both unstrained and strained semiconductors. Also shown are experimentally measured points ([5] for calculations; [6] for experimental results).

In the valence band, the effect of strain on the carrier mass is quite dramatic. The reasons for this were discussed in [1]. In unstrained alloy, the hole masses do not change much with alloying, since the density of state mass for the holes decreases only by ~ 8% as one goes from GaAs to InAs. However, the strain causes a much greater effect primarily due to the separation of the heavy hole and light hole bands. The hole bands are highly non-parabolic so that the carrier masses are quite sensitive to the position in the Brillouin zone where the mass is measured. At the Γ-point the hole masses in bulk GaAs have the dependence on strain shown in FIGURE 2 [8]. The strain is assumed to be pseudomorphic i.e.

$$\varepsilon_{xx} = \varepsilon_{yy} = \varepsilon; \ \varepsilon_{zz} = \frac{-2c_{12}}{c_{11}}\varepsilon$$

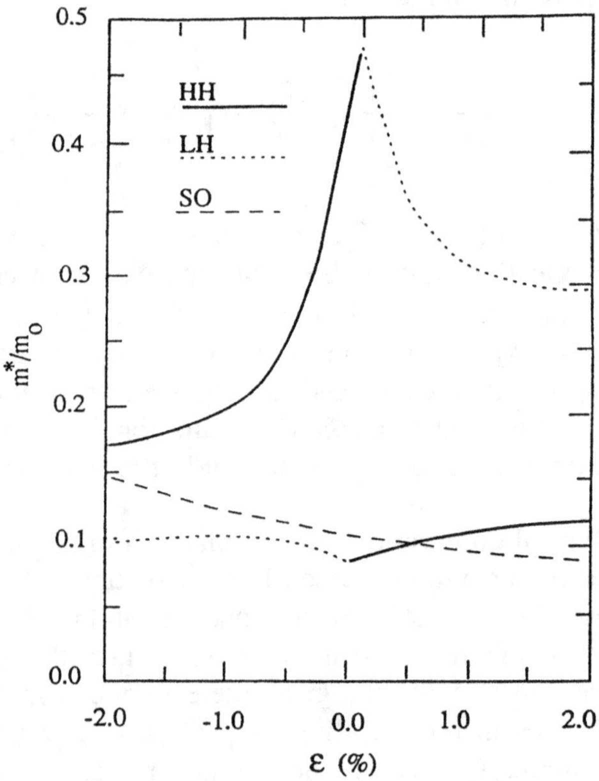

FIGURE 2 Effect of biaxial strain on hole masses in bulk GaAs [8].

A more relevant carrier mass is that averaged over the carrier occupation energies ($\sim k_BT$) and FIGURE 3 shows the results of such calculations along with experimental measurements [7,9]. Similar results have been obtained by magneto-optic techniques as well [10,11]. The effect of pseudomorphic strain is quite dramatic and hole masses can be reduced by up to a factor of 3. It is also interesting to note that the hole masses can be increased in a quantum well by choosing a small tensile strain so that the HH and LH states merge in a quantum well. This effect manifests itself directly in the enhanced acceptor binding energy [12].

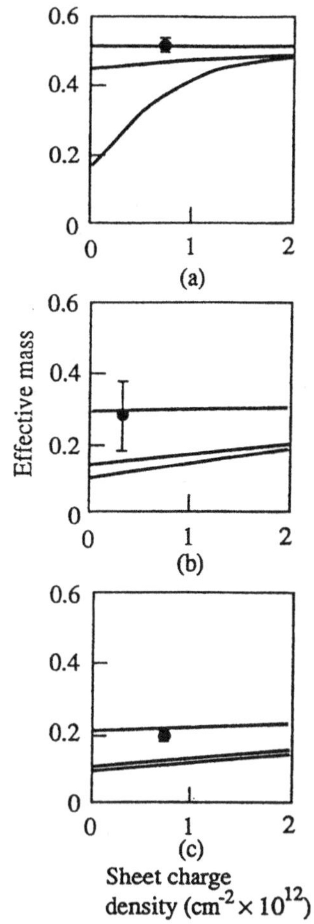

FIGURE 3 Calculated and measured (data points) hole effective mass for (a) lattice matched (b) 10% indium channel, and (c) 20% indium channel p-type MODFET as a function of sheet charge density. In each case, the upper curve is the average mass at 300 K, the middle curve is at 77 K and the bottom curve is at 0 K [7,9].

C CONCLUSION

In this Datareview we have seen how the carrier masses are affected by alloying in the InGaAs alloy. We see that the conduction band mass is affected only because of the changes in the bandgap that arise from the alloying or strain. On the other hand, the valence band mass is almost unaffected if there is no strain in the alloy. However, in the presence of strain the hole masses can be dramatically altered. In the case of compressive strain a mass reduction of up to a factor of 3 is possible.

ACKNOWLEDGEMENT

Discussions with and contributions of M. Jaffe and J. Loehr are gratefully acknowledged.

REFERENCES

[1] J. Singh [Datareview in this book: 3.1 *Energy gaps and bandstructure of lattice-matched and strained InGaAs*]

[2] E.O. Kane [*J. Phys. Chem. Solids (UK)* vol.8 (1959) p.38]

[3] E.O. Kane [*Semicond. Semimet. (USA)* Eds R.K. Willardson, A.C. Beer (Academic Press, New York, 1966) vol.1 p.81]

[4] P. Lawaetz [*Phys. Rev. B (USA)* vol.4 (1971) p.3460]

[5] M. Jaffe, J. Singh [*J. Appl. Phys. (USA)* vol.65 (1988) p.65]

[6] See, for example, Landolt-Bornstein [*Numerical Data and Functional Relations in Science and Technology* Eds O. Madelung, M. Schulz, H. Weiss (Springer-Verlag, New York, 1987) vol.22]; M. Neuberger [*Handbook of Electronic Materials* (IFI/Plenum, New York, 1971)]

[7] M. Jaffe [*Studies on the Electronic Properties and Applications of Coherently Strained Semiconductors* Ph.D. Thesis, The University of Michigan (1991)]

[8] J. Loehr [*Theoretical Studies of Pseudomorphic Quantum Well Optoelectronic Devices* Ph.D. Thesis, The University of Michigan (1991)]

[9] M. Jaffe, J.E. Oh, J. Pamulapati, J. Singh, P.K. Bhattacharya [*Appl. Phys. Lett. (USA)* vol.54 (1989) p.2345]

[10] T.E. Zipperian, C.R. Dawson, T.J. Drummond, J.E. Shiba, I.J. Fritz [*Appl. Phys. Lett. (USA)* vol.52 (1988) p.975]

[11] E.D. Jones, R.M. BieFeld, J.F. Klem, S.K. Kyo [*Gallium Arsenide and Related Compounds,* 1989, Inst. of Phys. Conf. Ser. no.106 (Institute of Physics, Bristol, 1989) p.435]

[12] J.P. Loehr, Y. Chen, D. Biswas, P. Bhattacharya, J. Singh [*Appl. Phys. Lett. (USA)* vol.57 (1990) p.180]

CHAPTER 4

TRANSPORT PROPERTIES

4.1 Low-field electron and hole mobilities in lattice-matched InGaAs on InP

L.F. Eastman

February 1993

A INTRODUCTION

The alloy $In_{0.53}Ga_{0.47}As$ has the same lattice constant, 5.87 Å, as InP. It has excellent electron transport properties, including low effective mass, 0.04 m_o, high mobility, up to 15,000 cm^2/V s at room temperature, and a large energy separation, 0.55 eV, between the lowest two valleys in the conduction band. There has been a considerable amount published on this material [1]. There also is another, large band gap material, $In_{0.52}Al_{0.48}As$, that is lattice-matched to $In_{0.53}Ga_{0.47}As$. As in other III-V compound semiconductors and their alloys, the valence band of InGaAs is complicated. Light and heavy hole bands are nearly degenerate at the valence band edge. The heavy holes dominate, with ~ 0.43 m_o effective mass and high state density, and the light holes have an effective mass of ~ 0.05 m_o. The mobility of holes is low, being ≤ 300 cm^2/V s for moderately pure material [1]. Modulation-doped field effect transistors (MODFETs) and heterojunction bipolar transistors (HBTs) made with $In_{0.53}Ga_{0.47}As$ channels and bases, respectively, give excellent high frequency performance. Photo-detectors and lasers are also made using $In_{0.53}Ga_{0.47}As$. This material can be made by liquid phase epitaxy (LPE), organometallic vapour phase epitaxy (OMVPE), or by molecular beam epitaxy (MBE). Semi-insulating InP substrates are doped with iron, and are available commercially.

B THEORETICAL MOBILITY LIMIT

Takeda has calculated the theoretical mobility for electrons in $In_{0.53}Ga_{0.47}As$ at 300 K and 77 K [2]. His limiting mobility levels are ~15,000 cm^2/V s and ~160,000 cm^2/V s, respectively, at these two temperatures for ion density levels ≤ $10^{14}/cm^3$. (See FIGURES 1(a) and 1(b).) His calculations include the effect of alloy scattering as well as phonon and ion scattering. The effect of acceptor compensation is also included. The limit of mobility, if no alloy scattering were present, has also been calculated. The results are ~ 17,500 cm^2/V s at 300 K and ~ 800,000 cm^2/V s at 77 K. It would require a very ordered structure to reach such levels of mobility, and that has not been accomplished. If there were only alloy scattering, without phonon or ion scattering, the mobility for electrons would be ~ 85,000 cm^2/V s at 300 K and 170,000 cm^2/V s at 77 K. This alloy scattering depends on $T^{-1/2}$.

Mobility equals $(e\tau)/m^*$, where e is the electronic charge, τ is the time between momentum-changing collisions, and m^* is the effective mass. In comparing room-temperature mobility levels in pure $In_{0.53}Ga_{0.47}As$ and GaAs, it is clear that the change in effective mass is the dominant reason for mobility enhancement, and that the time between momentum-changing collisions is nearly the same in both materials.

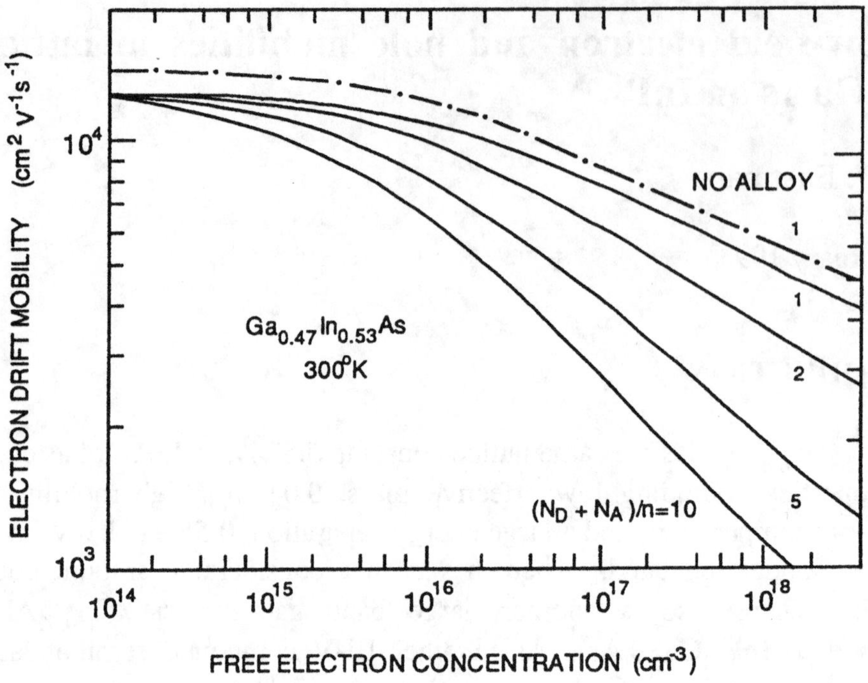

FIGURE 1(a) Calculated electron drift mobilities in $In_{0.53}Ga_{0.47}As$ at 300 K (after Y. Takeda [2]). The chain curve is the mobility with no alloy scattering. The parameter numbers are the compensation ratios $(N_D + N_A)/n$.

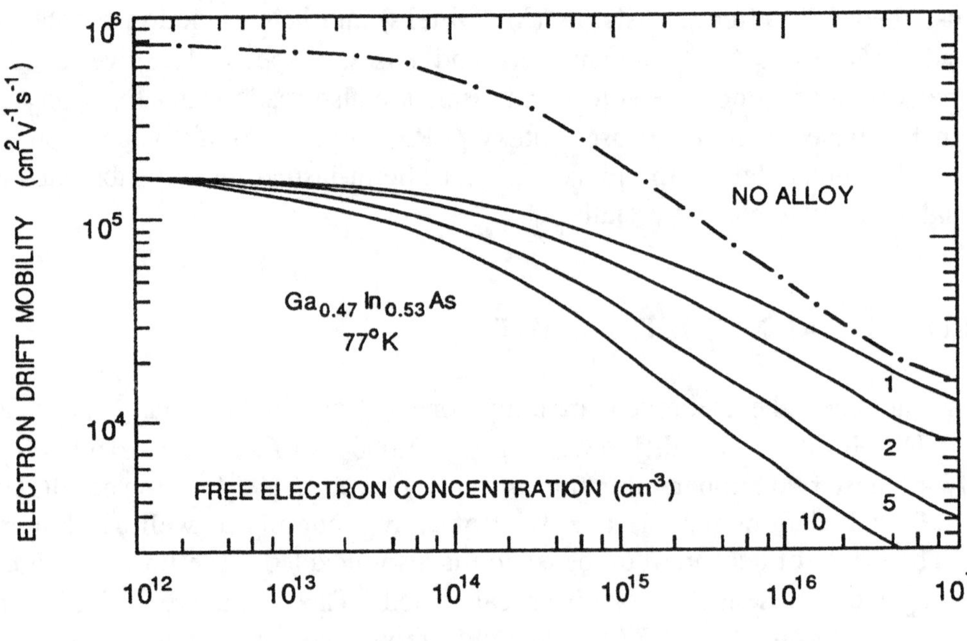

FIGURE 1(b) Calculated electron drift mobilities in $In_{0.53}Ga_{0.47}As$ at 77 K (after Y. Takeda [2]), with the same nomenclature as FIGURE 1(a).

The reciprocal of the electron mobility due to alloy scattering alone is dependent on y(1-y), in $In_yGa_{1-y}As$. This yields the lowest alloy-scattering electron mobility for $y = 0.5$. The overall electron mobility in the limit where $y \rightarrow 1$, for InAs, is about four times higher than that for the limit where $y \rightarrow 0$ for GaAs. As a result, the overall electron mobility for this alloy minimises in the range of $0.3 < y < 0.4$.

When holes are considered [1] many of the same physical phenomena are involved in limiting the mobility, in addition to the substantially higher effective mass. All this leads to hole mobility in bulk InGaAs that is limited to ~ 300 cm^2/V s.

C EXPERIMENTAL RESULTS

Early LPE results yielded electron mobility at 300 K and 77 K of 13,800 cm^2/V s and 70,000 cm^2/V s respectively [3]. In order to eliminate the effects of surface depletion, and the depletion or accumulation at the heterointerface with InP, it was necessary to make the layers several microns thick. These layers were compensated, with net electron density values of ~ 2 x $10^{15}/cm^3$ for the wafer yielding the 300 K data, and ~ 3 x $10^{14}/cm^3$ for the wafer yielding the 77 K data. These were different wafers due to optimisation of composition for a better lattice match at the two different measurement temperatures. InP and InGaAs have different coefficients of expansion, being 4.5 x 10^{-6}/K and ~ 5.25 x 10^{-6}/K respectively. The epitaxial layer thickness of ~ 6 μm did not allow pseudomorphic conditions to be achieved over a wide temperature range.

OMVPE growth yielded high electron mobility in a few cases, but has only been reported with relatively thin $In_{0.53}Ga_{0.47}As$ layers on InP. As a result, these results were very likely just those for a two-dimensional electron gas against the InP buffer layer. In an early result, 80,000 cm^2/V s was achieved at 77 K with ~ 3 x $10^{11}/cm^2$ electron sheet density [4], when the $In_{0.53}Ga_{0.47}As$ layer was grown on top of the InP. Shubnikov-de Haas measurements were made to verify the quantum Hall effects in this case. Later results yielded 77 K results as high as 155,000 cm^2/V s electron mobility [5], with only 0.6 μm $In_{0.53}Ga_{0.47}As$. Again this is not a bulk result, but a two-dimensional electron gas result, where the electron density is high enough to screen some of the impurity scattering.

MBE has been used to obtain the highest purity bulk sample to date [6]. These results are 15,200 cm^2/V s and 104,000 cm^2/V s, for 300 K and 77 K electron mobility respectively, for samples 3 μm thick grown on undoped $In_{0.52}Al_{0.48}As$ buffer layers. The latter prevented the outdiffusion of impurities from the iron-doped InP substrate. The effective net electron concentration was about 2 x $10^{13}/cm^3$, without adjusting for depletion on either side of the layer under test.

Some experiments on hole mobility in bulk InGaAs have yielded 300 and 600 cm^2/V s values at 300 and 77 K, respectively. Two-dimensional hole gas mobility values of up to nearly 15,000 cm^2/V s have been measured at 4.2 K [5].

D CONCLUSION

$In_{0.53}Ga_{0.47}As$ grown lattice-matched to InP has excellent electron transport properties, with rather limited hole transport properties. Experimental transport results agree closely with theoretical predictions that include alloy scattering along with phonon and ion scattering. MODFETs, HBTs, lasers and photo detectors will have high performance using this alloy.

REFERENCES

[1] See for example [*GaInAsP Alloy Semiconductors*, Ed T.P. Pearsall (J. Wiley & Sons, New York, 1982)]
[2] Ibid, ch.9, by Yoshikazu Takeda.
[3] J.D. Oliver, L.F. Eastman [*J. Electron. Mater. (USA)* vol.9 no.4 (1980) p.693-712]
[4] K.T. Chan, L.D. Zhu, J.M. Ballantyne [*Appl. Phys. Lett. (USA)* vol.47 no.1 (1985) p.44-6]
[5] M. Razeghi [*The MOCVD Challenge* (Adam Hilger, 1989) p.79]
[6] T. Mishima, M. Takahama, Y. Uchida, T. Tanoue, S. Takahashi [J. Electron. Mater. (USA) vol.20 no.1 (1991) p.113-6]

4.2 High-field transport in InGaAs and related heterostructures

M.A. Littlejohn, K.W. Kim and H. Tian

March 1993

A INTRODUCTION

High-field transport in InGaAs and related heterostructure materials is important in most electronic and optoelectronic devices fabricated with one or more reduced (submicron) dimensions in an active region [1]. This type of transport is determined by the energy band structure and by a large number of fundamental material properties in these regions. It is also influenced by the inherent geometric structures used in device fabrication and by externally-applied potentials and excitations. Here, we will examine some of the material parameters which control high field transport in InGaAs and related (In-based) material systems and heterostructures. We will also describe an empirical approach which allows determination of material parameters which have not been experimentally investigated beginning with available experimental data. Finally, we will discuss a predictive approach for determining properties of sparsely-studied materials systems and devices fabricated from them based on the use of the Monte Carlo method.

B ELECTRON VELOCITY-FIELD CHARACTERISTICS

The velocity-electric field (v - E) characteristic is one of the most important indicators of the potential for useful high-field transport properties in a material. Yet, v - E measurements are very difficult to perform and useful hot electron transport data for compound semiconductors, especially ternary materials such as InGaAs, therefore remain quite scarce, even after over twenty years of developments in the field. For example, since our original review paper was published in 1982 [1], we have found only a handful of experiments which report v - E data for $In_{0.53}Ga_{0.47}As$. The experimental difficulty is frequently attributed to contact and/or parasitic effects in test device structures. Thus, the standard 'modelling' approach is to use this limited number of direct or indirect measurements of velocity vs. electric field as a test domain to empirically determine a large space of material parameters which can fit the data in some relatively subjective 'best fit' approach. Often, some of the important material parameters have been previously measured in independent experiments, which aids the empirical study. Then, this data can be used to explore other properties and define experimental conditions for device related studies.

For example, FIGURE 1 shows the results of five experiments for $In_{0.53}Ga_{0.47}As$, and two calculated v - E curves obtained by the Monte Carlo simulation method. The calculated curves include our current work [2] and the recent theoretical work of Brennan and Park [3]. In chronological order, the experiments include the data of Marsh et al [4], Windhorn et al [5], Nag et al [6], Haase et al [7], and Shigekawa et al [8]. Nag [6] also presents Monte Carlo calculations which are in good agreement with those of Brennan and Park [3]. The data in FIGURE 1 was obtained using either Hall measurements [6, 7] or time-of-flight measurements [4,5,8]. One primary concern for comparing experiment with theory resides in the uncertainty of the actual doping level and the amount of carrier compensation in the samples. Also, it is important to note that Shigekawa's experiments were made in heterostructures fabricated from $In_{0.52}Al_{0.48}As/In_{0.53}Ga_{0.47}As/In_{0.52}Al_{0.48}As$ layers. Thus, the heterojunction band offsets play an important role in interpreting the experimental results and in correlating the results of the simulations [8] with the data. It is important to note that in spite of these uncertainties the actual trends between the theory and the data are quite good, and the basic agreement between the experimental values and the theoretical values are sufficient to place a good deal of confidence in the material parameters which can be deduced from the comparison. The only other pertinent point is that the disagreement between the measurements of Windhorn et al [5] and Shigekawa et al [8] for fields near 10 kV/cm cannot be resolved at this point. TABLE 1 presents a summary of the theoretical and experimental v - E values for $In_{0.53}Ga_{0.47}As$.

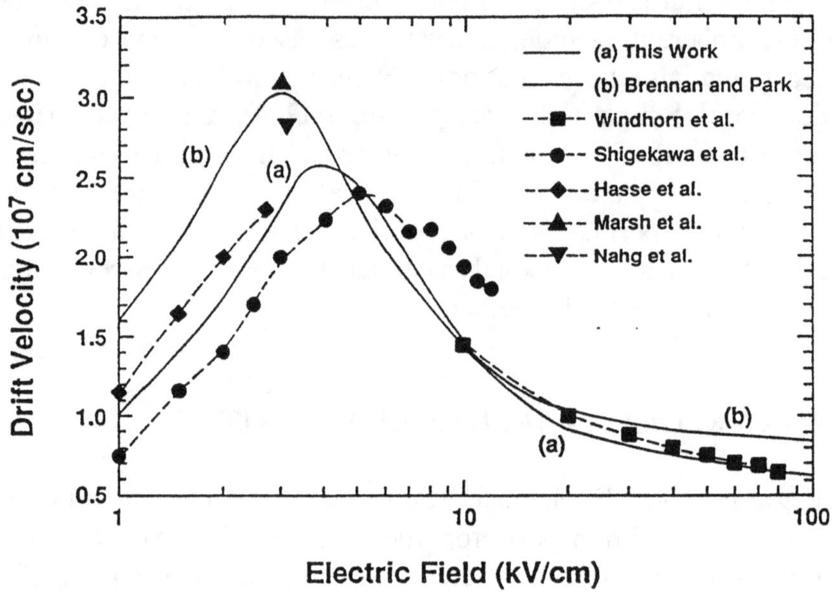

FIGURE 1 Electron drift velocity as a function of electric field for uncompensated n-type $In_{0.53}Ga_{0.47}As$ at 300 K. The solid curves are Monte Carlo calculations and the dashed curves with data points are experimental data.

Recently, a comprehensive treatment of Monte Carlo simulations for diamond and zinc-blende semiconductor structures, including $In_xGa_{1-x}As$, has been presented by Fischetti [9]. This work employs band structures obtained from local empirical psuedopotential calculations. Because of this fact, Fischetti's work should be considered as the current standard with which all other Monte Carlo calculations for compound semiconductors are compared. In this context, Fischetti's calculations agree with the experimental data of Marsh [4] and Windhorn [5], and they are in general good agreement with the theoretical data shown in TABLE 1 [2,3].

The theoretical values in TABLE 1 represent both computed points and values obtained from an interpolated least squared error curve fitting program. The differences between the data from [2] and [3] in this table arise from differences in material parameters used to obtain them. In particular, the doping level used to compute the values in [2] is 1×10^{16} cm^{-3} while [3] computes the values of undoped material. This difference is especially significant in the sub-threshold region. The material parameters employed will be discussed in the next section.

TABLE 1 Theoretical [2,3] and experimental [4-8] electron drift velocity values as a function of electric field intensity for $In_{0.53}Ga_{0.47}As$.

	Electron Drift Velocity (10^7 cm/s)						
Field (kV/cm)	Ref [2]	Ref [3]	Ref [4]	Ref [5]	Ref [6]	Ref [7]	Ref [8]
1.0	1.015	1.610				1.15	0.75
1.2	1.180	1.820					
1.5	1.400	2.102				1.65	1.16
2.0	1.735	2.591				2.00	1.40
2.5	2.036	2.914					1.70
2.7	2.150	2.990				2.30	
3.0	2.314	3.030	3.10				2.00
3.1	2.380	3.023			2.83		
3.3	2.465	2.985					
3.5	2.550	2.930					
4.0	2.570	2.730					2.23
5.0	2.458	2.330					2.40
6.0	2.230	2.030					2.32
7.0	1.989	1.846					2.16
8.0	1.759	1.690					2.17
9.0	1.598	1.542					2.06
10.0	1.465	1.421		1.45			1.94
11.0	1.350	1.360					1.85
12.0	1.262	1.302					1.80
15.0	1.080	1.150					
20.0	0.920	1.035		0.88			
30.0	0.808	0.960		0.80			
40.0	0.752	0.910		0.75			
50.0	0.710	0.895		0.71			
60.0	0.686	0.880		0.69			
70.0	0.651	0.865		0.65			
80.0	0.643	0.860					
90.0	0.634	0.855					
100.0	0.630	0.853					

C MATERIAL PARAMETERS FOR $In_{0.53}Ga_{0.47}As$

The calculation of the v - E curves shown in FIGURE 1 requires information for a large number of material parameters, most of which are unknown. In this case, one generally resorts to an interpolation scheme in order to estimate these unknown material parameters based on the best available data and theories. Several interpolation schemes have been discussed in the literature [1,2, 9-11] and found to be useful in predicting acceptable sets of material parameters. Most of these approaches begin with or are based on a knowledge of the material parameters for the binary constituents of the ternary (or quaternary) materials [12]. Some approaches use interpolations from known (experimentally-derived) material parameters for other ternary compositions [2,3,8]. For a general ternary material, these interpolation approaches include a linear form [1,2,11] and a nonlinear (or quadratic) form [1,10], in order to predict the 'bowing' effect which has been observed in many ternary material parameters, such as energy bandgap. An excellent discussion of both approaches is given by Adachi [10]. Also, the reader should refer to Fischetti's work for a detailed discussion of band structure effects and the important binary material band parameters which influence Monte Carlo simulations [9].

The Monte Carlo method has been used to calculate the v - E curves shown in FIGURE 1. Generally, this method comprises a model for the motion of sample electrons in a three-valley (Γ-L-X) semiconductor with nonparabolic analytical energy band structure under the influence of a uniform applied electric field [1-3,6,8-13]. A very detailed presentation of the Monte Carlo method is contained in the recent book by Singh [14]. The Monte Carlo method requires a knowledge of both bulk material parameters and energy band or valley-dependent parameters. TABLE 2 lists those material parameters which we used in the calculation of our theoretical curve in FIGURE 1. These parameters in TABLE 2 were obtained from a trial and error curve fitting approach using Windhorn's data [5] as the primary standard for the curve fit while considering both Haase's data [7] and Shigekawa's data [8] as secondary standards. The data of Haase and of Shigekawa (for electric fields up to 7 kV/cm) can both be fit with the parameters in TABLE 2 by varying either the total doping level or the ionised impurity scattering compensation ratio [15]. Several of the other authors who report the data presented in FIGURE 1 also present a similar set of material parameters based on their studies. The differences between their data and those shown in TABLE 2 are relatively minor. As additional experiments are performed in the future, and as additional data are extracted from device studies, there will be continual refinements to the material parameters for $In_{0.53}Ga_{0.47}As$ provided in TABLE 2.

Finally, Shigekawa has obtained a set of material parameters which fit his data quite well [8]. The deviations of these parameters from those shown in TABLE 2 are not significant, and are primarily in the values of the intervalley scattering parameters. These parameters (intervalley deformation potential and intervalley phonon energy) are very important in determining the high field region (above 10 kV/cm) of the v - E curve, and they greatly influence the saturation velocity and the shape of the v - E curve in the region of large negative differential resistance [9]. However, there is no unique set of material parameters for $In_{0.53}Ga_{0.47}As$ which will provide a good fit for both Shigekawa's data and Windhorn's data. Also, alloy scattering is an important mechanism in high field transport which affects v - E curves, especially in the velocity saturation region [1,2]. The magnitude of alloy scattering depends on a parameter which is empirically-derived with some degree of uncertainty [9,14]. Therefore, the value used for the alloy scattering potential in TABLE 2 will to a small extent influence the values of the intervalley scattering parameters deduced from the curve fits. In conclusion, we believe that the parameter set in TABLE 2 is a reliable current set of material parameters for this important ternary material which can be used with confidence for device studies, and we will elaborate on this point in the next section.

D PREDICTION OF OTHER TERNARY ($In_{0.52}Al_{0.48}As$) MATERIAL PARAMETERS

The lattice-matched $In_{0.52}Al_{0.48}As/In_{0.53}Ga_{0.47}As/InP$ system is an extremely promising set of materials for digital, microwave, and optical device applications. Impressive device performance has been achieved for the high electron mobility transistor (HEMT), the charge injection transistor (CHINT), and the semiconductor laser. However, certain material-related

TABLE 2 Values of bulk material parameters and energy band (valley)-dependent material parameters for $In_{0.53}Ga_{0.47}As$.

I. BULK MATERIAL PARAMETERS			
Material Parameter	Value		
Lattice constant (cm)	5.867×10^{-8}		
Longitudinal (LO) phonon energy (eV)	0.03271		
Velocity of sound (cm/s)	4.101×10^{5}		
Low frequency dielectric constant	13.10		
High frequency dielectric constant	11.09		
Density (g/cm^3)	5.482		
Alloy scattering potential (eV)	0.529		
II. ENERGY BAND (VALLEY)-DEPENDENT MATERIAL PARAMETERS			
Material Parameter	Energy Band		
	Γ	L	X
Effective mass ratio (m^*/m_o)			
Longitudinal	0.041	1.980	2.365
Transverse	0.041	0.162	0.355
Band nonparabolicity (eV^{-1})	1.167	0.588	0.649
Energy bandgap (wrt valence band, eV)	0.77	1.323	1.438
Valley separation (from Γ, eV)	0	0.553	0.668
Optical phonon energy (eV)	0	0.0327	0.0327
Number of equivalent valleys	1	4	3
Intervalley deformation potential (eV/cm)			
From Γ	0	7.0×10^{8}	7.0×10^{8}
From L	7.0×10^{8}	7.0×10^{8}	5.0×10^{8}
From X	7.0×10^{8}	5.0×10^{8}	5.8×10^{8}
Intervalley phonon energy (eV)			
From Γ	0	0.02276	0.02384
From L	0.02276	0.02312	0.02696
From X	0.02384	0.02696	0.02276

issues remain which merit careful study if potential applications of these devices are to be fully exploited. For example, high output conductance has been repeatedly observed in deep-submicron channel length HEMTs using this material system [2]. Unfortunately, there is very little available information on material parameters or transport properties of $In_{0.52}Al_{0.48}As$. Thus, we have achieved a set of physical parameters for this ternary material based on a linear interpolation of the best available data for the InAs and AlAs binary components and theoretical and experimental properties of the $In_{0.75}Al_{0.25}As$ ternary [2]. TABLE 3 provides a set of material parameters which we have used for the study of high field transport in $In_{0.52}Al_{0.48}As$. A perusal of this table shows some differences in precision and knowledge. For example, the satellite (L and X) energy bands are treated as spherical bands instead of ellipsoids as they were for $In_{0.53}Ga_{0.47}As$. Also, it is obvious that several of the parameters for $In_{0.52}Al_{0.48}As$ are more 'estimates' with greater uncertainty than those parameters listed in TABLE 2. However, as shown in FIGURE 2, these parameters predict low-field electron mobilities which are in good agreement with experimental mobilities for $In_{0.52}Al_{0.48}As$ [16-18]. While this is not a full test of the parameters, it is an important first test, and the theoretical curve in FIGURE 2 should be considered as an upper limit to the true mobility since the alloy scattering potential is only an estimate and the effects of imperfections and clustering have not been included in the calculations. As another example, FIGURE 3 shows the resultant v - E curves for $In_{0.52}Al_{0.48}As$ at 300 K as a function of doping concentration between 1×10^{16} and 1×10^{18} cm^{-3}. The v - E curves strongly depend on the doping level and the amount of carrier compensation, and FIGURE 3 gives an indication of this dependence. Usually, the $In_{0.52}Al_{0.48}As$ serves as a donor supply layer in modulation-doped devices and the ionised impurity concentration in these layers is in the upper range shown in FIGURE 3. Also, the negative differential resistance is almost eliminated at heavy doping levels, and the low-field mobility, peak velocity, and saturation velocity are reduced while the threshold field is shifted toward higher values. On the other hand, the saturation velocity is not significantly influenced by doping concentration. However, alloy scattering is very important in $In_{0.52}Al_{0.48}As$, as shown by the curve for S = 0 in FIGURE 3. At high electric fields alloy scattering becomes more important due to the rapid increase of the alloy scattering rate as electron energy increases. The calculated saturation velocities (2.5×10^6 - 4×10^6 cm/s) are in good agreement with reported values estimated for different doping concentrations obtained from experimental $In_{0.52}Al_{0.48}As$ metal-semiconductor field effect transistors [19], while the values without alloy scattering considerably overestimate the saturation velocity. Thus, the material parameters shown in TABLE 3 for this ternary do predict agreeable values with the sparse set of available experimental data which has been reported.

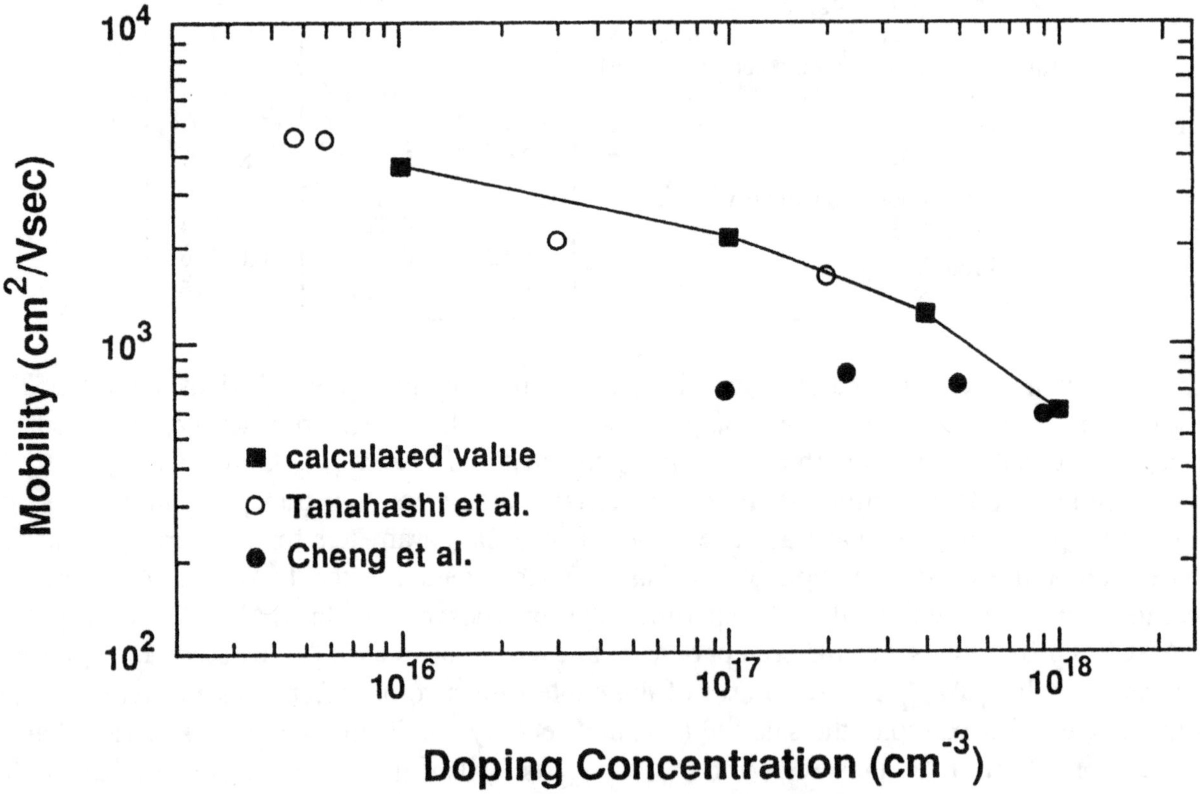

FIGURE 2 Low-field electron mobility as a function of doping concentration for uncompensated n-type $In_{0.52}Al_{0.48}As$ at 300 K. The solid curve is for calculated data and the experimental data points are shown for comparison.

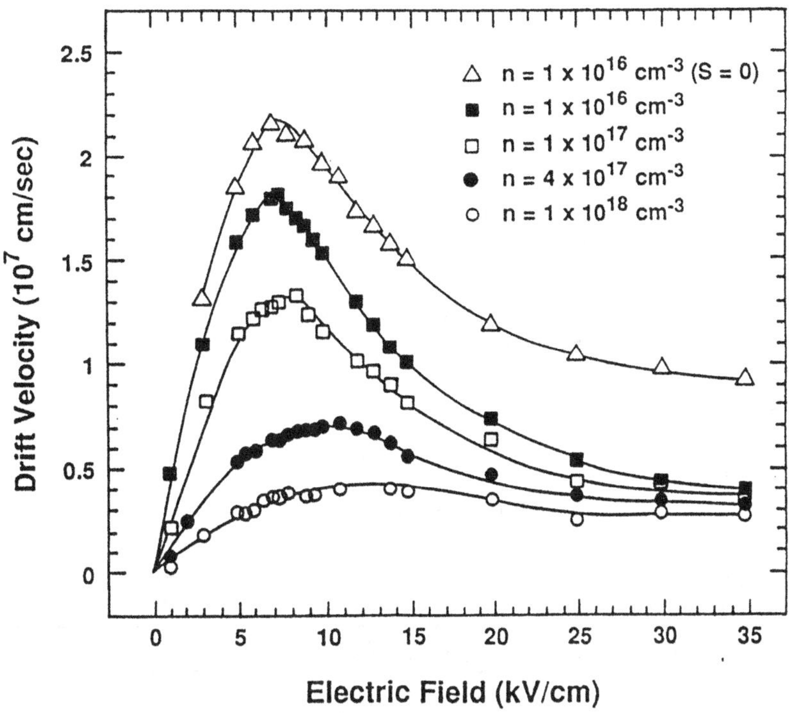

FIGURE 3 Electron drift velocity as a function of electric field for uncompensated n-type $In_{0.52}Al_{0.48}As$ at 300 K with doping concentration as a parameter. A random alloy is assumed for all cases except the one denoted as S = 0 for which a perfectly ordered alloy without alloy scattering is assumed.

We have used the parameter sets shown in TABLE 2 and TABLE 3 to model electron transport and device characteristics of $In_{0.52}Al_{0.48}As/In_{0.53}Ga_{0.47}As/InP$ high electron mobility transistors (HEMTs). These studies were done in order to interpret the effects of n or p channel doping on device behaviour reported in experimental efforts to reduce the problematic high output conductance consistently observed in these devices [20]. This is an ultimate test of both the material parameters and the role of high field transport in such complicated heterostructures. The simulation is accomplished with a two-dimensional, self consistent ensemble Monte Carlo model [21] which employs a three-valley (Γ-X-L) nonparabolic analytical band structure for both materials with all of the appropriate scattering mechanisms. FIGURE 4 shows the experimental and calculated transconductance values for a 0.25 μm HEMT as a function of gate voltage (V_g) with a drain voltage of 2.0 V and three different values of channel doping. The insert shows the vertical device configuration, which is a complicated set of heterostructure layers. As can be seen, there is a good overall agreement between the measured (data points with connecting lines) and calculated (lines) data. Also, the calculated average channel velocities are 2.1 x 10^7, 2.4 x 10^7, and 2.6 x 10^7 cm/s for the n-type, undoped, and p-type channel HEMTS shown in FIGURE 4. This demonstrates that high field dynamics strongly influence the device characteristics. The observed differences between the experimental and simulation results may be from uncertainties in the experimental device parameters, such as barrier layer thickness and doping. FIGURE 5 shows the experimental and calculated output conductance as a function of channel doping density for HEMTs with 0.25 μm and 1.0 μm gate lengths. The results clearly indicate that employing different channel doping can change output conductance significantly, with calculated trends that are consistent with both DC and RF (10 GHz) data. We have further used these calculations to interpret the performance enhancement of the HEMT with p-type channel doping as arising from improved carrier confinement in the potential well of the channel and from a reduced spread in the channel electric field profile.

TABLE 3 Values of bulk material parameters and energy band (valley)-dependent material parameters for $In_{0.52}Al_{0.48}As$.

I. BULK MATERIAL PARAMETERS			
Material Parameter	Value		
Lattice constant (cm)			
Longitudinal (LO) phonon energy (eV)	0.0395		
Velocity of sound (cm/s)			
Low frequency (static) dielectric constant	12.46		
High frequency dielectric constant	9.84		
Density (g/cm^3)	4.90		
Alloy scattering potential (eV)	0.47		
II. ENERGY BAND (VALLEY)-DEPENDENT MATERIAL PARAMETERS			
Material Parameter	Energy Band		
	Γ	L	X
Effective mass ratio (m^*/m_o)	0.070	0.390	0.602
Band nonparabolicity (eV^{-1})	0.843	0.552	0.588
Energy bandgap (wrt valence band, eV)	1.44	1.780	2.040
Valley separation (from Γ, eV)	0.0	0.340	0.60
Optical phonon energy (eV)			
Number of equivalent valleys	1	4	3
Intervalley deformation potential (eV/cm)			
From Γ	0	5×10^8	7.4×10^8
From L	5×10^8	5×10^8	5×10^8
From X	7.4×10^8	5×10^8	7.4×10^8
Intervalley phonon energy (eV)			
From Γ	0	0.0308	0.0291
From L	0.0308	0.0311	0.0360
From X	0.0291	0.0360	0.0291

In addition to the review of material parameters for $In_{0.53}Ga_{0.47}As$ and related heterostructures, we have also attempted to present a justification for use of the Monte Carlo method as a predictive tool for device performance. Recently, Fischetti and Laux [22] have discussed this point using submicrometer MOSFETs on Ge, Si, GaAs, InP and $In_{0.53}Ga_{0.47}As$ as a predictive vehicle. They suggest that this approach is both a conservative and realistic approach for device predictions while pointing out areas of uncertainty and reasonableness of assumptions related to doping, biases, temperature of operation, and device design and scaling. We believe that our work supports their conclusions in general and specifically related to the role of In-based materials technology in the search for improved device speed and performance.

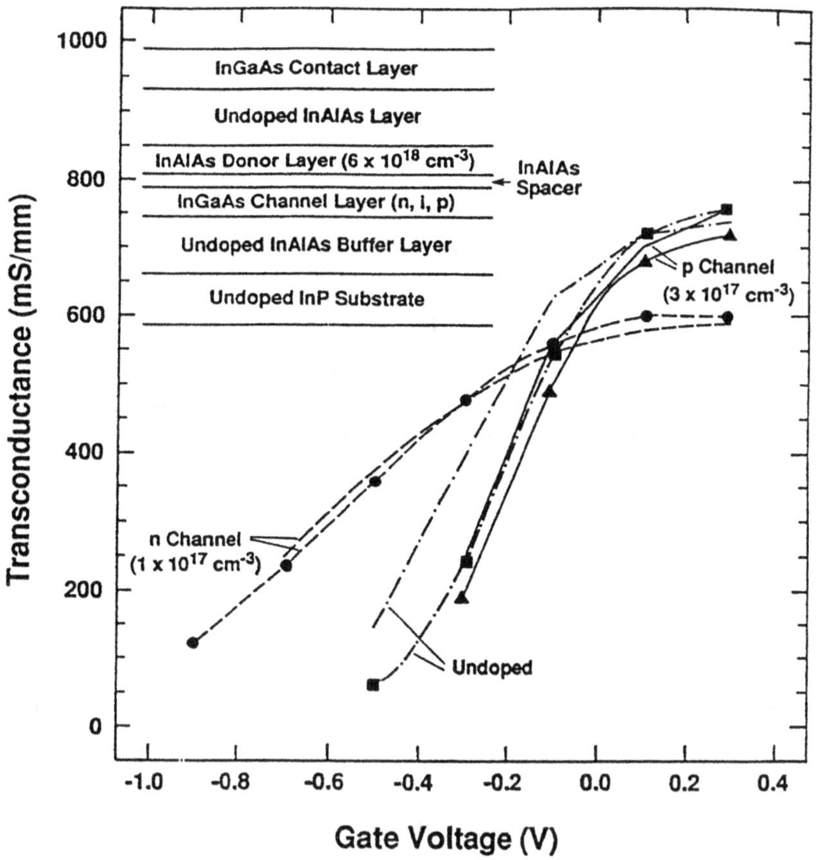

FIGURE 4 Transconductance (G_m) as a function of gate voltage (V_g) for 0.25 μm gate length HEMT devices with n-type, undoped, and p-type channel doping and a drain voltage of 2.0 V. Measured results and calculated results are shown as data points (with connecting lines) and lines, respectively. The inset illustrates the material layer configuration for fabricated devices.

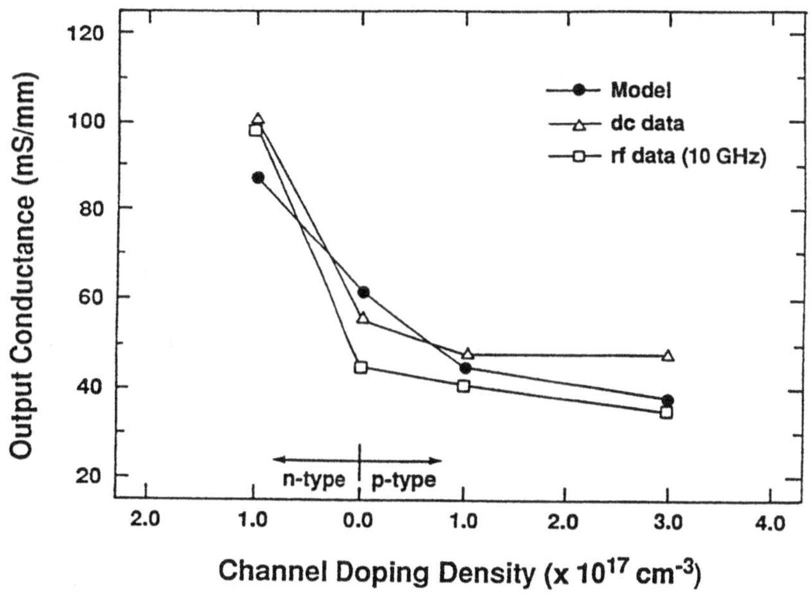

FIGURE 5 Output conductance as a function of channel doping for HEMT devices with 0.25 μm gate length and a gate voltage of 0 V.

E CONCLUSION

We discussed and summarised the experimental high field and related transport properties of $In_{0.53}Ga_{0.47}As$ and $In_{0.52}Al_{0.48}As$ ternary materials. We then discussed a scheme based on the Monte Carlo method and the use of available experimental data to calculate and predict the important bulk and band-related material properties of these important materials. Finally, we summarised the results of a study of a complicated heterostructure material system in the form of a HEMT device which showed that the material parameters obtained from this approach can lead to agreement with experimental device data and can result in an interpretation of physical phenomena in these devices which are related to high field transport in the ternary heterostructure material system. This approach can be continually refined and will become more powerful as more experimental studies are conducted which lead to new material data.

REFERENCES

[1] M.A. Littlejohn, T.H. Glisson, J.R. Hauser [in *GaInAsP Alloy Semiconductors* Ed. T.P. Pearsall (John Wiley, 1982) ch.10 p.243-74]

[2] H.S. Kim, H. Tian, K.W. Kim, M.A. Littlejohn [*Appl. Phys. Lett. (USA)* vol.61 (1992) p.1202-4]

[3] K.F. Brennan, D.H. Park [*J. Appl. Phys. (USA)* vol.65 (1989) p.1156-63]

[4] J.H. Marsh, P.A. Houston, P.N. Robson [in *Gallium Arsenide and Related Compounds* Ed. H.W. Thim (Institute of Physics, 1980) p.621-30]

[5] T.H. Windhorn, L.W. Cook, G.E. Stillman [*IEEE Electron Device Lett. (USA)* vol.3 (1982) p.18-20]

[6] B.R. Nag, S.R. Ahmed, M.D. Roy [*Physica (Netherlands)* vol.134B (1985) p.519-22]

[7] M.A. Haase, V.M. Robbins, N. Tabatabaie, G.E. Stillman [*J. Appl. Phys. (USA)* vol.57 (1985) p.2295-8]

[8] N. Shigekawa, T. Furata, K. Arai [*J. Appl. Phys. (USA)* vol.69 (1991) p.4003-10]

[9] M.V. Fischetti [*IEEE Trans. Electron Devices (USA)* vol.38 (1991) p.634-49]

[10] S. Adachi [*J. Appl. Phys. (USA)* vol.58 (1985) p.R1-R29]

[11] A.P. Long, P.H. Beton, M.J. Kelly [*J. Appl. Phys. (USA)* vol.62 (1987) p.1842-49]

[12] S. Adachi [*J. Appl. Phys. (USA)* vol.53 (1982) p.8775-92]

[13] K.W. Kim, K. Hess [*J. Appl. Phys. (USA)* vol.64 (1988) p.3057-64]

[14] J. Singh [*Physics of Semiconductors and Their Heterostructures* (McGraw Hill, 1993) p.397-471]

[15] T.G. van der Roer, F. Widdershoven [*J. Appl. Phys. (USA)* vol.59 (1986) p.813-20]

[16] T. Tanahashi, K. Nakajima, A. Yamaguchi, I. Umebu [*Appl. Phys. Lett. (USA)* vol.43 (1983) p.1030-32]

[17] K.Y. Cheng, A.Y. Cho, W.R. Wagner [*J. Appl. Phys. (USA)* vol.52 (1981) p.6328-34]

[18] W.-P. Hong et al [*J. Vac. Sci. Technol. B (USA)* vol.5 (1987) p.800-7]

[19] G.G. Zhou et al [*IEDM Tech. Digest (USA)* (1991) p.247-9]

[20] U.K. Mishra, L.M. Jelloian, M. Liu, M. Thompson, S.E. Rosenbaum, K.W. Kim [*Inst. Phys. Conf. Ser. (UK)* vol.120 (1991) p.287-92]

[21] K.W. Kim, H. Tian, M.A. Littlejohn [*IEEE Trans. Electron Devices (USA)* vol.38 (1991) p.1737-42]

[22] M.V. Fischetti, S.E. Laux [*IEEE Trans. Electron Devices (USA)* vol.38 (1991) p.650-60]

4.3 Low- and high-field transport in pseudomorphic InGaAs based heterostructures

P.K. Bhattacharya

February 1993

A INTRODUCTION

Pseudomorphic $In_xGa_{1-x}As/AlGaAs$ [1] ($x > 0$) and $In_xGa_{1-x}As/In_{0.52}Al_{0.48}As$ [2] modulation doped (MD) heterostructures have emerged as excellent material systems for the design and fabrication of high-frequency and low-noise field-effect transistors. A lesser known and potentially important heterostructure system is $InAs_xP_{1-x}/InP$ ($x \leq 1.0$). Vast improvements in mobilities and device performance are observed in modulation-doped field-effect transistors (MODFETs) made with pseudomorphic heterostructures, compared with lattice-matched ones. Some of the improvement in device performance results from a larger conduction band discontinuity, compared to those in lattice-matched heterostructures, leading to more efficient modulation doping and better confinement. The improvement also results from the enhancement in low- and high-field carrier transport properties with strain. This Datareview describes the measured low- and high-field carrier transport properties in pseudomorphic MD heterostructures. The low- and high-field transport properties in lattice-matched modulation doped heterostructures have been described in the Datareviews by L.F. Eastman and M.A. Littlejohn in this volume. It is important to point out that the data presented here are for steady-state conditions and do not consider transient effects.

B LOW- AND HIGH-FIELD ELECTRON TRANSPORT IN PSEUDOMORPHIC InGaAs/GaAs/AlGaAs MD HETEROSTRUCTURES

For the reasons mentioned above, steady improvements in the electron transport properties have been observed with increase of compressive strain in InGaAs-channel MD heterostructures. From high-field magnetoresistance measurements Masselink et al [3] measured electron velocities in $In_{0.15}Ga_{0.85}As/Al_{0.15}Ga_{0.85}As$ MODFETs which were at least 15% larger than equivalent $GaAs/Al_{0.15}Ga_{0.85}As$ MODFETs. The electron velocity in the pseudomorphic $In_{0.15}Ga_{0.85}As/ Al_{0.15}Ga_{0.85}As$ heterostructure at 77 K was 1.7×10^7 cm/s at a field of 3.5 kV/cm. The velocity is expected to peak at 2×10^7 cm/s. Effective saturation velocities in GaAs and $In_{0.15}Ga_{0.85}As$ channel MODFETs with 1 μm gate lengths were measured to be 1.75×10^7 and 2×10^7 cm/s, respectively [4]. The improvements in low- and high-field transport properties in GaAs- and InP-based pseudomorphic InGaAs are very well summarised in FIGURE 1 from Ballingall et al [5].

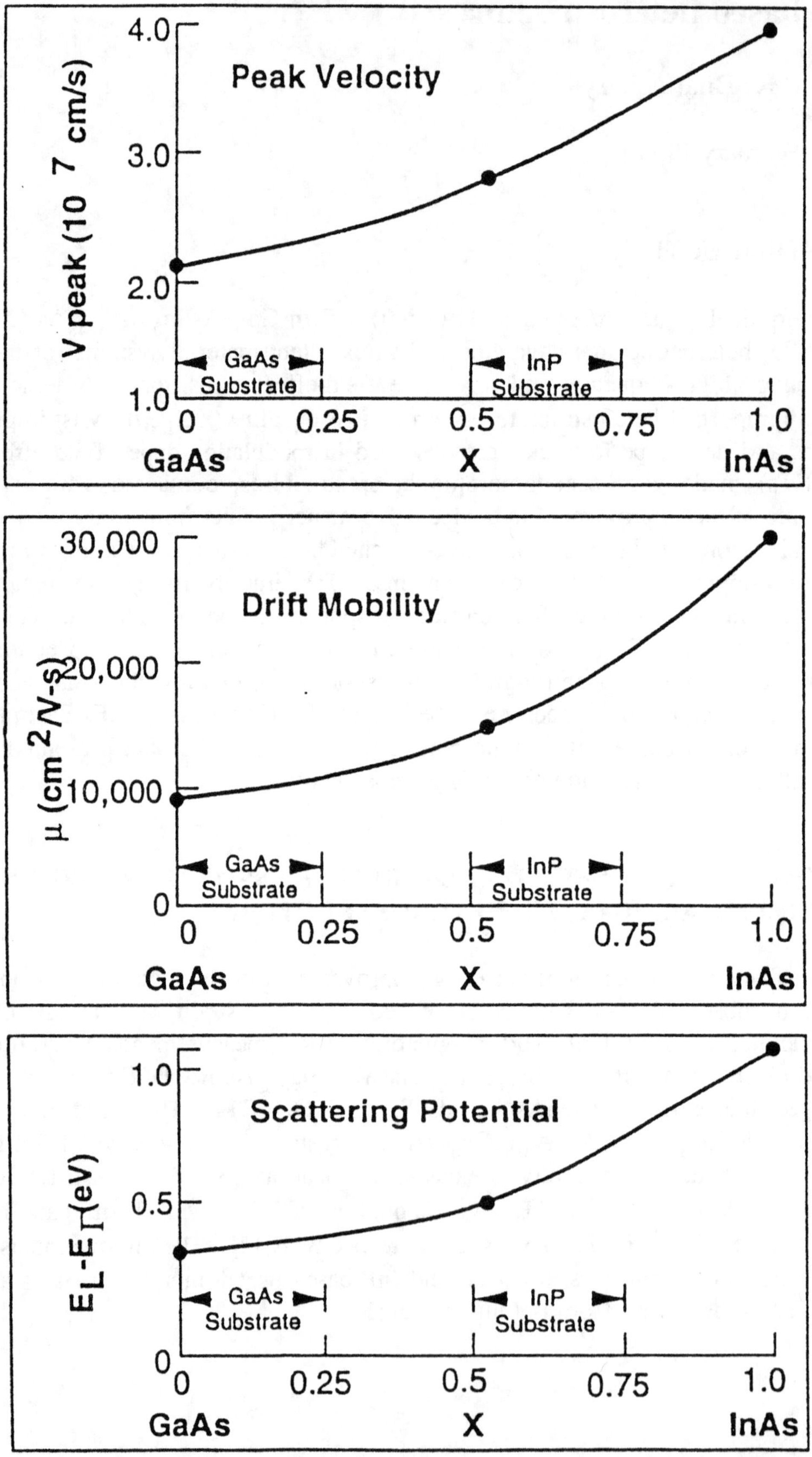

FIGURE 1 Transport properties of InGaAs/GaAs and InGaAs/InP [5].

C LOW- AND HIGH-FIELD TRANSPORT IN PSEUDOMORPHIC $In_xGa_{1-x}As/In_{0.52}Al_{0.48}As$ MODULATION-DOPED HETEROSTRUCTURES

A typical MD heterostructure grown by molecular beam epitaxy (MBE) is shown in FIGURE 2. Mobilities and carrier effective masses in these heterostructures have been measured by Hong et al [6] using the van der Pauw and Shubnikov-de Haas techniques, respectively. The data are shown in FIGURE 3 and TABLES 1 and 2. It is believed that the steady improvement in mobility with In composition x in the channel results from a combination of improved confinement (reduced intersubband scattering) and reduced alloy scattering.

TABLE 1 Measured Hall data from $In_xGa_{1-x}As/In_{0.52}Al_{0.48}As$ heterostructures.

Sample	Channel composition	Mobility (cm^2/V s)		2DEG density (10^{12} cm^{-2})	
		300 K	77 K	300 K	77 K
A	0.53	11500	50100	1.65	1.60
B	0.60	12300	65200	1.79	1.74
C	0.65	13900	74000	1.82	1.79

TABLE 2 Transport data obtained from Hall and Shubnikov-de Haas (SDH) measurements.

Sample	Hall data at 4.2 K		m^*/m_o	N_0 [a] (10^{12} cm^{-2})	N_1 [b]	$E_1 - E_0$ [c] (meV)
	μ_H (cm^2/V s)	n_H (10^{12} cm^{-2})				
A (0.53)	67900	1.46	0.046 ± 0.002	1.12	0.26	43.2
B (0.60)	95000	1.65	0.046 ± 0.002	1.53	0.04[d]	76.1
C (0.65)	13400	1.65	0.046 ± 0.002	1.65	0.04[d]	84.6

[a]Lowest subband occupation at 4.2 K determined from SDH measurements.
[b]Occupation in first excited subband at 4.2 K.
[c]Energy separation between ground state and first excited state in the 2DEG.
[d]These values are approximate.

The high-field electron transport characteristics [6,7] in these heterostructures are shown in FIGURES 4 and 5. The reduction in high-field mobility at high fields is believed to be due to optical phonon scattering. The pre-threshold velocity-field characteristics were obtained from pulsed current-voltage measurements on planar H-shaped devices. The improvement in the velocities at both 300 and 77 K are again due to reduced alloy scattering and reduced intersubband scattering.

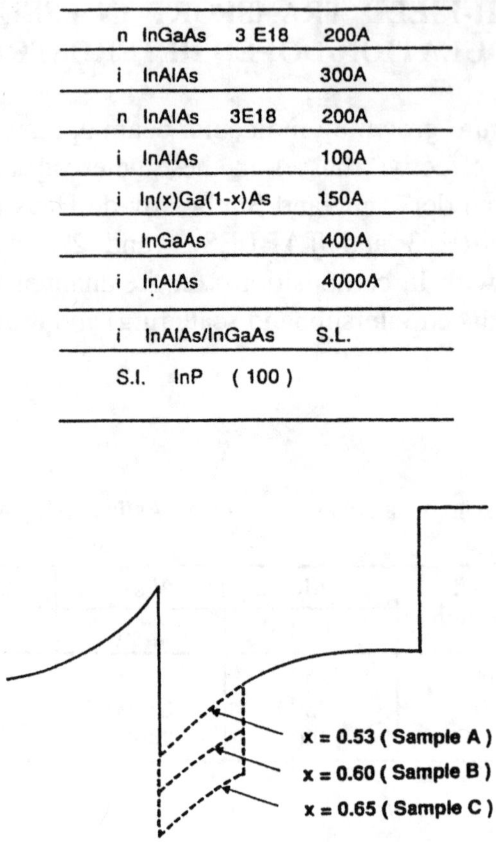

FIGURE 2 Schematic of pseudomorphic modulation doped heterostructures and corresponding band diagrams.

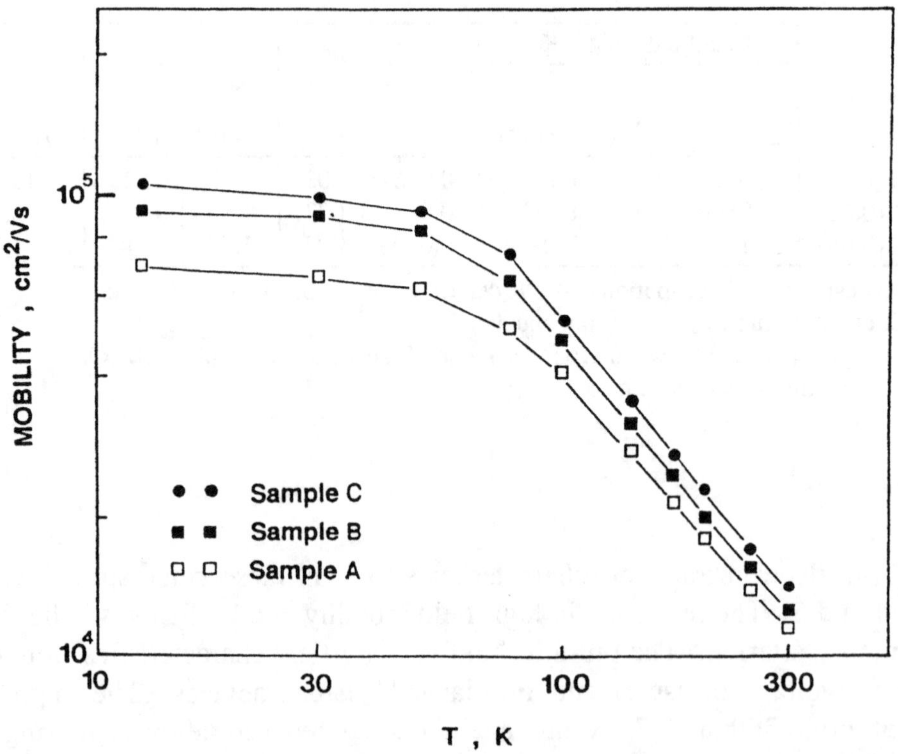

FIGURE 3 Temperature-dependent Hall mobilities as measured in samples A, B, and C.

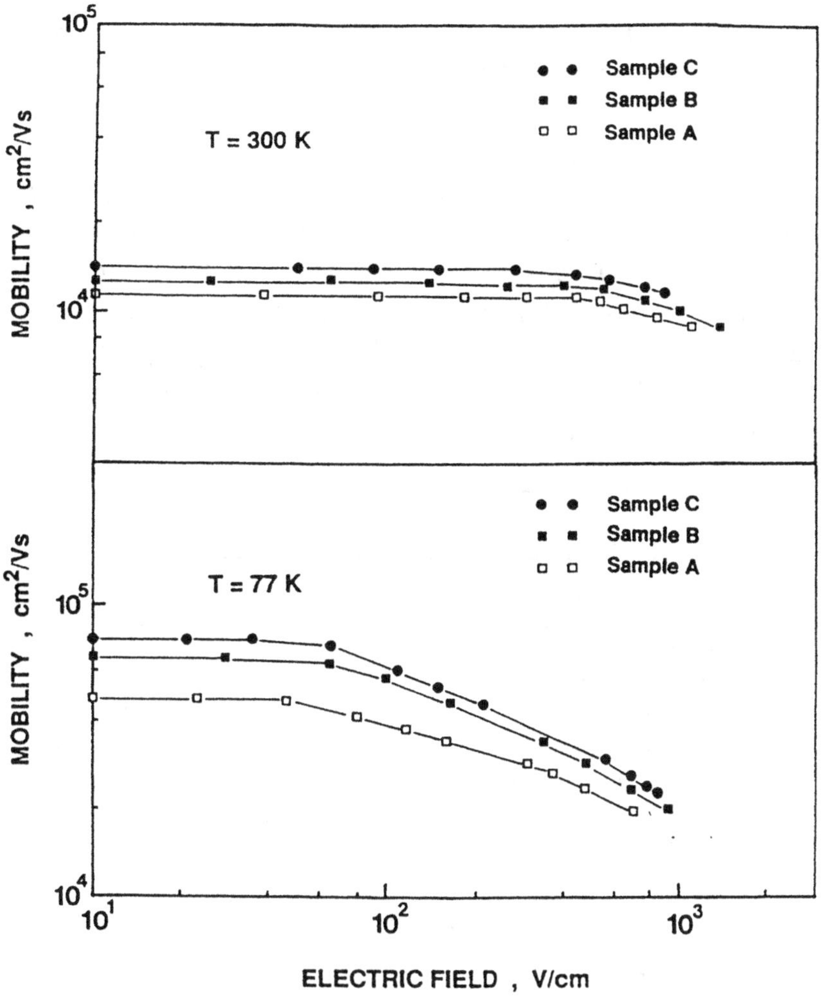

FIGURE 4 300 and 77 K mobility-field characteristics obtained from pulsed Hall measurements.

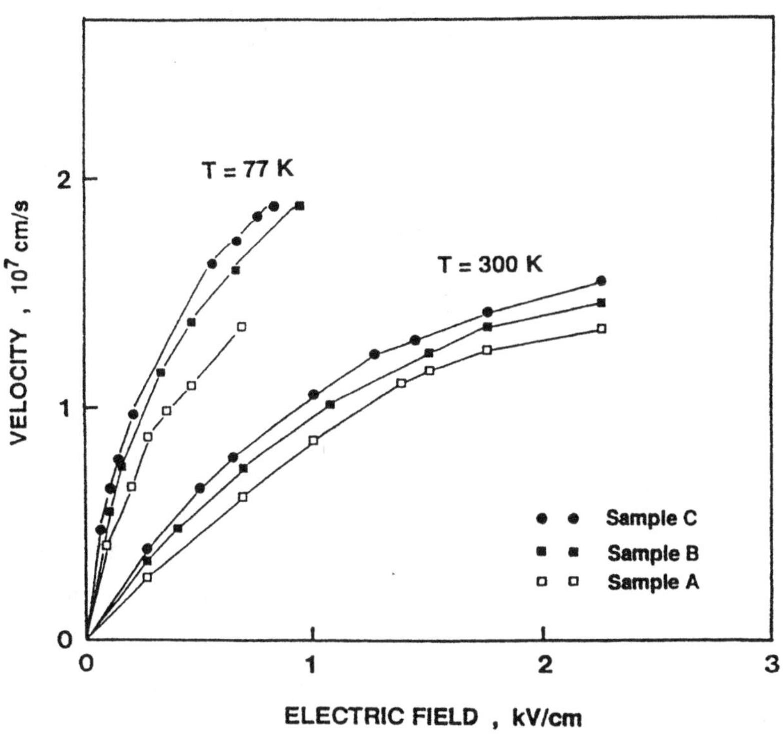

FIGURE 5 Velocity-field characteristics obtained from pulsed current-voltage measurements and pulsed Hall measurements at 300 and 77 K, respectively.

D LOW- AND HIGH-FIELD TRANSPORT IN $InAs_xP_{1-x}$/InP ($0.3 \le x \le 1.0$) MODULATION DOPED HETEROSTRUCTURES

The InAsP/InP pseudomorphic heterostructure system is potentially useful for long-wavelength optoelectronic device applications. It is also a heterostructure of choice for high-performance heterostructure FETs. The low- and high-field electron transport properties of this heterostructure system have been measured by Yang et al [8] in the structure shown in FIGURE 6. The low-field transport properties of typical pseudomorphic heterostructures are listed in TABLE 3. Note that the values of the Γ-L intervalley separation and the conduction band discontinuity are for relaxed InAsP alloys. Again, a steady improvement in the mobility is observed as the channel material approaches the binary InAs.

Undoped	InP		200 Å
S - doped	InP	3×10^{18} cm^{-3}	200 Å
Undoped	InP		100 Å
Undoped	$InAs_xP_{1-x}$ or $In_{0.53}Ga_{0.47}As$		y Å
Undoped	InP		5000 Å
S. I. (100)	InP	Substrate	

FIGURE 6 Schematic of $InAs_xP_{1-x}$/InP and $In_{0.53}Ga_{0.47}As$/InP modulation doped heterostructures grown by low-pressure OMCVD.

TABLE 3 Transport-parameters of $InAs_xP_{1-x}$/InP and $In_{0.53}Ga_{0.47}As$/InP modulation doped heterostructures.

Sample	$InAs_xP_{1-x}$ channel composition (x)	Channel thickness (nm)	Channel bandgap at 300 K (eV)	$\Delta E_{\Gamma-L}$ in InAsP (eV) [9]	$\Delta E_c = 0.65\ \Delta E_g$ for InAsP/InP (eV)	Hall data for E = 0 at 300 K: $n_s \times 10^{12}$ (cm^{-2})	Hall data for E = 0 at 300 K: μ (cm^2/V s)
A	0.3	10	1.053	0.743	0.193	1.88	3100
B	0.6	10	0.756	0.887	0.386	2.29	6100
C	0.8	10	0.558	0.982	0.515	2.39	7700
D	1.0	5	0.36	1.078	0.644	5.75	6900
E	$In_{0.53}Ga_{0.47}As$/InP	10	0.74	0.55	0.308	1.55	8500

The measured high-field electron mobilities in this heterostructure are depicted in FIGURE 7. The measured velocity-field characteristics at 300 and 77 K are shown in FIGURE 8. Note the steady improvement in the velocity with increase of x. The cross-over in the velocities of samples B and D in FIGURE 8(b) occurs due to an increase in the field-dependent alloy scattering in sample B ($InAl_{0.6}P_{0.4}$). It may be noted that the velocity of 3×10^7 cm/s in the InAs channel MD heterostructure at 77 K at $E = 2 \times 10^3$ V/cm is the highest measured. Considering only Γ and L valleys and insignificant real space transfer of electrons, a Monte

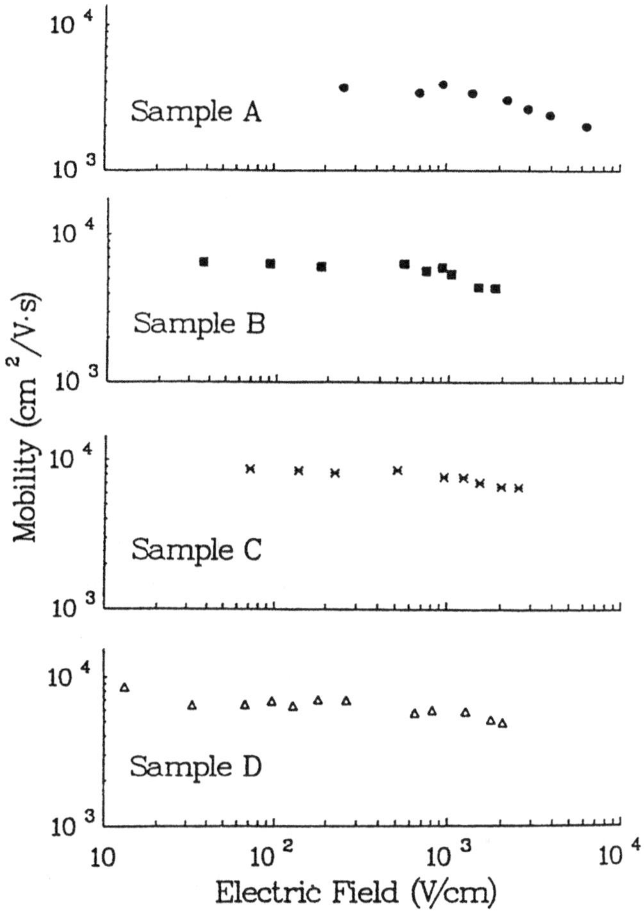

FIGURE 7 Measured mobility-field characteristics of modulation doped $InAs_xP_{1-x}$/InP with x = 0.3, 0.6, 0.8 and 1.0 at room temperature.

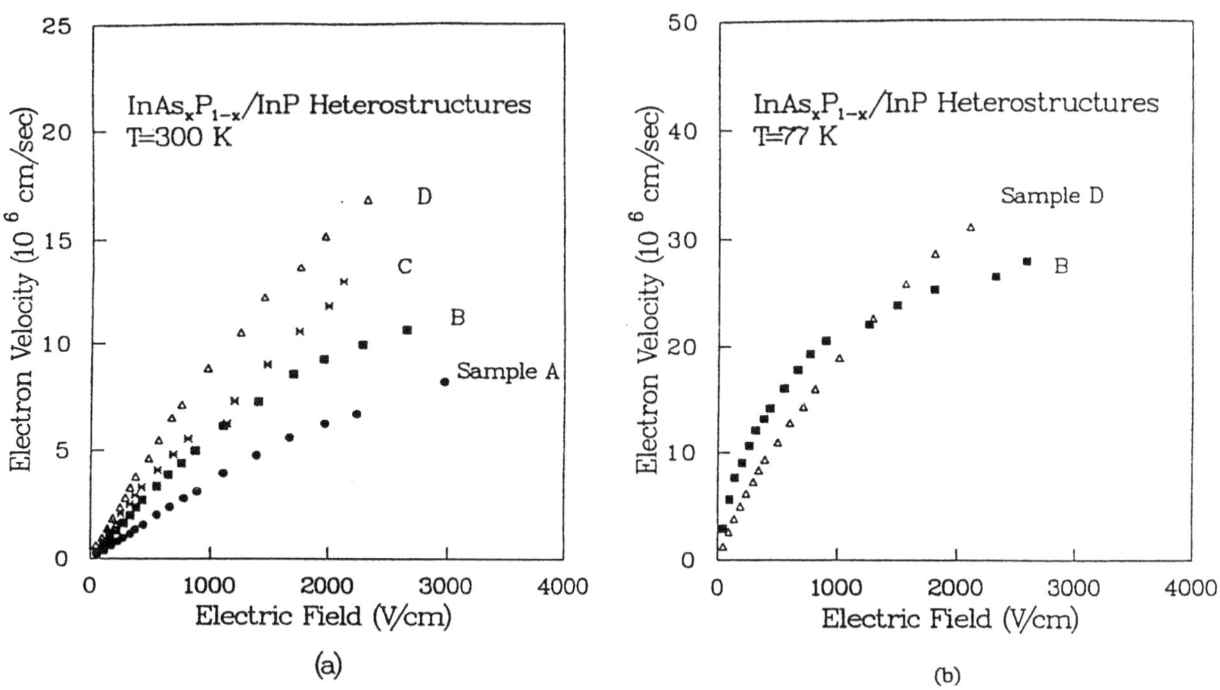

FIGURE 8 (a) Measured velocity-field characteristics of modulation doped $InAs_xP_{1-x}$/InP and $In_{0.53}Ga_{0.47}As$, and (b) Velocity-field characteristics of $InAs_xP_{1-x}$/InP modulation doped heterostructure (x = 0.6 and 1.0) at 77 K.

Carlo analysis for bulk $InAs_xP_{1-x}$ was performed by Yang et al [8]. All the relevant scattering mechanisms were included and for the sake of comparison with the measured data, it is assumed that the transport properties in a bulk channel and in a 2DEG are similar. The results are shown in FIGURES 9(a) and 9(b). The calculated electron velocities are in excellent agreement with the measured data of FIGURE 8. The data of FIGURE 9b indicate that even in the InAs/InP channel, the average electron energy at the highest field (~ 0.1 eV) is much smaller than $\Delta E_c \cong 0.64$ eV, or the Γ-L separation of 1.08 eV [9] in InAs. Therefore, even at high fields, most of the carriers remain confined in the channel.

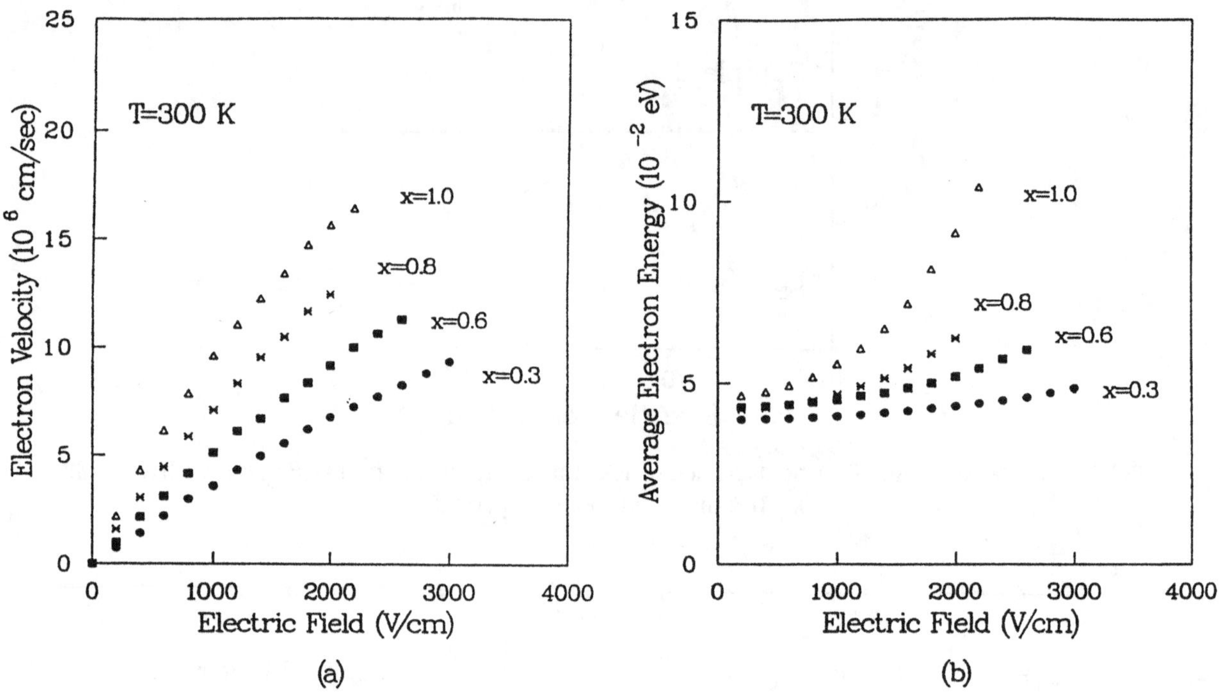

FIGURE 9 (a) Velocity-field characteristics and (b) average electron energy in the channel of $InAs_xP_{1-x}$/InP modulation doped heterostructures at 300 K obtained from steady-state Monte Carlo calculations.

E HOLE TRANSPORT IN InGaAs/GaAs STRAINED HETEROSTRUCTURES

Theoretical calculations, as outlined by J. Singh in this volume, indicate that the changes in the hole band structure and effective masses are more dramatic than those in the conduction band. Biaxial strain splits the light and heavy-hole subbands. As a result, the in-plane hole effective masses and hole transport are significantly altered. Measured hole mobilities in InGaAs/GaAs strained layer superlattices [10] are shown in FIGURES 10(a) and 10(b). The higher mobility in FIGURE 10(a) is attributed to a higher valence band offset for a larger compressive strain, and consequently a more efficient transfer doping.

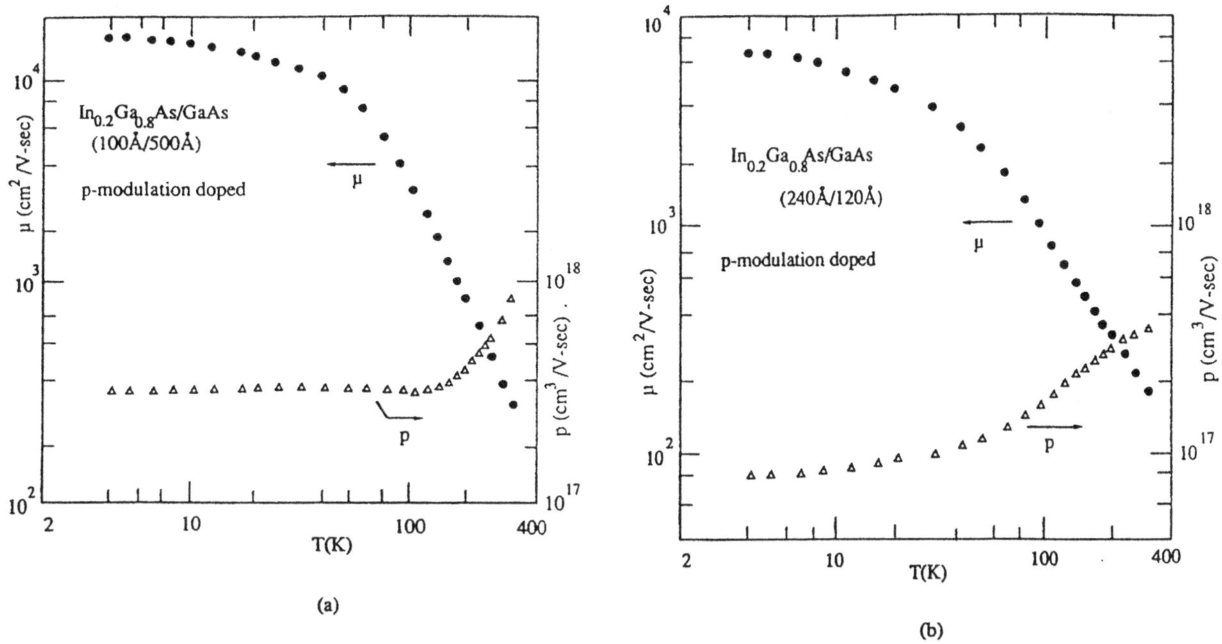

FIGURE 10 Variation of μ_{th} and $p = (eR_H)^{-1}$ versus T for $In_{0.2}Ga_{0.8}As/GaAs$ p-modulation doped structures for the cases of (a) large strain (misfit = 1.2%) and (b) relatively small strain (misfit = 0.5%) [10].

F CONCLUSION

The measured steady-state low- and high-field transport properties of some important pseudomorphic modulation doped heterostructures are reported here. The steady improvement of transport properties with compressive strain (accompanied by a lowering of the bandgap in the channel material) is principally due to an increase of the band offset. For the same reason (higher band offset) strain-relaxed heterostructures have demonstrated [11] higher electron mobilities than similar pseudomorphic structures with the same sheet electron concentration.

REFERENCES

[1] J.J. Rosenberg, M. Benlamri, P.D. Kirchner, J.M. Woodall, G.D. Pettit [*IEEE Electron Device Lett. (USA)* vol.6 (1985) p.491]

[2] U.K. Mishra, A.S. Brown, S.E. Rosenbaum, M.J. Delaney, S. Vaughn, K. White [*IEEE Trans. Electron Devices (USA)* vol.35 (1988) p.2441]

[3] W.T. Masselink et al [*International Electron Devices Meeting Techn. Dig.* (1985) p.755]

[4] P.R. de la Houssaye, D.R. Allee, Y-C. Pao, D.G. Schlom, J.S. Harris, R.F.W. Pease [*IEEE Electron Device Lett. (USA)* vol.9 (1988) p.150]

[5] J.M. Ballingall [private communications]

[6] W-P. Hong, G.I. Ng, P.K. Bhattacharya, D. Pavlidis, S. Willing, B. Das [*J. Appl. Phys. (USA)* vol.64 (1988) p.1945]

[7] W.-P. Hong, P.K. Bhattacharya [*IEEE Trans. Electron Devices (USA)* vol.34 (1987) p.1491]

[8] D. Yang, P.K. Bhattacharya, W-P. Hong, R. Bhat, J.R. Hayes [*J. Appl. Phys. (USA)* vol.72 (1992) p.174]

[9] M.V. Fischetti [*IEEE Trans. Electron Devices (USA)* vol.38 (1991) p.634]

[10] I.J. Fritz, L.P. Dawson, T.J. Drummond J.E. Schirber, R.M. Biefeld [*Appl. Phys. Lett. (USA)* vol.48 (1986) p.139]

[11] J. Chen, J.M. Fernandez, H.H. Wieder [*Appl. Phys. Lett. (USA)* vol.61 (1992) p.1116]

4.4 Carrier diffusion lengths and lifetimes in InGaAs

N. Chand

March 1993

A INTRODUCTION

Whenever the carrier concentrations are disturbed from their equilibrium values, they attempt to return to equilibrium. In case of injection of excess carriers, introduced by light or by a forward-biased pn junction, return to equilibrium is through recombination of the injected minority carriers with the majority carriers. In case of extraction of carriers, as in the space-charge region (SCR) of a reverse-biased diode, return to equilibrium is through the process of generation of electron-hole pairs. The parameters which characterise the rate of return to equilibrium are the minority carrier lifetime (τ) and the minority carrier diffusion length (L) which are related by

$$L = \sqrt{D\tau} \tag{1}$$

where D is the diffusion coefficient of the minority carrier. Obviously, lifetimes fall into two primary categories: recombination lifetimes τ_r and generation lifetimes τ_g. During recombination an electron-hole (eh) pair ceases to exist on average after a time τ_r. The generation lifetime, by analogy, is the time that it takes on average to generate an eh pair. Because devices consist of bulk regions and surfaces, both bulk and surface recombination or generation take place simultaneously, and their separation is sometimes quite difficult.

The minority carrier recombination occurs through the competing nonradiative and radiative processes. The nonradiative process includes recombination at deep traps or other defects, recombination at the surface, Auger recombination, and leakage of minority carriers to the ohmic contacts. Thus, both L and τ are very sensitive to the bulk and surface properties of the material, and are often used to characterise the material quality. A detailed description of carrier lifetimes and methods of their determination is given in [1,2].

B CARRIER LIFETIME

Very little information is currently available on τ and L in both lattice matched and strained InGaAs. Henry and coworkers [3] reported both radiative and nonradiative lifetimes in n- and p-type $In_{0.53}Ga_{0.47}As$. They deduced the values of lifetimes from data on luminescence decay time and luminescence efficiency. Their data are shown in FIGURE 1 as a function of majority carrier density. The nonradiative electron lifetime at $p = 2 \times 10^{18}$ cm^{-3} is 7 ns. At $p = 1 \times 10^{18}$ cm^{-3}, the radiative and nonradiative electron lifetimes are approximately equal. In n-$In_{0.53}Ga_{0.47}As$, the nonradiative hole lifetime at $n = 2 \times 10^{18}$ cm^{-3} is about 50 ns and the carrier density for equal radiative and nonradiative rates is about 3×10^{18} cm^{-3}.

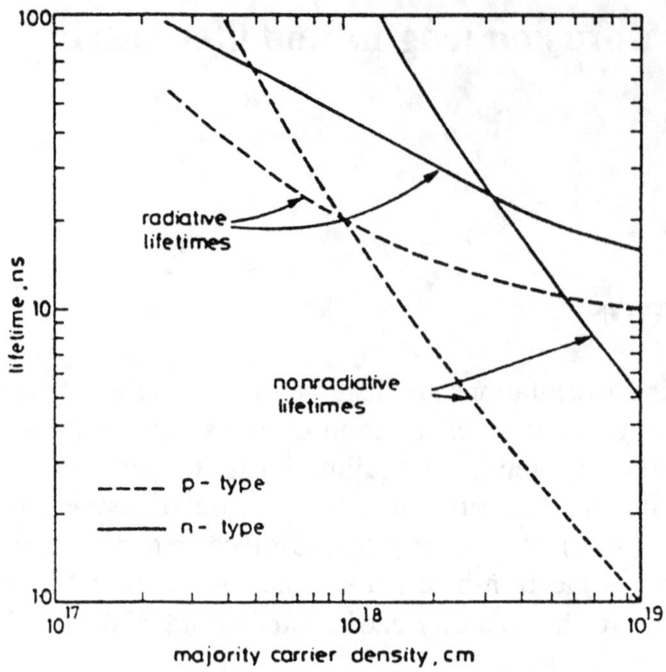

FIGURE 1 Radiative and nonradiative lifetimes in InGaAs [3].

Gallant and Zemel [4] reported a minority hole lifetime of 18.5 μs at 300 K in n-InGaAs with n ~1 x 10^{15} cm^{-3}. At 77 K, these authors estimated a lifetime of 1.2 μs. Furthermore, from theoretical considerations they expected the radiative recombination lifetime to follow a $T^{3/2}$ temperature dependence. Trommer and Hoffmann [5] reported a minority-hole lifetime of 6 μs in n-InGaAs with n = 1 - 3 x 10^{15} cm^{-3}. Sugiura et al [6] reported that there was no discernible difference in the minority carrier lifetime in strained $In_{0.7}Ga_{0.29}As$ and unstrained $In_{0.53}Ga_{0.4}As$ MQW lasers.

C DIFFUSION LENGTH

From the measured gain of InGaAs/InP HBTs or the quantum efficiency of InGaAs/InP photodetectors, several groups [4,5,7-12] have reported minority carrier diffusion lengths in $In_{0.53}Ga_{0.47}As$. The results are summarised in TABLE 1. In strained p-$In_{0.14}Ga_{0.86}As$ on GaAs substrates [13] with p ~ 2 x 10^{18} cm^{-3}, L_e was measured as 1.3 μm in LPE grown layers and ~ 1.0 μm in VPE grown material.

TABLE 1 Minority carrier diffusion length in $In_{0.53}Ga_{0.47}As$.

Hole diffusion length (L_h, μm)	Electron diffusion length (L_e, μm)
2.0 (n ~5 x 10^{15} cm^{-3}) [7]	2.5 (p ~1.4 x 10^{16} cm^{-3}) [9]
0.4 (n ~2 x 10^{16} cm^{-3}) [8]	0.8 (p ~5 x 10^{18} cm^{-3}) [9]
140 at 300 K (n ~1 x 10^{15} cm^{-3}) [5]	0.4-0.6 (p ~2-4 x 10^{19} cm^{-3}) [10]
47 at 77 K (n ~1 x 10^{15} cm^{-3}) [5]	7.5 (p ~5 x 10^{18} cm^{-3}) [11]
52-74 (n = 1-3 x 10^{15} cm^{-3}) [5]	4.0 (1 x 10^{19} cm^{3}) [12]

D CONCLUSION

The carrier diffusion lengths and lifetimes are important transport parameters in semiconductor materials. The values of these parameters in lattice-matched and strained InGaAs, as measured and reported by different authors, are summarised in this Datareview.

REFERENCES

[1] D.K. Schroder [*Semiconductor Material and Device Characterisation* (John Wiley & Sons, 1990) p.359-447]

[2] J.W. Orton, P. Blood [*The Electrical Characterisation of Semiconductors: Measurement of Minority Carrier Properties* (Academic Press, 1990)]

[3] C.H. Henry, R.A. Logan, F.R. Merrit, C.G. Bethea [*Electron. Lett. (UK)* vol.20 no.9 (1984) p.358-9]

[4] M. Gallant, A. Zemel [*Appl. Phys. Lett. (USA)* vol.52 (1988) p.1686-8]

[5] R. Trommer, L. Hoffmann [*Electron. Lett. (UK)* vol.22 (1986) p.360-2]

[6] H. Sugiura et al [*Appl. Phys. Lett. (USA)* vol.61 (1992) p.318-20]

[7] K. Alavi, A.N.M.N. Choudhury, J. Vleek, N.J. Slater, C.G. Fonstad, A.Y. Cho [*IEEE Electron Device Lett. (USA)* vol.3 (1982) p.379]

[8] Y. Takeda, M. Kuzuhara, A. Sasaki [*Jpn. J. Appl. Phys. (Japan)* vol.19 (1980) p.899-903]

[9] M.M. Tashima, L.W. Cook, G.E. Stillman [*J. Electron. Mater. (USA)* vol.11 (1982) p.831]

[10] P. Ambree, B. Gruska, K. Wandel [*Semicond. Sci. Technol. (UK)* vol.7 (1992) p.858-60]

[11] R. Bhat, J.R. Hayes, H. Schumacher, M.A. Koza, D.M. Hwang, M.H. Meynadier [*J. Cryst. Growth (Netherlands)* vol.93 (1988) p.919-23]

[12] I. Tonai, T. Yano [*Proc. SPIE, Materials and Technologies for Optical Communications (USA)* vol.866 (1988) p.63-8]

[13] T. Ota, K. Oe, M. Yamaguchi [*J. Appl. Phys. (USA)* vol.46 (1975) p.3674]

CHAPTER 5

SURFACE PROPERTIES

5.1 Surface structure and bonding configurations of InGaAs

P.N. Fawcett

February 1993

A INTRODUCTION

The surface structures of InAs and (In,Ga)As alloys have been studied in much less detail than the corresponding GaAs surfaces. Most recent surface studies of III-V semiconductors have concentrated on MBE grown material and in particular on the GaAs(001) surface while older studies tended to concentrate on the GaAs(110) surface which is the natural cleavage plane. In recent years, greater understanding of these surface structures has been obtained by the use of the Scanning Tunnelling Microscope (STM). Though these reconstructions have been primarily studied for GaAs, due to the similar chemical nature of the group III species, many of the same reconstructions are also observed for the (In,Ga)As system using electron diffraction techniques and it is to be expected that these are essentially the same as those of GaAs [1].

B InAs(001)

Ludeke [2] prepared (In,Ga)As alloys on InAs(001) and GaAs(001) surfaces using MBE and investigated the surface structure using reflection high energy electron diffraction (RHEED). Both 2x4 and 4x2 surface reconstructions were observed for (In,Ga)As over the entire alloy range. Electron energy loss spectroscopy revealed the presence of two empty dangling bond derived surface states. The In derived surface state was 0.4 - 0.5 eV below that of Ga and moved from above the conduction band edge into the bandgap for $x \geq 0.6$.

From previous work on the GaAs(001) surface [3], it is known that the 2x4 and c(2x8) reconstruction which is more commonly observed by low energy electron diffraction (LEED) can both coexist on the same surface, the c(2x8) unit cell being a special case of the 2x4 surface with anti-phase domain boundaries between the 2x4 unit cells. The two-fold periodicity was generally accepted to arise from dimerisation of the surface arsenic atoms [4] but the origin of the four-fold periodicity was the subject of much controversy until direct observation of the surface by an STM revealed it to consist of a regular arrangement of missing dimers in which every fourth dimer on the surface was missing [5]. This was in agreement with earlier tight-binding total energy calculations by Chadi [6]. Subsequent observations also demonstrated that unit cells containing two dimer vacancies can exist resulting in a surface with an arsenic coverage variable from 50 - 75% [7].

The origin of the 4x2 reconstruction is less certain, due to technical difficulties in imaging the surface at the temperatures needed to maintain the reconstruction. However, a study of the 4x2 reconstruction by Biegelsen [8] using an STM found it to consist of unit cells containing two adjacent dimers and two dimer vacancies in which the dimer bond direction was rotated by 90° with respect to the arsenic dimer bond direction of the 2x4 reconstruction. This is consistent with the presence of Ga-Ga dimers on the surface. In addition, anti-phase boundaries were observed creating a larger c(8x2) unit cell coexistent with the 4x2 unit cell, in a similar manner to the 2x4 - c(2x8) reconstruction. Though the same reconstructions are observed for both InAs and GaAs, the conditions needed to maintain the reconstructions are quite different, the lower thermal stability of InAs requiring a lower temperature to maintain the As-stable reconstructions.

The kinetics of surface processes involved in the growth of (In,Ga)As alloys from beams of the constituent elements was investigated by Foxon and Joyce [9] using modulated beam techniques, RHEED and AES. The principal limitation to growth of these alloy films is the thermal stability of InAs which is the less stable of the two III-V elements of which the alloy may be considered to be composed. This leads to the preferential desorption of As_2 which produces a temperature dependent sticking coefficient of As_4. This attains its maximum value of $S_{As_4} = 0.5$ as the reconstruction changes from the 2x4 to the 4x2 reconstruction and an excess population of indium on the surface is created as free metal. At higher temperatures, preferential desorption of indium compared to gallium can occur. Normally, for the growth of (In,Ga)As alloys a temperature of $\leq 500°C$ is used to prevent the reconstruction changing from the 2x4 to the 4x2 reconstruction and to prevent the associated formation of free indium on the surface.

On the static surface, at temperatures < 380°C which are much lower than those normally used to grow (In,Ga)As, a c(4x4) reconstruction is observed. Recently, Biegelsen [8] demonstrated that the surface unit cell consists of three As-As dimers aligned perpendicularly to the arsenic bond direction of the bulk terminated surface. This is in agreement with studies using X-ray photoelectron spectroscopy which showed that the reconstruction is not an intrinsic reconstruction but is instead due to trigonally bonded arsenic chemisorbed on the GaAs surface [10].

C InAs(110)

The InAs(110) surface is the natural cleavage plane and being a non-polar surface, structural rearrangement is likely to be less pronounced than on other surfaces. Measurements using medium energy ion blocking [11] and LEED [12] have indicated that while the 1x1 symmetry of the surface is preserved, a structural rearrangement of the surface does occur with the bonds connecting first layer atoms rotating and the bond lengths remaining essentially unchanged. Smit [11] obtained a value for this rotation angle of $30° \pm 2.4°$ while Duke and Paton [12] deduced a value of $31° \pm 3°$ with the In being displaced upwards by 0.07 ± 1 Å relative to its bulk position and the As being displaced downwards by an equal amount. Feenstra et al [13] performed a more direct investigation of the GaAs(110) surface using an STM and found a similar rearrangement with a buckling angle of 29 - 31°.

D InAs(111)

An early study of InAs(111) and $(\bar{1}\bar{1}\bar{1})$ surface structures was performed by Grant and Haas [14] who used LEED to study surfaces prepared by ion bombardment. On the InAs(111) surface a 2x2 reconstruction was observed while on the $(\bar{1}\bar{1}\bar{1})$ surface, faint streaks were observed between the integral order beams and on heating to higher temperatures facets began to form. More recent work by Sugiyama [15] on MBE grown films showed that the InAs

$(\bar{1}\bar{1}\bar{1})$ surface exhibits the As-stable 2x2 and In-stable ($\sqrt{19}$ x $\sqrt{19}$) R23.4° reconstructions in the temperature range 250 - 470°C. The change in reconstruction due to the variation in the As_4/In flux ratio and substrate temperature was compared with that of the InAs(001) 2x4 and 4x2 reconstructions. Above 470°C, fractional order streaks were no longer observed from the $(\bar{1}\bar{1}\bar{1})$ surface. On the InAs(111) surface only the As-stable 2x2 structure was observed for a substrate temperature between 450 and 500°C. A study of the GaAs(111) surface performed by an STM suggested that the 2x2 reconstruction is due to As adatom trimers bonded to the underlying surface while the ($\sqrt{19}$ x $\sqrt{19}$) R23.4° reconstruction is free from As trimers and consists of two-layer hexagonal rings [16].

E OTHER SURFACES

Sugiyama [15] also investigated the surface structure exhibited by InAs in the (211) and $(\bar{2}\bar{1}\bar{1})$ orientations both of which showed a 1x1 unreconstructed surface. During epitaxial growth on the $(\bar{2}\bar{1}\bar{1})$ surface using MBE, the RHEED pattern became weak and spotty which was associated with the formation of inverted pyramidal hollows in the layer.

F CONCLUSION

The (In,Ga)As(001) surface exhibits 2x4 and 4x2 reconstructions over the entire composition range. The origin of these has been discussed, together with the surface structure of InAs(110) and InAs(111), in terms of previous studies of the GaAs(001) surface.

REFERENCES

[1] J. Zhang, P.J. Dobson [in *Properties of Gallium Arsenide, 2nd Edition,* EMIS Datareviews Series no.2 (INSPEC, IEE, 1990) ch.15 p.363-8]

[2] R. Ludeke [*IBM J. Res. Dev. (USA)* vol.22 (1978) p.304-14]

[3] B.A. Joyce, J.H. Neave, P.J. Dobson, P.K. Larsen [*Phys. Rev. (USA)* vol.29 (1984) p.814-9]

[4] A.Y. Cho [*J. Appl. Phys. (USA)* vol.47 (1976) p.2841-3]

[5] M.D. Pashley, K.W. Haberern, W. Friday, J.M. Woodall, P.D. Kirchener [*Phys. Rev. Lett. (USA)* vol.60 (1988) p.2176-9]

[6] D.J. Chadi [*J. Vac. Sci. Technol. (USA)* vol.5 (1987) p.834-7]

[7] M.D. Pashley, K.W. Haberern, J.M. Gaines [*Surf. Sci. (Netherlands)* vol.267 (1992) p.153-60]

[8] D.K. Biegelsen, R.D. Bringans, J.E. Northrup, L.E. Swartz [*Phys. Rev. B (USA)* vol.41 (1990) p.5701-6]

[9] C.T. Foxon, B.A. Joyce [*J. Cryst. Growth (Netherlands)* vol.44 (1978) p.75-83]

[10] P.K. Larsen, J.H. Neave, J.F. van der Veen, P.J. Dobson, B.A. Joyce [*Phys. Rev. B (USA)* vol.27 (1983) p.4966-77]

[11] L. Smit, J.F. van der Veen [*Surf. Sci. (Netherlands)* vol.166 (1986) p.183-205]

[12] C.B. Duke, A. Paton [*Phys. Rev. (USA)* vol.27 (1983) p.6189-98]

[13] R.M. Feenstra, J.A. Stroscio, J. Tersoff, A.P. Fein [*Phys. Rev. Lett. (USA)* vol.58 (1987) p.1192-5]

[14] J.T. Grant, T.W. Haas [*Surf. Sci. (Netherlands)* vol.26 (1971) p.669-76]
[15] K. Sugiyama [*J. Cryst. Growth (Netherlands)* vol.75 (1986) p.435-40]
[16] D.K. Biegelsen, R.D. Bringans, J.E. Northrup, L.E. Swartz [*Phys. Rev. Lett. (USA)* vol.65 (1990) p.452-5]

5.2 Surfaces and dielectric/InGaAs interfaces

H.H. Wieder

January 1993

A INTRODUCTION

Technological control of the surfaces and interfaces of the ternary $In_xGa_{1-x}As$ alloys is required for a large number of present day and future electronic and electro-optic device and circuit applications where edge and surface passivation represent important considerations. The surfaces of $In_{0.53}Ga_{0.47}As$ lattice matched to InP and its interfaces with metals, insulators and other semiconductors are of particular importance.

B FREE SURFACES AND METAL/SEMICONDUCTOR INTERFACES

$In_xGa_{1-x}As$ surfaces exposed to the ambient environment develop complex overlayers [1] which consist of indium, gallium and arsenic oxides as well as elemental arsenic. A major advantage of $In_{0.53}Ga_{0.47}As$ surfaces is their considerably lower surface recombination velocity, SRV = 5 x 10^3 cm/s, compared to that of GaAs, SRV = 10^6 cm/s. Their SRV can be further reduced to ~ 20 cm/s by etching in a dilute bromine-methanol solution or in a sulphuric acid-based solution, (SRV = 50 cm/s), as shown by Yablonovitch et al [2]. Epitaxial regrowth of $In_{0.53}Ga_{0.47}As$ layers following their removal from molecular beam epitaxial (MBE) or organometallic chemical vapour phase deposition (OMCVD) reactors requires special chemical processing of their surfaces prior to their reinsertion. Frei et al [3] have described the advantages of using, for this purpose, a sulphuric acid-based etching solution. Regrowth of p-n junctions made in this manner exhibits ideality factors indistinguishable from those grown by uninterrupted procedures and symmetric as well as asymmetric $In_{0.53}Ga_{0.47}As$/InP heterojunctions exhibit an upper limit SRV = 200 cm/s. The surface barrier of metal covered, clean (100)-oriented $In_xGa_{1-x}As$ structures was investigated by Brillson et al [4] by X-ray photoemission spectroscopy (XPS). They found that the Fermi level, E_F, is unpinned and that the equilibrium bandbending depends on the interfacial chemical reactions. The situation is quite different for air or for oxygen-exposed $In_xGa_{1-x}As$ surfaces or for metal insulator semiconductor (MIS) structures. While oxidised GaAs surfaces are pinned near the middle of the bandgap Baier et al [5] found E_F to be pinned 0.13 eV above the conduction band edge on oxidised and (110)-cleaved n- or p-type InAs. The surfaces of air-exposed n-type InAs are, therefore, accumulated and those of p-type InAs are inverted.

Kajiyama et al [6] measured the Schottky barrier height, F_{bn}, as a function of composition of air exposed, n-type $In_xGa_{1-x}As$ alloys, shown in FIGURE 1. Also shown on the same figure is the composition dependence of the surface Fermi level $E_F^* = E_{cs} - E_F$ and where E_{cs} is the conduction band edge at the surface. The data were derived from capacitance vs. voltage, (C-V), measurements [7], made on MIS structures with 0.1 μm thick SiO_2 insulating layers deposited by means of a low temperature plasma assisted pyrolysis of silane in the presence of

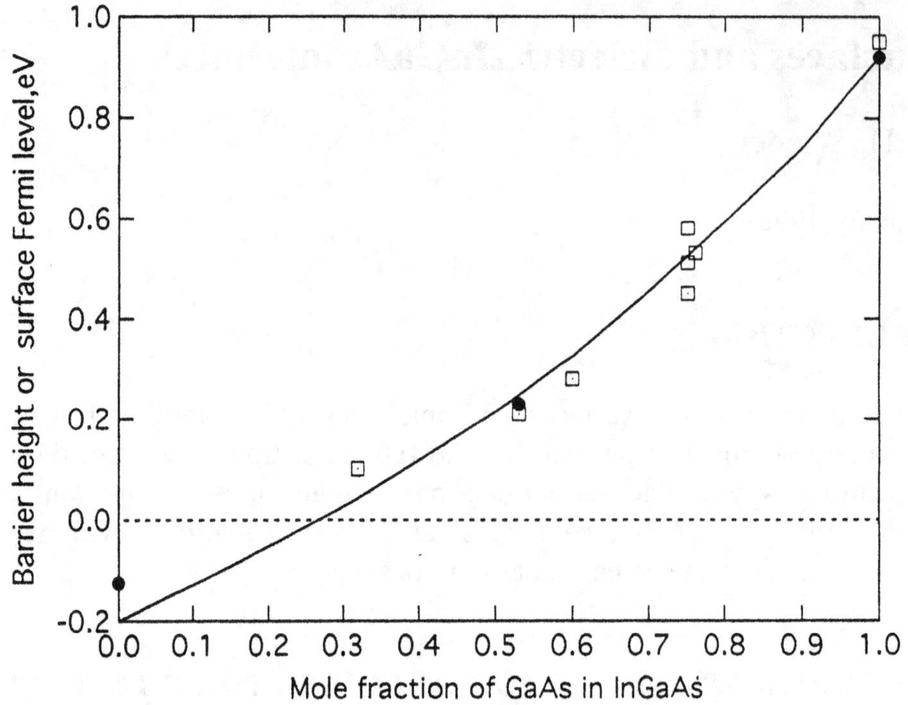

FIGURE 1 Equilibrium surface Fermi level dependence on the fractional GaAs concentration of $In_xGa_{1-x}As$ MIS structures in comparison with the barrier height of metal covered, air-exposed $In_xGa_{1-x}As$ surfaces [6,7].

nitrous oxide. Decreasing the fractional indium concentration of $In_xGa_{1-x}As$ from x = 1 to x = 0 moves E_F^* from ~ 0.122 eV, above the conduction band minimum (CBM) of InAs, to ~ 0.21 eV below the CBM for $In_{0.53}Ga_{0.47}As$ and thence to 0.91 eV below the CBM for GaAs. In the case of $In_{0.53}Ga_{0.47}As$ Morgan and Frey [8], using a thin interfacial SiO_2 layer, found, from current vs. voltage (I-V) measurements made on n-type 'Schottky barrier' diodes, that F_{bn} = 0.19 eV. The complementary barrier height of air-exposed, p-type $In_{0.53}Ga_{0.47}As$ was derived by Veteran et al [9] from C-V measurements, F_{bp} = 0.55 eV. All of these results confirm the significance of Fermi level pinning of air exposed 'Schottky barrier' structures and that of the interfacial surface chemistry of this compound.

C BARRIER MODIFICATION

Concerted efforts have been made to increase F_{bn} of $In_{0.53}Ga_{0.47}As$ by chemical procedures. Loualiche et al [10] have described such a method which consists of removing the native oxide in 10% HF solution followed by immersion in 100°C deionised water and subsequent oxidation in (100:5:1, H_2O:NH_4:H_2O_2). Schottky diodes made by such procedures have F_{bn} = 0.65 eV, an ideality factor n = 1.10, and a breakdown voltage, V_{bd} = 2.5 V. A similar procedure for barrier enhancement was described by Hwang et al [11]. After removal of their native oxide the $In_{0.53}Ga_{0.47}As$ specimens were immersed in a $P_2S_5/(NH_4)_2S$ solution which produced a stable interfacial layer assumed to be some form of P oxide. Such diodes are reported to have F = 0.54 eV but a comparatively low ideality factor, n = 1.53. It appears, therefore, that the barrier modification is due to a complex interfacial layer which may have some of the attributes of a low resistivity (leaky) insulator. Such structures cannot be described adequately in terms of a Schottky barrier nor are they insulated gate structures.

D MIS AND MISFET STRUCTURES

The potentially advantageous properties of insulator/$In_{0.53}Ga_{0.47}As$ interfaces have focused attention on the use of this alloy for metal insulator semiconductor field effect transistors (MISFET). A wide variety of insulating layers have been used in such structures in order to emulate, as nearly as possible, the favourable properties of silicon/silicon oxide MOSFETs. Mullin and Wieder [12] have used such MIS structures with Al_2O_3 insulating layers grown by a low temperature plasma assisted CVD process to make field effect-controlled electrical and galvanomagnetic measurements. The data was used to derive N_{ss}, the density of the surface states within the fundamental bandgap of $In_{0.53}Ga_{0.47}As$. Curve (a) in FIGURE 2 shows that a high density of surface states prevents the displacement of the Fermi level from the CBM below E_c - 0.2 eV, in good agreement with the Fermi level pinning position determined by means of C-V measurements [13,14] and with F_{bn} measured on oxidised surfaces. Curve (b) shows $N_{ss}(E)$ following annealing of such an MIS specimen in an ambient environment at a temperature of 160°C for 16 hr. It suggests that even low temperature annealing can reduce N_{ss} and prevent pinning of the Fermi level and allows its displacement over a substantial portion of the fundamental bandgap. Deep depletion as well as surface inversion is thus feasible. Inversion mode MISFETs with CVD-deposited SiO_2 on p-type $In_{0.53}Ga_{0.47}As$ heteroepitaxial layers grown on semi-insulating InP were first described by Wieder et al [15]. However, the field effect mobility and the transconductance, g_m, of the n-channel electrons was lower by more than an order of magnitude compared to n-type $In_{0.53}Ga_{0.47}As$ or that of depletion-mode devices and is attributed to severe scattering of the electrons in the two-dimensional inversion channel quantum well by defects located at the insulator interface.

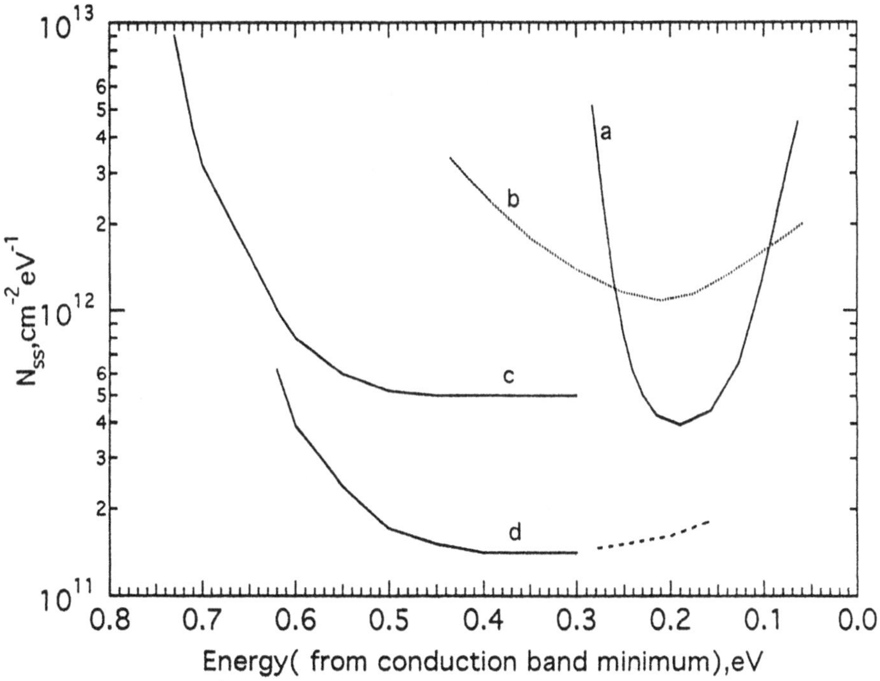

FIGURE 2 Energy dependence of surface state density of n-doped and p-doped $In_{0.53}Ga_{0.47}As$ MIS structures: (a) virgin n-type specimen [12]; (b) n-type specimen after isothermal annealing at low temperatures [12]; (c) with sputtered SiO_2 insulating layer [16]; (d) with CVD-grown SiO_2 insulating layer [16].

Kaumans et al [16] have confirmed inversion-mode operation of a similar MISFET but the g_m is still small, in spite of the low density of $N_{ss}(E)$, shown in FIGURE 2(c) and (d), derived from C-V measurements made on MIS structures with SiO_2 insulating layers grown by sputtering or CVD procedures. They also found that thermal annealing in an $AsH_3 + N_2$ or $PH_3 + N_2$ environment produces a further reduction and a broad minimum in $N_{ss}(E)$ in the $10^{11}/cm^2$-eV range. Considerably better results were obtained by Liao et al [17] using a plasma oxidation technique [18] to grow a 45 nm thick gate insulator on acceptor-doped, p ~ $2 \times 10^{17}/cm^3$, $In_{0.53}Ga_{0.47}As$ heteroepitaxial layers, with n^+ source and drain junctions. On an inversion-mode MISFET with a 4.5 μm long gate they obtained a g_m = 20 mS/mm and a saturated electron drift velocity, $v_s = 2.6 \times 10^6$ cm/s. However a major impediment of such devices is the slow temporal drift of the transistor characteristics associated with deep level traps at the native oxide interface, in a manner similar to InP MISFETs [19,20]. Better results were obtained by O'Connor et al [21] with depletion mode MISFETs using 10 to 25 nm thick layers of silicon nitride deposited by plasma assisted low temperature CVD on donor doped, $n = 2\text{-}8 \times 10^{16}/cm^3$ $In_{0.53}Ga_{0.47}As$ heteroepitaxial layers. With gate lengths of 1.2 μm they obtained a peak g_m = 130 mS/mm and $v_s = 2 \times 10^7$ cm/s. Further improvements with such MISFETs were obtained by Cheng et al [22] by the use of self-alignment techniques to reduce the gate length to 0.5 μm and reduce the source series resistance so that peak g_m = 250 mS/mm and $v_s = 24 \times 10^7$ cm/s. This suggests that interfacial scattering, shielded by the depletion region, is less effective than in inversion layers where it reduces the saturated velocity by a factor of 5 in comparison with $v_s = 2.95 \times 10^7$ cm/s in an n-type channel, determined by Bandy et al [23]. The microwave properties of depletion mode MISFETs with CVD-grown SiO_2 insulating layers and with 1 μm gate were investigated by Gardner et al [24,25]. They obtained a small signal gain of 5.8 dB and developed 110 mW with 3 dB power gain and 10.5% power added efficiency at 12 GHz.

A serious and, as yet, only a partially resolved problem is the instability and time-dependent drift of $In_{0.53}Ga_{0.47}As$ MISFET characteristics. Selders and Beneking [26] investigated the drift behaviour of n-channel inversion mode MISFETs made with a CVD grown SiO_2 insulating layer and found that rapid thermal annealing decreases the short term drift, as shown in FIGURE 3. The long term drift is essentially unaffected. The near term drift is a strong function of the oxide deposition temperature. If the depletion width of a depletion mode $In_{0.53}Ga_{0.47}As$ MISFET is greater than the thickness of its source drain channel then, as shown by Ishii et al [27], such a normally 'cut off' transistor can be driven into conduction by a positive gate bias. A different type of enhancement mode MISFET can be made [28] by using a closely compensated, n-type, non-conducting, $In_{0.53}Ga_{0.47}As$ layer for the transistor channel; a positive gate voltage is then used to induce surface accumulation in such normally off devices. However, their g_m and v_s values are low; at best they are comparable to inversion mode MISFETs with the electron velocity limited by severe interfacial scattering.

It is evident from work performed on MIS structures and on MISFETs that technological control is required not only of the $In_{0.53}Ga_{0.47}As$ surface but also of the insulator and its interfaces. Selders et al [29] performed secondary ion mass spectroscopic profiling of $In_{0.53}Ga_{0.47}As$ specimens coated with SiO_2 layers using a variety of annealing procedures in H_2/AsH_3 environments. They found that for such layers deposited at temperatures between 200 and 400°C Ga and In signals are observed far into the oxide. Even after the deposition of 60 to 70 nm of the oxide Ga, In and As are detected in concentrations of the order $10^{18}/cm^3$. The investigations of Taillepied et al [30] on Si_3N_4-coated $In_{0.53}Ga_{0.47}As$ indicate lower

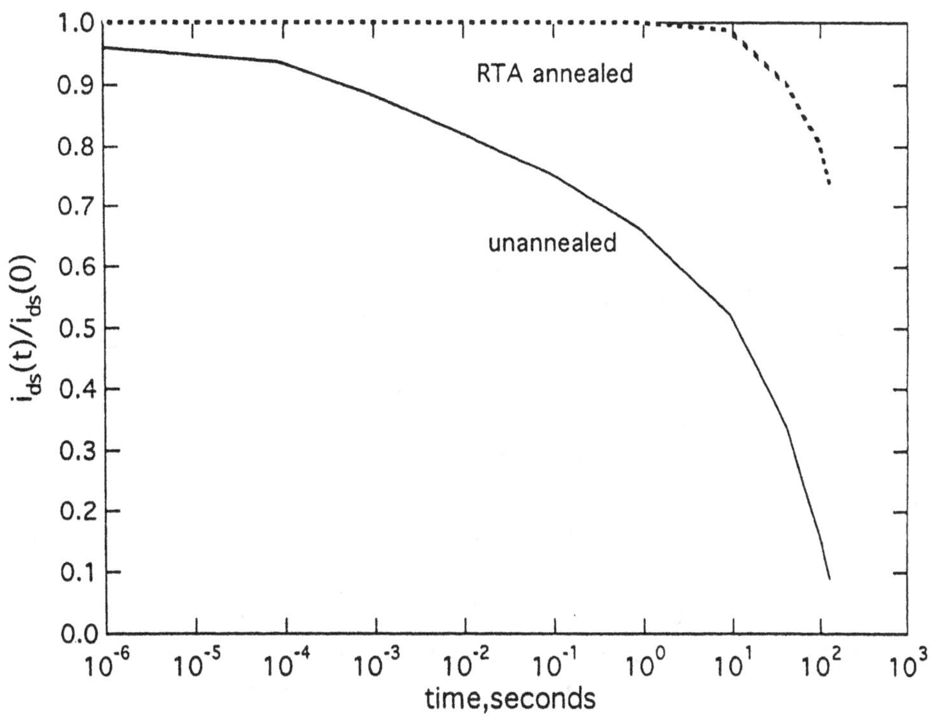

FIGURE 3 Effect of rapid thermal annealing (700°C) on the time-dependent drift of the drain current of $In_{0.53}Ga_{0.47}As$ MISFET normalised to its initial value at t = 0. After Selders and Beneking [26].

concentrations of In and Ga in the insulator consistent with its higher diffusion barrier compared to the oxide. In either case preservation of the stoichiometry of the semiconductor surface must be considered to be a prime consideration in the choice of appropriate insulators for device purposes.

E HETEROJUNCTION-ASSISTED INTERFACES

Lattice matched or pseudomorphic heterojunctions with large band edge discontinuities may represent a possible solution to the impediments associated with homomorphic or heteromorphic insulators. Lee and Forrest [31] have shown that $In_xGa_{1-x}As/InP$ heterojunctions lattice mismatched over a range from -0.24% to 0.26% have interface state densities < $10^{10}/cm^2$. Comparable interface state densities might be inferred from the data of Wada et al [32] obtained on $In_xGa_{1-x}As/In_yAl_{1-y}As/InP$ photodiodes. However, lattice matching alone is not a sufficient condition for low interface state densities. The congruent MBE deposition of the crystalline insulator SrF_2 destroys the surface stoichiometry of the heterojunction and is qualitatively worse than SiO_x, as shown by Leonard et al [33]. Considerably better results have been obtained by Akazawa et al [34] by MBE grown at 500°C of 1 to 1.5 nm thick silicon layers on MBE grown $In_{0.53}Ga_{0.47}As$ followed by deposition of SiO_2 using a low temperature photo CVD process. For effective surface passivation it is necessary for the Si CBM to be above that of the $In_{0.53}Ga_{0.47}As$ channel in accordance with considerations advanced by Hasegawa et al [35]. The MIS and MISFET structures made in this manner were annealed in hydrogen at 350°C. Such annealing reduces the $N_{ss}(E)$ by nearly an order of magnitude with a minimum $N_{ss} = 1.6 \times 10^{11}/cm^2$-eV at 0.28 eV below the CBM. However it is not clear whether equilibrium surface inversion can be achieved with such structures although the drain current drift of n-channel depletion mode MISFETs seems to be negligible.

F CONCLUSION

The interfacial chemistry of insulator/$In_{0.53}Ga_{0.47}As$ interfaces determines the energy, density and the distribution of the interface states. Assuming that fast interface states reside on the surface of this semiconductor and are related to lattice defects produced by its exposure to various adsorbates then the model of Fermi level pinning [28] might be applicable to $In_{0.53}Ga_{0.47}As$. Surface traps located at 0.55 eV relative to the valence band edge were calculated to account for the equilibrium position of the Fermi level in terms of a surface donor density, $N_{dss} = 5 \times 10^{12}/\text{cm}^2$ and surface acceptors, $N_{ass} = 10^{11}/\text{cm}^2$; these appear to be at least one order of magnitude smaller than those which cause Fermi level pinning in GaAs.

Hasegawa et al [36] have summarised the main features of insulator/semiconductor interfaces, revealed experimentally by a variety of techniques, including C-V, admittance measurements, photocapacitance spectroscopy and deep level transient spectroscopic measurements: (a) the U-shaped $N_{ss}(E)$ distribution of III-V semiconductors is defined in terms of its curvature, the energy location of its minimum, E_{HO}, and its density, N_{ss} min; (b) the curvature and the magnitude of N_{ss} min depend on the insulator type and on its processing condition; (c) for any given semiconductor E_{HO} is invariant and represents the charge neutrality point of the interface state distribution; (d) the states in the vicinity of the insulator/semiconductor interface cover a wide range of time constants from a fraction of a microsecond to many seconds. They propose that at non-ideal interfaces a distortion of the local bonds which may be caused by chemical disproportionation or strain relaxation can produce an incomplete separation of bonding and anti-bonding states forming a continuum around the charge neutrality level. They found that for $In_xGa_{1-x}As$ with x = 0.75, 0.47 and 0.35 MIS structures the calculated $E_{HO} - E_V$ is in fair agreement with experiment and suggest that there is a strong tendency for the Fermi level to be pinned at E_{HO} if a large amount of disorder is introduced at the interfaces.

A number of troublesome issues remain unsolved. Among the most important of these is understanding the origin and technological control of the time dependent drift of $In_xGa_{1-x}As$ MISFET parameters while simultaneously improving the electron velocity in inversion layers to values comparable to those of depletion mode devices.

REFERENCES

[1] G.P. Schwartz [*Thin Solid Films (Switzerland)* vol.103 (1983) p.3]

[2] E. Yablonovitch, R. Bhat, C.E. Zah, T.J. Gmitter, M.A. Koza [*Appl. Phys. Lett. (USA)* vol.60 (1992) p.37]

[3] M.R. Frei, J.R. Hayes, H.F. Shirokmann, C. Caneau [*J. Appl. Phys. (USA)* vol.70 (1991) p.3967]

[4] L.J. Brillson et al [*Appl. Phys. Lett. (USA)* vol.48 (1986) p.1458]

[5] H.-U. Baier, L. Koenders, W. Muench [*J. Vac. Sci. Technol. B (USA)* vol.4 (1986) p.1095]

[6] K. Kajiyama, Y. Mizushima, S. Sakata [*Appl. Phys. Lett. (USA)* vol.23 (1973) p.458]

[7] H.H. Wieder [*Appl. Phys. Lett. (USA)* vol.38 (1981) p.170]
[8] D.V. Morgan, J. Frey [*Electron. Lett. (UK)* vol.14 (1978) p.1737]
[9] J.L. Veteran, D.P. Mullin, D.I. Elder [*Thin Solid Films (Switzerland)* vol.97 (1982) p.187]
[10] S. Loualiche, H. L'Haridon, A. Le Corre, D. Lecrosnier, M. Salvi, P.N. Favennec [*Appl. Phys. Lett. (USA)* vol.52 (1988) p.540]
[11] K.C. Hwang, S.S. Li, C. Park, T.J. Anderson [*Appl. Phys. Lett. (USA)* vol.67 (1990) p.6571]
[12] D.P. Mullin, H.H. Wieder [*J. Vac. Sci. Technol. B (USA)* vol.1 (1983) p.782]
[13] T. Kobayashi, Y. Shinoda [*J. Appl. Phys. (USA)* vol.53 (1982) p.3339]
[14] K. Ishii, T. Sawada, H. Ohno, H. Hasegawa [*Electron. Lett. (UK)* vol.18 (1982) p.1034]
[15] H.H. Wieder, A.R. Clawson, D.I. Elder, D.A. Collins [*IEEE Electron Device Lett. (USA)* vol.2 (1981) p.73]
[16] R. Kaumans, J. Selders, H. Beneking [*Inst. Phys. Conf. Ser. (UK)* no.63 (1982) p.329]
[17] A.S.H. Liao, B. Tell. R.F. Leheny, T.Y. Chang [*Appl. Phys. Lett. (USA)* vol.41 (1982) p.280]
[18] B. Tell, R.E. Nahory, R.F. Leheny, J.D. DeWinter [*Appl. Phys. Lett. (USA)* vol.39 (1981) p.744]
[19] M. Okamura, T. Kobayashi [*Jpn. J. Appl. Phys. (Japan)* vol.19 (1980) p.2143]
[20] D.L. Lile, M.J. Taylor [*J. Appl. Phys. (USA)* vol.54 (1983) p.260]
[21] P. O'Connor, T.P. Pearsall, K.Y. Cheng, A.Y. Cho, J.C.M. Hwang, K. Alavi [*IEEE Electron Device Lett. (USA)* vol.3 (1982) p.64]
[22] C.L. Cheng, A.S.H. Lao, T.Y. Chang, R.F. Leheny, L.A. Coldren, B. Lalevic [*IEEE Electron Device Lett. (USA)* vol.5 (1984) p.169]
[23] S. Bandy, C. Nishimoto, S. Hyder, C. Cooper [*Appl. Phys. Lett. (USA)* vol.38 (1981) p.817]
[24] P.D. Gardner et al [*Inst. Phys. Conf. Ser. (UK)* no.65 (1982) p.399]
[25] P.D. Gardner, S.Y. Narayan, Yong-Hoon Yun [*RCA Rev. (USA)* vol.42 (1981) p.542]
[26] J. Selders, H. Beneking [*IEEE Electron Device Lett. (USA)* vol.7 (1986) p.434]
[27] K. Ishii, T. Sawada, H. Ohno, H. Hasegawa [*Electron. Lett. (UK)* vol.18 (1982) p.1034]
[28] H.H. Wieder, J.L. Veteran, A.R. Clawson, D.P. Mullin [*Appl. Phys. Lett. (USA)* vol.43 (1983) p.287]
[29] J. Selders, M. Maier, H. Beneking [*Proc. Int. Conf. on Insulating Films on Semiconductors*, Toulouse, Eds J.J. Simonne, J. Buxo (North Holland, Amsterdam, 1986) p.81-4]
[30] M. Taillepied, S. Gourier, J.P. Chane [*Proc. Int. Conf. on Insulating Films on Semiconductors*, Toulouse, Eds J.J. Simonne, J. Buxo (North Holland, Amsterdam, 1986) p.85-8]
[31] C.D. Lee, S.R. Forrest [*Appl. Phys. Lett. (USA)* vol.57 (1990) p.469]
[32] O. Wada, H. Nobuhara, H. Hamaguchi, T. Mikawa, A. Tackeuchi, T. Fujii [*Appl. Phys. Lett. (USA)* vol.54 (1989) p.16]
[33] T.P. Leonard, J. Bregman, M. Pepper, G. Davies, E.G. Scott [*J. Appl. Phys. (USA)* vol.70 (1991) p.5090]
[34] M. Akazawa, E. Ohue, H. Ishii, H. Iwadate, H. Hasegawa [*Solid-State Electron. (UK)* vol.33 (1990) p.345]
[35] H. Hasegawa, M. Akazawa, H. Ishii, K. Matsuzaki, H. Ohno [*Jpn. J. Appl. Phys. (Japan)* vol.27 (1988) p.L2265]

[36] H. Hasegawa, H. Ohno, H. Ishii, T. Haga, Y. Abe, H. Takahashi [*Appl. Surf. Sci. (Netherlands)* vol.41/42 (1989) p.372]

5.3 Schottky barriers and ohmic contacts on InGaAs

P. Kordoš and M. Marso

December 1992

A INTRODUCTION

Investigations of Schottky barriers and ohmic contacts are important in obtaining more information about the properties of metal-semiconductor interfaces, and optimising the performance of electronic devices. At Schottky diodes and gates of field-effect transistors the barrier height should be as high as possible. For low-resistance ohmic contacts we need the opposite situation. In addition to the optimal barrier height, a good surface morphology to allow further lithographic processing, no excessive stress, as well as thermal stability of the interface properties, are other requirements for the device applications of Schottky and ohmic contacts.

On metal/n-InGaAs contacts ($In_{0.53}Ga_{0.47}As$ lattice-matched to InP hereafter referred to only as InGaAs) a barrier height of about 0.25 eV was predicted by the intrinsic interface state theory [1]. On p-InGaAs a barrier height of 0.5 eV will be expected according to the relation $\phi_{B(n)} + \phi_{B(p)} = E_g = 0.75$ eV. Barrier heights near to these values were experimentally determined [2,3]. From this it follows, that on n-InGaAs it should not be very difficult to prepare good ohmic contacts, but the barrier is too low for rectifying contacts. The properties of ohmic contacts on p-InGaAs should be similar to those on p-GaAs due to their similar barrier heights. In general, the barrier height can be enhanced (reduced) by two different effects, by band bending due to space charge in the highly counter-doped (highly alike-doped) interlayer and/or by conduction-band discontinuity at the heterointerface between an active layer and an interlayer of different composition with higher (lower) bandgap energy [4]. Highly doped interlayers can be produced by interdiffusion during the thermal processing (used in the preparation of low-resistance ohmic contacts) or by ion implantation. In the latter it is more convenient to grow a thin highly doped contact layer or a thin barrier enhanced heterolayer on top of InGaAs.

B SCHOTTKY BARRIERS

Schottky barrier investigations are made on samples with circular or square metal dots of different areas which are prepared by deposition of single- or multi-layer metallisation and subsequent lift-off process. For the device applications usually sub-m mushroom-like contacts (T-gates) and microwave adapted interconnections are formed. The barrier height is mostly evaluated [5] from the I-V characteristic by using the thermionic-emission equation with the saturation current density $J_s = A^*T^2\exp(-q\phi_B/kT)$, where A* is the effective Richardson constant; A* = 5.04 and 61.9 A/cm^2K^2 for n-InGaAs [6] and p-InGaAs [3], respectively. Another approach for determining the barrier height is based on the thermal activation energy method, i.e. by measurement of I-V characteristics at different temperatures and using a plot of $\ln(J_s/T^2)$ as a function of the reciprocal temperature. The C-V measurements, as well as the photoelectric measurements can be also used to evaluate the barrier height.

B1 Barriers Prepared Directly on InGaAs

From the first study of the Schottky diodes on n-type $In_xGa_{1-x}As$ of different compositions grown on GaAs substrates it follows that the barrier height on Au/n-InGaAs contacts will be $\phi_B{}^o = 0.2$ eV [7]. Barrier heights around this value were published later by applying different metals on n-InGaAs; Nb contact - $\phi_B{}^o = 0.08$ eV [8], Ti - 0.15 eV [6], Au - 0.2 eV [2], Ni - 0.2 eV [9] and Al in situ deposited on MBE layer - 0.2 eV [8].

On Al/p-InGaAs contacts the barrier height of 0.55 eV was reported [3]. From the more detailed study of metal/p-InGaAs contacts it follows that the barrier heights obtained by depositing various metals (Au, Al, Ag, Ti, Ni, and Pt) on wet chemically etched surfaces of p-InGaAs are in the range between 0.49 and 0.62 eV [10]. On sputter etched surfaces higher barriers, $\phi_B = 0.7$ eV, independent of the metal used were obtained [11]. Oxide-like interfacial layers or high density of surface states are probably responsible for the barrier enhancement.

B2 Barriers Enhanced by Thin Interlayers

Many attempts have been made to enhance the Schottky barrier on n-InGaAs by thin interlayers of different compositions, which are situated between the InGaAs layer and the barrier metal [4,12-31]. A relatively convenient procedure is the preparation of thin counter-doped p^+-InGaAs interlayers [4,12-17]. Barrier heights $\phi_B = \phi_B{}^o + \Delta\phi_B$ of 0.6 eV can be achieved with p^+-InGaAs for a thickness d = 30 nm and a dopant density $p = 1.5 \times 10^{18}$ cm^{-3} [15]. The barrier enhancement can be expressed as $\Delta\phi_B = (q/\varepsilon_s)pd^2$ [5], but according to the two-carrier calculation [32] at higher dopant densities and thicknesses the effective barrier height will saturate near the bandgap energy. This behaviour was recently confirmed [4] and on n-InGaAs Schottky diodes with 60 nm thick p^+-InGaAs interlayers doped to $p = 8 \times 10^{18}$ cm^{-3} barrier heights up to 0.68 eV have been measured ($J_{R(1V)} = 4.5 \times 10^{-6}$ A/cm^2, n = 1.15) [17]. Unfortunately, some arguments can be found against the use of an acceptor in the growth equipment where high purity layers should be prepared.

Another approach to barrier enhancement consists of the preparation of an interfacial oxide-like layer [18-21]. The barrier enhancement described by the tunnelling model [33] is proportional to the interlayer thickness $\Delta\phi_B = Dd$, where D depends on the carrier effective mass in the semiconductor (in n-InGaAs $D \simeq 1 \times 10^6$ V/cm). A Cd-based organic complex as thin dielectric layers [19,21] or SiO_x layers [18] were used and barrier heights in the range from 0.38 to 0.55 eV are reported. However, the application of dielectric interlayers to high-speed devices might be accompanied by interface-state formation and possible degradation of the device performance.

Thin layers of higher bandgap semiconductors such as lattice-matched InAlAs [22-24] and InP [4,16,25,26] and lattice-mismatched GaAs [4,27,28], AlGaAs [29] and InGaP [4,30,31] can also be used to enhance the barrier on n-InGaAs due to the conduction band discontinuity at the heterointerface, $\Delta\phi_B \simeq \Delta E_c$ [34]. InAlAs layers are used at the preparation of n-InGaAs devices [35] because of high ΔE_c at the InAlAs/InGaAs interface, $\Delta E_c \simeq 0.52$ eV [36]. On the other hand, barrier heights of only 0.50 - 0.55 eV were measured on n-InGaAs Schottky diodes with 30 nm thick InAlAs interlayers [23,24] in contradiction to values between 0.55 and 0.85 eV reported on thick n-InAlAs diodes by using various metals [37-39]. Unfortunately, deep traps were observed at the InAlAs/InGaAs interfaces [40] and together with an air-oxidation of Al-rich materials these can be responsible for the limited performance of InAlAs/InGaAs devices [41].

Barrier heights up to 0.55 eV were measured on n-InGaAs diodes with n-InP interlayers of different thicknesses [4,16]. This is in agreement with expected values considering $\Delta E_c = 0.24 - 0.36$ eV at the InP/InGaAs interfaces [42], or with barrier heights of 0.48 - 0.55 eV measured directly on n-InP diodes [43]. Additional barrier height enhancement can be achieved by counterdoping of the InP interlayer. Barrier heights up to 0.66 eV have been reported on n-InGaAs Schottky diodes with 30 nm thick p-InP doped to 2×10^{18} cm^{-3} [4].

In the past, GaAs and AlGaAs barrier-enhanced layers were also used on n-InGaAs [27-29], but due to the high lattice-mismatch and following critical layer thickness for pseudomorphic growth of only $d_c \simeq 4$ nm it is difficult to prepare Schottky diodes with reasonable parameters [4]. By using InGaP layers [30,31] one can adjust both the bandgap (i.e. ΔE_c) and the lattice-mismatch (i.e. d_c) by changing the composition of InGaP; e.g. the bandgap of InGaP with $x_{GaP} = 0.11$ is the same as that of GaAs but $d_c = 12$ nm. Unfortunately, barrier heights not exceeding values reported for InAlAs and InP interlayers have been obtained with InGaP on n-InGaAs [4].

C OHMIC CONTACTS

Results obtained for the preparation of ohmic contacts on GaAs [44] are mostly transferred to InGaAs [45,60]. The ohmic contacts are usually prepared by e-beam evaporation of a metallic multilayer structure in one pump down cycle (e.g. layers of 5 nm Ni, 70 nm 88:12 wt% Au:Ge and 30 nm Ni are successively deposited on the semiconductor surface) and usually they are subsequently sintered by means of rapid thermal processing under N_2 + (5-15%)H_2 forming gas ambient. For the device preparation a Au layer of about 0.5 μm thickness is finally electroplated upon the ohmic contact to improve the intercontact resistivity. The properties of the ohmic contacts are evaluated by electrical measurements using the transmission line method [46] and as a characteristic measure the specific contact resistance ρ_c (Ωcm^2) is used (unfortunately some authors present only the contact resistivity $R_c = \rho_c/L_T$, where L_T is the zero resistance distance in the R-L plot used in TLM). In general ρ_c increases with decreased dopant density in InGaAs [45]. At the preparation of ohmic contacts on low doped InGaAs the barrier height can be reduced by thin highly doped interlayers, which are usually produced during an alloying process, by ion implantation [47], or by intentional growth of a cap-layer. Another possibility to lower the barrier height, and thus the contact resistance, is by a heterointerlayer with lower bandgap energy, which can also be produced during thermal processing of a contact or by an intentional growth (e.g. an InAs layer between a metal and InGaAs [48]).

C1 Contacts to n-InGaAs

For the preparation of ohmic contacts on n-InGaAs different metallisations were used, but no priority for a specific one can be given. As it follows from summarised published data given in TABLE 1, the contact resistance ρ_c decreases from about 10^{-6} to 10^{-8} Ωcm^2 with increased dopant density from 10^{16} to 10^{19} cm^{-3}. When thermally untreated contacts are used $\rho_c = 10^{-7} - 10^{-8}$ Ωcm^2 can be obtained on highly doped n-InGaAs layers [49,50].

TABLE 1 Specific contact resistances of ohmic contacts on n-InGaAs.

n (cm^{-3})	ρ_c (Ωcm^2)	Thermal process T (°C) - t (s)	Metallisation	Ref
1×10^{16}	5×10^{-5}	350 - 120	Ni-AuSn-Ni	[51]
1.9×10^{16}	2×10^{-6}	400 - 120	Ni-AuGe-Ni	[51]
1.0×10^{17}	5×10^{-7}	450 - 25	AuGe-Ni-Au	[52]
1.4×10^{17}	3×10^{-7}	420 - 120	Ni-AuGe-Ni	[51]
1.9×10^{17}	3×10^{-7}	400 - 120	Ni-AuSn-Ni	[51]
2.2×10^{18}	7×10^{-8}	400 - 120	Ni-AuGe-Ni	[51]
2.7×10^{18}	4×10^{-8}	400 - 120	Ni-AuSn-Ni	[51]
2.7×10^{18}	1.6×10^{-6}	none	Au-Sn-Au	[49]
8×10^{18}	5×10^{-8}	420 - 15	AuGe-Ag-Au	[53]
1×10^{19}	6×10^{-7}	350	Pd-Ge	[54]
1×10^{19}	5.5×10^{-7}	none	Ti-Pt-Au	[49]
2.9×10^{19}	2.9×10^{-7}	none	Au-Sn-Au	[49]
2.9×10^{19}	2.1×10^{-7}	none	Ti-Pt-Au	[49]
5×10^{19}	1.7×10^{-8}	none	Cr-Au	[50]

C2 Contacts to p-InGaAs

In the primary work concerning the ohmic contacts on p-InGaAs a 90:10 wt% Au:Zn alloy source was used to deposit a 150 nm thick AuZn layer on the InGaAs surface ($p \simeq 5 \times 10^{18}$ cm^{-3}). After sintering at 430 - 460°C for 20 s the specific contact resistance $\rho_c = 1.6 \times 10^{-5}$ Ωcm^2 was obtained [55]. Different kinds of metallisations on highly doped p-InGaAs ($p = 10^{18}$ - 10^{19} cm^{-3}) were later used, e.g. Au-Zn, Ni-AgZn-Ni, Ni-AuZn-Ni, Ni-AuZn-Ti, Ni-AuMn-Ni, Pd-Ge, Ti-Au, generally alloyed at 350 - 500°C for 2 min (for more details see Tab.3.3.2 in [45]), but only similar or even higher values of ρ_c were obtained. Only on contacts prepared by alloying of 120 nm thick 95:5 wt% Au:Zn at 400°C for 2 min on high dose implanted p-InGaAs surfaces was much lower specific contact resistance obtained, with $\rho_c = 1.3 \times 10^{-6}$ Ωcm^2 [47]. Similar values of 2.3×10^{-6} Ωcm^2 have been obtained on 50 nm Pd - 115 nm Ge alloyed contacts on InGaAs with $p = 1.8 \times 10^{18}$ cm^{-3} [54].

Recently Ti-Pt(-Au) metallisation on p-InGaAs was systematically studied [48, 56-59]. On Ti-Pt ohmic contacts on highly doped p-InGaAs (1.5×10^{19} cm^{-3}) processed by RTA at 300 - 600°C for 30 s an extreme decrease of the specific contact resistance (near two decades) at temperatures between 400 and 470°C was found with $\rho_c = 3.4 \times 10^{-8}$ Ωcm^2 at 450°C [56]. Unfortunately, in the later works of the same authors on InGaAs with $p = 5 \times 10^{18}$ cm^{-3} only $\rho_c = 9 \times 10^{-6}$ Ωcm^2 is observed for Ti-Pt contacts [48] and 5.5×10^{-7} Ωcm^2 for Ti-Pt-Au contacts [57], in both cases processed by RTA at 450°C for 30 s. On the other hand, on Ti-Pt-Au contacts on p-InGaAs ($p = 2 \times 10^{19}$ cm^{-3}) without using thermal processing the specific contact resistances of $\rho_c = 1 \times 10^{-5}$ Ωcm^2 were measured [49].

D CONCLUSION

The Schottky barrier height on n-InGaAs is too low for device applications, $\phi_b^o \simeq 0.2$ eV. Different kinds of barrier-enhanced interlayers can be used, e.g. counter-doped p^+-InGaAs, undoped n-InP or n-InAlAs, and barrier heights of 0.5 - 0.6 eV can be obtained. For

InGaAs-based MODFET preparation 25 - 30 nm thick undoped InAlAs layers are used. On p-InGaAs the barrier height is about 0.55 eV.

The specific contact resistance on n-InGaAs decreases with increased dopant density - from $\rho_c \simeq 10^{-6}$ Ωcm^2 at $n \simeq 10^{16}$ cm^{-3} to about 10^{-8} Ωcm^2 at $n \simeq 10^{19}$ cm^{-3}. On p-InGaAs the contact resistance is higher because of higher Schottky barrier and values of $\rho_c = 10^{-5}$ - 10^{-7} Ωcm^2 can be obtained on highly doped layers with $p \simeq 10^{18}$ - 10^{19} cm^{-3}. For device preparation Au-Ge-Ni based ohmic contacts are preferred on MODFET structures and Ti-Pt-Au metallisation is used to contact both n- and p-InGaAs heterojunction bipolar transistors.

REFERENCES

[1] J. Tersoff [*Phys. Rev. B (USA)* vol.32 (1985) p.6968-71]

[2] P.K. Bhattacharya, M.D. Yeaman [*Solid-State Electron. (UK)* vol.24 (1981) p.297-300]

[3] J.L. Veteran, D.P. Mullin, D.I. Elder [*Thin Solid Films (Switzerland)* vol.97 (1982) p.187-90]

[4] P. Kordoš, M. Marso, R. Meyer, H. Lüth [*J. Appl. Phys. (USA)* vol.72 (1992) p.2347-55]

[5] S.M. Sze [*Physics of Semiconductor Devices* (Wiley-Interscience, 1981) p.279-97]

[6] U. Kunze, W. Kowalsky [*J. Appl. Phys. (USA)* vol.63 (1988) p.1597-602]

[7] K. Kajiyama, Y. Mizushima, S. Sakata [*Appl. Phys. Lett. (USA)* vol.23 (1973) p.458-9]

[8] H. Tamura, A. Yoshida, S. Muto, S. Hasuo [*Jpn. J. Appl. Phys. (Japan)* vol.24 (1987) p.L7-9]

[9] H. Kräutle, E. Woelk, J. Selders, H. Beneking [*IEEE Trans. Electron Devices (USA)* vol.32 (1985) p.1119-22]

[10] L. Malacky, P. Kordoš, J. Novák [*Solid-State Electron. (UK)* vol.33 (1990) p.273-8]

[11] J. Selders, N. Emeis, H. Beneking [*IEEE Trans. Electron Devices (USA)* vol.32 (1985) p.605-9]

[12] C.Y. Chen, A.Y. Cho, K.Y. Cheng, P.A. Garbinski [*Appl. Phys. Lett. (USA)* vol.40 (1982) p.401-3]

[13] J.H.Kim, S.S.Li, L.Figueroa [Electron. Lett. (UK) vol.24 (1988) p.687-9]

[14] J.H. Kim, S.S. Li, L. Figueroa, T.F. Carruthers, R.S. Wagner [*J. Appl. Phys. (USA)* vol.64 (1988) p.6536-40]

[15] P. Kordoš, M. Marso, R. Meyer, H. Lüth [*Electron. Lett. (UK)* vol.27 (1991) p.1759-61]

[16] P. Kordoš, M. Marso, R. Meyer, H. Lüth [*Proc. 4th Int. Conf. on Indium Phosphide and Related Materials*, Newport, USA, 20-24 April 1992 (IEEE LEOS, USA, 1992) p.230-3]

[17] P. Kordoš, M. Marso, R. Meyer, H. Lüth [*IEEE Trans. Electron Devices (USA)* vol.39 (1992) p.1970-2]

[18] D.V. Morgan, J. Frey [*Electron. Lett. (UK)* vol.14 (1978) p.737-8]

[19] W.K. Chan, H.M. Cox, J.H. Abeles, S.P. Kelty [*Electron. Lett. (UK)* vol.23 (1987) p.1346-8]

[20] K.C. Hwang, S.S. Li, C. Park, T.J. Anderson [*J. Appl. Phys. (USA)* vol.67 (1990) p.6571-3]

[21] T.J. Licata et al [*J. Electron. Mater. (USA)* vol.19 (1990) p.1239-46]

[22] J. Barnard, H. Ohno, C.E.C. Wood, L.F. Eastman [*IEEE Electron Device Lett. (USA)* vol.1 (1980) p.154-6]

[23] D.V. Morgan, H. Ohno, C.E.C. Wood, W.J. Schaff, K. Board, L.F. Eastman [*IEEE Proc. (USA)* vol.128 (1981) p.141-3]

[24] S.R. Bahl, M.H. Leary, J.A. del Alamo [*IEEE Trans. Electron Devices (USA)* vol.39 (1992) p.2037-43]

[25] D. Kuhl, F. Hieronymi, E.M. Böttcher, T. Wolf, A. Krost, D. Bimberg [*Electron. Lett. (UK)* vol.26 (1990) p.2107-9]

[26] Ch.X. Shi, D. Grützmacher, M. Stollenwerk, Q.K. Wang, K. Heime [*IEEE Trans. Electron Devices (USA)* vol.39 (1992) p.1028-31]

[27] C.Y. Chen, S.N.G. Chu, A.Y. Cho [*Appl. Phys. Lett. (USA)* vol.46 (1985) p.1145-7]

[28] J. Selders, P. Roentgen, H. Beneking [*Electron. Lett. (UK)* vol.22 (1986) p.14-16]

[29] T. Kikuchi, H. Ohno, H. Hasegawa [*Electron. Lett. (UK)* vol.24 (1988) p.1208-10]

[30] A. Hosseini Terani, D. Decoster, J.P. Vilcot, M. Razeghi [*J. Appl. Phys. (USA)* vol.64 (1988) p.2215-8]

[31] P. Kordoš, J. Novák, O. Kayser, K. Heime [*Phys. Status Solidi A (Germany)* vol.127 (1991) p.K25-7]

[32] G.P. Schwartz, G.J. Gualtieri [*J. Electrochem. Soc. (USA)* vol.133 (1986) p.1266-8]

[33] D.V. Morgan, J. Frey [*Phys. Status Solidi A (Germany)* vol.51 (1979) p.K29-33]

[34] M. Eizenberg, M. Heiblum, M.I. Nathan, N. Braslau, P.M. Mooney [*J. Appl. Phys. (USA)* vol.61 (1987) p.1516-22]

[35] L.D. Nguyen, A.S. Brown, M.A. Thompson, L.M. Jelloian [*IEEE Trans. Electron Devices (USA)* vol.39 (1992) p.2007-14]

[36] J.H. Huang, T.Y. Chang, B. Lalevic [*Appl. Phys. Lett. (USA)* vol.60 (1992) p.733-5]

[37] N. Harada, S. Kuroda, T. Katakami, K. Hikosaka, T. Mimura, M. Abe [*Proc. 3rd Int. Conf. on Indium Phosphide and Related Materials*, Cardiff, UK, 1991 (IEEE LEOS, USA,1991) p.377-80]

[38] L.P. Sadwick, C.W. Kim, K.L. Tan, D.C. Streit [*IEEE Electron Device Lett. (USA)* vol.12 (1991) p.626-8]

[39] F. Gueissaz, M. Gailhanou, R. Houdré, M. Ilegems [*Appl. Phys. Lett. (USA)* vol.60 (1992) p.1099-101]

[40] J.K. Luo, H. Thomas, I.L. Morris [*Electron. Lett. (UK)* vol.28 (1992) p.797-9]

[41] K. Inoue [*Proc. 4th Int. Conf. on Indium Phosphide and Related Materials*, Newport, USA, 20-24 April 1992 (IEEE LEOS, USA, 1992) p.10-13]

[42] J.D. Lambkin [in *Properties of Indium Phosphide*, EMIS Datareviews Series no.6 (INSPEC, IEE, 1991), ch.14 p.318-22]

[43] C. Gaonach, S. Cassette, M.A. Di Forte-Poisson, C. Brylinski, M. Champagne, A. Tardella [*Semicond. Sci. Technol. (UK)* vol.5 (1990) p.322-7]

[44] M. Murakami [*Mater. Sci. Rep. (Netherlands)* vol.5 (1990) p.273-317]

[45] K. Heime [*InGaAs field-effect transistors* (Taunton, Research Studies Press, 1989) p.52-8]

[46] H.H. Berger [*J. Electrochem. Soc. (USA)* vol.119 (1972) p.507-14]

[47] A.D. Huelsman, C.G. Fonstad [*IEEE Trans. Electron Devices (USA)* vol.33 (1986) p.294-7]

[48] S.N.G. Chu et al [*J. Appl. Phys. (USA)* vol.67 (1990) p.3754-60]

[49] L.G. Shantharama, H. Schumacher, H.P. Leblanc, R. Esagui, R. Bhat, M. Koza [*Electron. Lett. (UK)* vol.26 (1990) p.1127-9]
[50] T. Fujii, T. Inata, K. Ishii, S. Hiyamizu [*Electron. Lett. (UK)* vol.22 (1986) p.191-2]
[51] H. Kräutle, E. Woelk, J. Selders, H. Beneking [*IEEE Trans. Electron Devices (USA)* vol.32 (1985) p.1119-23]
[52] H. Morkoc, T.J. Drummond, C.M. Stanchak [*IEEE Trans. Electron Devices (USA)* vol.28 (1981) p.1-3]
[53] P. O'Connor, T.P. Pearsall, K.Y. Cheng, A.Y. Cho, J.C.M. Hwang, K. Alavi [*IEEE Electron Device Lett. (USA)* vol.3 (1982) p.64-6]
[54] W.X. Chen, S.C. Hsueh, P.K.L. Yu, S.S. Lau [*IEEE Electron Device Lett. (USA)* vol.7 (1986) p.471-3]
[55] Y. Nakano, S. Takahashi, Y. Toyoshima [*Jpn. J. Appl. Phys. (Japan)* vol.19 (1980) p.L495-7]
[56] A. Katz et al [*Appl. Phys. Lett. (USA)* vol.54 (1989) p.2306-8]
[57] A. Katz, B.E. Weir, W.C. Dautremont-Smith [*J. Appl. Phys. (USA)* vol.68 (1990) p.1123-8]
[58] K.L. Jiao, A.L. Soltyka, W.A. Anderson, A. Katz [*Appl. Phys. Lett. (USA)* vol.57 (1990) p.1913-5]
[59] A. Katz et al [*Mater. Res. Soc. Symp. Proc. (USA)* vol.216 (1991) p.305-11]
[60] T.C. Shen, G.B. Gao, H. Morkoc [*J. Vac. Sci. Technol. B (USA)* vol.10 (1992) p.2113-32]

5.4 Oxides of InGaAs

C.W. Wilmsen

January 1993

A INTRODUCTION

Oxides readily form on the surface of InGaAs when exposed to air or any other oxidising ambient. These oxides are generally a mixture of In, Ga and As oxides; the exact composition depends strongly on the conditions of oxide growth and any subsequent annealing. The thermodynamics, chemical composition, growth rate and properties of these oxides are summarised here.

B THERMODYNAMICS

The growth of an oxide layer on a semiconductor is the result of the chemical reaction between the surface atoms of the semiconductor and an oxidant. The chemical compounds formed by this reaction are those which yield the lowest possible energy for the oxidation conditions. The equilibrium phase diagram serves as a guide to the composition of the oxide layer and any changes that may occur during annealing. FIGURE 1 illustrates the phase diagram for the quaternary, In-Ga-As-O [1]. The phase diagram indicates that the equilibrium oxidation products are In_2O_3, Ga_2O_3 and elemental As. However, the compounds, As_2O_3, As_2O_5, $InAsO_4$ and $GaAsO_4$, may also form, but these compounds are not stable when in proximity with the substrate and can decompose to form In_2O_3, Ga_2O_3 and elemental As through reactions such as:

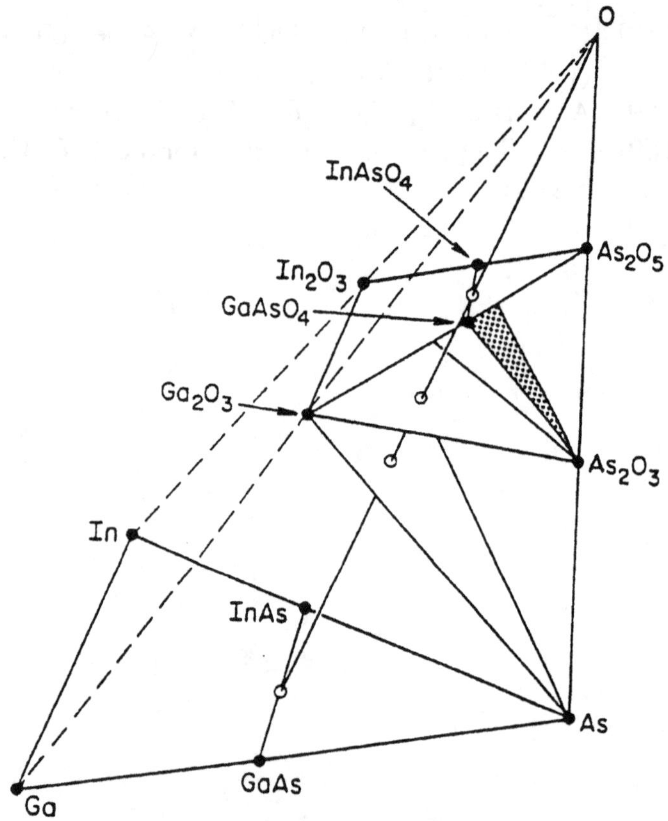

FIGURE 1 Equilibrium phase diagram for In-Ga-As-O [1].

$$2In_xGa_{(1-x)}As + As_2O_3 \rightarrow xIn_2O_3 + (1-x)Ga_2O_3 + 4As.$$

The enthalpies and Gibbs free energies of formation at room temperature for the oxides are given in TABLE 1.

TABLE 1 Enthalpies and Gibbs energies of formation [2-4].

Phase	ΔH^f(kcal/mol)	ΔG^f(kcal/mol)
In	0	0
Ga	0	0
As	0	0
InAs	-14.0	-12.8
GaAs	-17.0	-16.2
In_2O_3	-221.3	-198.6
As_2O_3	-157.0	-137.7
As_2O_5	-221.1	-187.0
$InAsO_4$	-233.8	-209.4
Ga_2O_3	-260.3	-238.6
$GaAsO_4$	-240.6	-212.8

C PROPERTIES OF THE OXIDES

TABLE 2 Properties of single phase oxides [5].

Phase	Melting Point, °C	Solubility	Bandgap, eV	Index of Refraction
In_2O_3	~2000	conc. acid & base	3.55	1.9
Ga_2O_3	~1800	conc. acid & base	4.7	1.93
As_2O_3	315	water	4.0	2.01
As_2O_5	decomposes @315	water		

D CHEMICAL COMPOSITION AND GROWTH RATE OF THE OXIDES

Very little has been published on the chemical composition of the oxides grown on InGaAs, and while there is sufficient data to determine the primary composition, the details of the oxide composition will require additional experiments.

D1 Thermal Oxides

The composition of oxides grown on $In_{0.53}Ga_{0.47}As$ substrates in dry O_2 [5-6] or air [1] in the temperature range 350 to 550°C has been determined by X-ray photoelectron and/or Raman spectroscopies. These oxides are primarily composed of In_2O_3 and Ga_2O_3 in approximately equal proportions. The concentration of elemental As observed at the oxide/substrate interfaces varies with growth temperature. Very low concentrations of As_2O_3 were found at the surface of oxides grown at $T < 400$°C. There is no direct evidence for the presence of $InAsO_4$, $GaAsO_4$ or As_2O_5, although oxidising the substrate for long periods may result in small surface concentrations of these compounds as is the case with the thermal oxides of InAs [2] and GaAs [7].

The thermal oxide growth rate on $In_{0.53}Ga_{0.47}As$ [5] lies between that of InAs and GaAs as shown in FIGURE 2, which indicates that very little oxide grows when $T < 300$°C and that rapid growth occurs for $T > 450$°C.

D2 Anodic Oxides

Raman spectra indicate that the as grown anodic oxide of InGaAs is a mixture of In_2O_3, Ga_2O_3 and As_2O_3 [1]. The ratio of these oxides and their distribution are not known and these will depend on the pH and composition of the electrolyte as is observed with anodic oxides on other compound semiconductors [8,9]. Annealing in vacuo for 8 hrs at 400°C causes the As_2O_3 to decompose and form elemental As. The oxide grew at the rate of ~ 20 Å/volt, which is about the same as that of InAs [1] and GaAs [10].

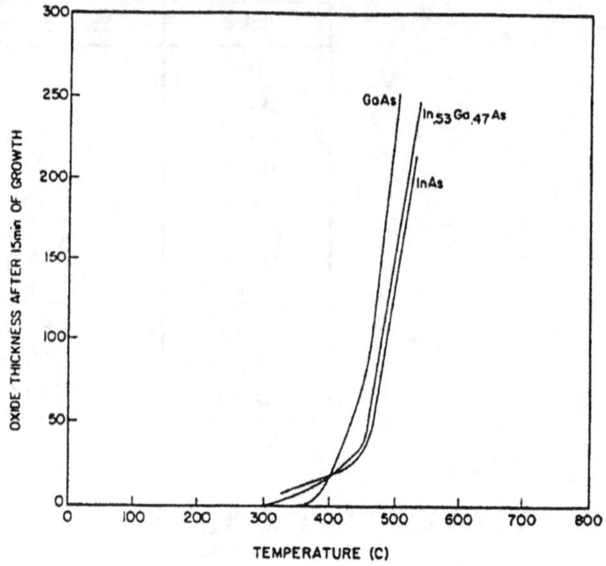

FIGURE 2 Oxide thickness after 15 min. in dry O_2 as a function of oxidation temperature [5].

D3 Chemically Grown Oxides

Oxides grown in 30% H_2O_2 are a porous mixture of In_2O_3 and Ga_2O_3 of about 7 - 10 Å in thickness [11]. The pores result from the dissolution of the As_2O_3 which is grown along with the In and Ga oxides. Dilution of the H_2O_2 and/or lowering the pH leads to the incorporation of a small concentration of elemental As. Lowering the pH below 3 dissolves the oxides and leaves a thick porous layer of elemental As.

D4 UV/Ozone Oxides

Exposing an $In_{0.53}Ga_{0.47}As$ surface to ozone and UV radiation from a low pressure Hg lamp for ~ 10 minutes, results in the growth of a 35 Å thick oxide layer composed of In_2O_3, Ga_2O_3 and As_2O_3 in the ratio ≈1:1:3 [12]. Annealing at 354°C causes the As_2O_3 to decompose, leaving a residue of elemental As.

E CONCLUSION

All of the oxide layers which grow on InGaAs are primarily composed of In_2O_3 and Ga_2O_3. Elemental As is often found at the oxide/substrate interface, particularly if the oxide has been annealed. As_2O_3 is usually only present in low concentrations because it is volatile, soluble in water and it decomposes in the presence of the substrate. The oxidation rate appears to be similar to that of InAs and GaAs.

REFERENCES

[1] G.P. Schwartz, J.E. Griffiths, G.J. Gualtieri [*Thin Solid Films (Switzerland)* vol.94 (1982) p.213-22]

[2] G.P. Schwartz, W.A. Sunder, J.E. Griffiths, G.J. Gualtieri [*Thin Solid Films (Switzerland)* vol.94 (1982) p.205-12]

[3] C.D. Thurmond, G.P. Schwartz, G.W. Kammlott, B. Schwartz [*J. Electrochem. Soc. (USA)* vol.127 (1980) p.1366-71]

[4] D.D. Wagman, W.H. Evans, V.B. Parker, I. Halow, S.M. Bailey, R.H. Schumm [*Natl. Bur. Stand. Tech. Note (USA)* (1969) p.270-3]

[5] C.W. Wilmsen [in *Physics and Chemistry of III-V Compound Semiconductor Interfaces*, Ed C.W. Wilmsen (Plenum Press, New York, 1985) p.428-30]

[6] C.W. Wilmsen, D.H. Laughlin [unpublished]

[7] G.P. Schwartz, G.J. Gualtieri, G.W. Kammlott, B. Schwartz [*J. Electrochem. Soc. (USA)* vol.126 (1979) p.1737]

[8] D.H. Laughlin, C.W. Wilmsen [*Appl. Phys. Lett. (USA)* vol.37 (1980) p.915]

[9] K.M. Geib, C.W. Wilmsen [*J. Vac. Sci. Technol. (USA)* vol.17 (1980) p.952]

[10] C.R. Zeisse, L.J. Messick, D.L. Lile [*J. Vac. Sci. Technol. (USA)* vol.14 (1977) p.957]

[11] D.E. Aspnes, H.J. Stocker [*J. Vac. Sci. Technol. (USA)* vol.21 (1982) p.413]

[12] S.I.J. Ingrey, W.M. Lau, R.N.S. Sodhi [*J. Vac. Sci. Technol. A (USA)* vol.7 (1989) p.1554]

CHAPTER 6

RADIATIVE AND NON-RADIATIVE RECOMBINATION

6.1 Photoluminescence of pure InGaAs alloys

D. Bimberg and B. Srocka

November 1992

A INTRODUCTION

Luminescence is a widely-used tool for examining intrinsic electronic properties of semiconductors, crystal perfection [1-3], homogeneity [4] and impurity incorporation [5,6]. In the field of semiconductor alloys, composition determination is an additional important goal. This review will give some insight into a number of important properties of three-dimensional InGaAs layers on various substrates revealed by photoluminescence. Quantum confinement effects as well as interface problems (for a review see [7]) are discussed elsewhere in this volume. For an excellent treatment of the theoretical basis of this technique see [8].

B GENERAL PROBLEMS

B1 Alloy Broadening

The random distribution of atoms in ternary alloys implies statistical potential fluctuations which change the microscopic energy gap. Photoluminescence (PL) lines are therefore a superposition of signals belonging to different gap energies and are considerably broadened. Treating the potential fluctuations within the characteristic volume V of a transition, Schubert et al [9] found a relation for the standard deviation σ_E of the bandgap energy seen by the exciton:

$$\sigma_E = \frac{dE_g}{dx}\sqrt{\frac{x\,(1-x)}{K\,V_{ex}}} \tag{1}$$

where x, K, and V_{ex} are the composition, cation density, and exciton volume, respectively. Analogous relations hold for transitions including shallow bound levels (with appropriate characteristic volumes V). For lattice matched InGaAs Charreaux et al [10] calculated with Eqn (1) a minimum inhomogeneous broadening of 1.6 meV for a gaussian shaped excitonic line and of 3.3 meV for a donor-acceptor-pair recombination. Singh and Bajaj derived a similar value with a quantum mechanical approach [11,12]. Thus at temperatures below 20 K (40 K) the alloy broadening exceeds thermal broadening (FWHM of the order of kT) for the excitonic (shallow impurity related) lines. It should be noted, that:

- according to Eqn (1) the alloy broadening is largest for x = 0.5 (like in lattice matched InGaAs) and goes to zero for the binaries
- these considerations are based on the assumption of randomly distributed cations (no clustering or ordering effects).

As a consequence, the different bound exciton lines are not resolved in the spectra, since the localisation energies of the excitons are very similar (see Section C). Analogous problems arise for the determination of ionisation energies from PL-transitions involving shallow impurities [5].

Another result of the composition fluctuations is the localisation of free excitons in local potential minima, leading to a relaxation of the momentum selection rule at low temperatures and hence to an exponential Boltzmann tail at the high energy side of the free exciton line [13,14].

B2 Composition Dependence

Using the known bandgap values of GaAs and InAs and measuring the photoluminescence of close to lattice matched layers, Goetz et al [5] determined the composition dependence of PL-peak emission E_{ex}, and hence the bowing parameter and the gap energy E_g at 2 K:

$$E_g(x_{Ga}) = E_{ex}(x_{Ga}) + 2.1 \text{ meV} = \{410.5 + 633.7\, x_{Ga} + 475\, x_{Ga}^2\} \text{ meV} \qquad (2)$$

For the very Ga-rich end ($x_{Ga} \geq 0.98$) Ye et al [15] found:

$$E_g(x_{Ga}) = E_{gGaAs} - 1590\,(1-x_{Ga}) \text{ meV} \qquad (3)$$

in agreement with the relation above.

It must be noted that Eqn (2) describes the behaviour of strain free material. Epitaxial layers grown on a substrate of different chemical composition are usually strained by some extent. A strain induced shift in E_g of more then 2 meV was found for a lattice mismatch $\Delta a/a$ as low as -2×10^{-4} (see Section B3 and [16]). The variation in substrate (InP) lattice constant caused by doping may explain some scatter of E_g-data found in the literature [17]. Note that for a 1 meV shift of E_g to occur, a composition change of less than 0.1% is sufficient.

Attempts to determine the bowing of $E_g(x)$ at higher temperatures have resulted in widely varying values of the bowing parameter in Eqn (2). It varies from 420 to 530 meV [18-22]. For room temperature we suggest as the most reliable relation:

$$E_g(x_{Ga}) = \{360 + 629\, x_{Ga} + 436\, x_{Ga}^2\} \text{ meV} \qquad (4)$$

given by Nahory et al [19]. The intensity maximum of PL emission is then at an energy given by:

$$E_{PLmax} = E_g + 1/2\, kT \qquad (5)$$

B3 Strain Effects

Strain alters the PL-spectra by changing the bandgap parameters in a threefold manner [16,17,23-25]:

- bandgap shift due to the hydrostatic term
- splitting of the J = 3/2 valence band states at k = 0 due to biaxial strain
- change of the spin orbit splitting.

Thus, for tensile strain (Ga-rich layers) the gap decreases with respect to that in unstrained material of the same composition. The minimum separation is now between conduction and light hole valence band (J = 3/2, $m_J = \pm 1/2$). In contrast, the minimum separation of the band edges increases relative to unstrained layers for compression (In-rich layers), and the luminescence transition is between the conduction and heavy hole band (J = 3/2, $m_J = \pm 3/2$). Of course one has to add the composition (Eqn (2)) and the strain-effect in order to calculate the absolute gap shift. For a Ga-rich epilayer one obtains an enhanced total shift (composition and strain alter the gap in the same direction), while for In-rich layers the shift is reduced (composition and strain influence on the gap have opposite sign) [23]. For clarity it should be noted that in [16,24] the heavy hole band was ascribed to the (J = 3/2, $m_J = \pm 1/2$) state.

A question of fundamental importance is how thick a strained layer can be grown before dislocation formation starts. Matthews and Blakeslee evaluated this critical thickness, d_c, assuming a force balance at the interface [26], whereas People and Bean suggested an energy balance model, predicting larger values of d_c [27]. At the present time, it is generally accepted that the force-balance model yields the correct value for the onset of dislocation formation [28-32] depending on growth technique and conditions. Although there will be no immediate impact on overall quantum efficiency or halfwidths of the PL-lines and the X-ray rocking curve at low dislocation densities, one can observe these degradation effects if the layer thickness considerably exceeds d_c and the dislocation density becomes larger [29,32]. In this case, PL and cathodoluminescence microscopy have significant advantages over simple PL-measurements. Dislocations are directly imaged as dark-lines. This is a much more sensitive [32-35] technique which is unaffected by inhomogeneous strain relaxation [35].

On the other hand, layers far exceeding d_c look similar to dislocated bulk material with no (or little residual) strain [36] if the thermal expansion coefficients of layer and substrate are not too different. In Sections C and D we will present typical results for luminescence from strained and unstrained InGaAs. One should keep in mind that the PL-spectra from strained InGaAs considered here are often masked by quantum confinement effects at the heterointerfaces and within thin layers.

C InGaAs/InP

This section focuses on layers on InP substrates. In Section D results for layers on other substrates, such as GaAs, Si, and InGaAs, are presented. C1 and C2 deal with lattice matched layers, whereas in C3 strained systems are reported.

C1 Excitonic and Free Carrier Recombination

While Eqn (2) gives $E_g = 808.1 \pm 0.1$ meV as the position of the excitonic PL-peak for the lattice matched composition at low temperatures (2 K), Bassignana et al [17] observed this peak at 801 ± 0.2 meV. They attribute the shift to the difference in substrates. Using a hydrogenic model, Haynes' rule, and experimentally obtained acceptor binding energies, Goetz et al [5] calculated the separation of several excitonic lines from the fundamental gap, which are listed in TABLE 1. All these lines lie within a range of 3 meV, comparable to the normally observed halfwidth. From intensity and temperature dependent PL it was shown that the PL-peak is dominated by bound excitons. The binding energy of free excitons was determined to be 2.1 ± 0.1 meV.

TABLE 1 Predicted spectral position of several excitonic lines (E_X, E_D and E_A are the binding energy of the free exciton, donor and several acceptors, respectively).

Recombination	Theoretical emission energy α, α', β are set to 0.1	Separation from E_g (meV)
(X)	$E_g - E_X$	2.0
(D^o,X)	$E_g - (E_X + \alpha E_D)$	2.3
$D^+,X)$	$E_g - (E_X + \alpha' E_D)$	3.3
(A^o,X)	$E_g - (E_X + \beta E_A)$	3.3 .. 4.5
(D^o,h)	$E_g - E_D$	3.3

As the temperature of the sample is increased, the electrons and holes are thermally ionised. The photoluminescence then results from electron-hole-recombination [37,38]. The variation of the luminescence peak position with temperature was investigated by several authors. Towe [37] found for the range from 77 to 300 K the empirical relation:

$$E_g(T) = \{821 + 0.230\, T + 2.39 \times 10^{-4} T^2\} \text{ meV} \qquad (6)$$

Pearsall [38], Louati [6], Yu [39] and Goetz [18] used a theoretical dependence claimed by Varshni [40]:

$$E_g = E_{g0} - \frac{\alpha\, T^2}{(T + \beta)} \qquad (7)$$

and obtained for the temperature range of 4-300 K:

$\alpha = 0.4500$ meV/K	$\beta = 327$ K	$E_{g0} = 803$ meV	[38]
$\alpha = 0.4700$ meV/K	$\beta = 327$ K	$E_{g0} = 810$ meV	[6]
$\alpha = 0.4906$ meV/K	$\beta = 301$ K	$E_{g0} = 814$ meV	[39]
$\alpha = 0.5700$ meV/K	$\beta = 323$ K	$E_{g0} = 811$ meV	[18]

in rough agreement with each other. Eqn (5) holds for the PL maximum. The derivation of Goetz et al is based on a comparative PL and X-ray study of a large number of samples grown by LPE, MOCVD and VPE. The E_g-value calculated from Goetz's data for 300 K is 727 meV. The scatter in the data for E_{g0} - the gap at 0 K - may be explained by composition differences, whereas the variation of α and β should be caused by uncertainties. Since Eqn (7) takes only free carrier - lattice interactions into account but no excitonic effects, the PL-peak position is overestimated by the exciton localisation energy at low temperatures, as can be seen in [18,39].

The hydrostatic pressure dependence of the band edge photoluminescence was investigated by Lambkin and Dunstan [41]. They found a sublinear shift of the peak position with pressure and a linear correlation to the lattice constant. They measured at 80 K:

$$\Delta E_g(P) = \{E_{g0} + 10.9 \pm 0.2\ P - 30 \pm 2 \times 10^{-3}\ P^2\}\ \text{meV} \quad (8)$$

and at 300 K:

$$\Delta E_g(P) = \{E_{g0} + 11.0 \pm 0.2\ P - 27 \pm 2 \times 10^{-3}\ P^2\}\ \text{meV} \quad (9)$$

where P is the pressure in kbar. By using the InP-values for the bulk modulus a deformation potential for E_g of -8.25 ± 0.1 eV was obtained in the 80 - 300 K temperature range.

Correlation between PL-halfwidth and impurity concentration, as well as its temperature evolution, was treated by Towe [37] and Pearsall et al [38]. They used the van Roosbroeck-Shockley model [8] and obtained reasonable agreement. Pearsall pointed out that the FWHM can be used to estimate the impurity content with an error of about ±50%.

C2 PL Related to Impurities and Defects

Depending on impurity concentrations, a set of PL-lines approximately 10 to 30 meV below the exciton line is observable at low temperatures. As a result of the alloy broadening, the participating (D°,A°) and (e,A°) components are merged and the lines are Gaussian and slightly shifted to higher energies [5]. However for low enough temperatures, the (D°,A°) part should dominate, whereas with increasing temperature the electrons are thermally activated from the neutral donors and the (e,A°) transition dominates. In TABLE 2 we summarise the PL peak separation between these lines and the excitonic one.

For increasing temperatures these lines also vanish due to the thermal activation of the holes from the neutral acceptors participating in the transitions. These activation energies are always smaller than the binding energies of the holes to the different acceptors. They equal the $1S_{3/2}$ - $2P_{3/2}$ energy differences of the impurities [42,5]. Activation and binding energy are, however, sometimes confused in the literature [6,43,44]. Low-temperature excitation-intensity dependence of these lines was investigated by Yu and co-workers [45].

TABLE 2 Peak separation of impurity bands from excitonic PL-lines. Precision is ±2 meV if not otherwise stated.

Acceptor	Energetic difference from E_{ex} (meV)	Temperature (K)	Refs
C	8 ± 1	2.4	[5,46,47]
Mg	13	4	[6]
Cd	22	2	[48]
Zn	17 ± 1	2	[5]
	19	4	[47]
	26 ± 1	4	[46]
Si	20 ± 1	2	[5]
	30	4	[46]
Hg	62 ± 5	4	[6]
Be	22	4	[49]

TABLE 3 lists the photoluminescence lines transmitting from transition metal (TM) impurity related transitions. Only the iron related ones are rather well investigated so far [50]. A set of four zero-phonon lines is located at ≈ 350 meV, representing an internal transition between excited and ground state of the Fe^{2+} level. The orbital degeneracy of this ^5D level is lifted by crystal field splitting [51]. For a discussion of TM levels see also [52].

TABLE 3 Photoluminescence due to transition metals (TM) in InGaAs (E_{ex} stands for the position of the exciton line).

TM	Line position (meV)	Identification	Temperature (K)	Refs
Cr	E_{ex} - 14	Cr-related	4	[53]
Mn	E_{ex} - 49	(D^o,A^o_{Mn})	6	[54]
Fe	350 (4 lines)	$Fe^{2+}:(^5T_2) \rightarrow (^5E)$	4	[50,55]
	660	Fe-related	2.4	[55,56]

A structure at about 700 meV occurs in low temperature measurements, which is particularly weak in intensity in layers with high crystallographic perfection [56]. The same line was probably observed by Yagi et al [57] at 77 K in LPE layers at 690 meV. They found that its intensity has a sharp minimum for lattice matched conditions at growth temperature. Thus, its origin should be some defect introduced during the crystal growth caused by mismatch effects.

Another feature at 600 meV found at 2 K is of unknown nature [55,56].

C3 Mismatched Layers on InP

Temkin et al examined the critical layer thickness, d_c, over the whole composition range and found good agreement with the force-balance predictions [28]. The critical thickness decreases rapidly with lattice mismatch and becomes smaller than 100 nm already for a composition mismatch of ±3%. For layers thicker than d_c, Salokatve and Hovinen showed that one has to deal with inhomogeneous strain relaxation in the growth direction through the whole layer [58]. Thus PL-signals integrating over a considerable thickness are additionally broadened.

In [17,24] the bandgap shift due to the lattice mismatch relative to the InP-substrate is investigated experimentally. Bassignana et al [17] pointed out that for epilayers the composition induced bandgap shift is masked by the strain effects, and therefore the slopes of the shift are different for layers under tension and compression even in the vicinity of the lattice matching composition. They determined (at 7 K) for the Ga-rich side:

$$\Delta E_g(eV) = -5 \times \Delta a/a \qquad (10)$$

and for the In-rich side:

$$\Delta E_g(eV) = -11.38 \times \Delta a/a \qquad (11)$$

where ΔE_g is the deviation from the gap-energy at zero mismatch. This is in excellent agreement with the calculations from People [23] resulting in two curves for the gap-energy variation intersecting at the lattice matched composition (see also Section B3). They also found good agreement at room temperature.

Reihlen et al observed directly the heavy hole light hole band splitting in an $In_{0.474}Ga_{0.526}As/InP$ layer [59]. They found that at 150 K the free carrier recombination line consists of two parts separated from each other by 28 meV due to the splitting of the valence band.

D InGaAs ON GaAs- AND Si-SUBSTRATES

D1 InGaAs/GaAs

Isoelectronic doping of GaAs with In (up to 10^{+19} cm^{-3}) was found to improve the layer quality considerably. The etch pitch density as well as the trap concentration are lowered [60,61] and the near band edge luminescence intensity of MOCVD-grown layers is increased by up to a factor of twenty [62]. This effect is thought to be caused by strain around the In atoms which lowers the vacancy supersaturation and thus suppresses vacancy formation [63]. These layers show GaAs-like PL-spectra [64].

The critical layer thickness of ternary InGaAs on GaAs was investigated by many authors on exactly oriented (001) as well as on slightly off cut substrates [29-34]. Andersson et al [31] examined d_c for the whole alloy region and found good agreement with the predictions by Matthews and Blakeslee [26]. For $x_{Ga} = 92.6\%$, d_c is reduced to 50 nm. Elman et al reported that for MBE the critical thickness is considerably larger in samples grown at low temperatures (460°C) [65].

For layers thicker than d_c grown on 2° off GaAs, a drastically enhanced PL-intensity was observed, due to the strain relaxation concentrated near the substrate/epilayer interface [36,66]. Also, considerable luminescence intensity improvement was achieved by confining the misfit dislocations in a superlattice buffer. Uppal and co-workers report on layers with x_{Ga}

down to 0.8 grown on top of such a buffer, and found 150 times higher photoluminescence intensity compared to structures without the buffer [67].

D2 InGaAs/Si

Oe and Takeuchi [68] prepared InGaAs epitaxial layers on a Si-substrate by MBE. The FWHM of the room temperature photoluminescence from these layers increases from 38 nm to 166 nm for x_{Ga} = 100% to x_{Ga} = 51%. The peak position shifts from 0.88 to 1.55 mm.

E InGaAs - BULK CRYSTALS

Several authors [69-71] have begun to close the gap between GaAs and InP substrates by growing InGaAs bulk crystals with a Ga-content of 91 - 97%. PL excitonic lines with FWHM of 5 meV could be obtained for this composition [70]. Donor-acceptor-pair transitions due to accidentally introduced impurities are also observed. Roth et al [71] have grown an InGaAs-layer with x_{Ga} = 92% on a substrate of the same composition. They found a 500 times higher intensity of the excitonic line compared to that from the substrate bulk material with few residual impurity-related signals.

F CONCLUSION

Typical features of the luminescence and information obtained from the spectra of InGaAs on various substrates or from bulk were presented here. Although general consensus is reached for the most important parameters, some values such as the temperature dependence of the PL-peak position show considerable scatter. Further systematic investigations of the relationship between material properties and growth conditions are required to overcome this problem.

REFERENCES

[1] S.G. Hummel et al [*Appl. Phys. Lett. (USA)* vol.60 no.12 (1992) p.1483-5]

[2] V. Swaminathan, R.A. Stall, A.T. Macrander, R.J. Wunder [*J. Vac. Sci. Technol. B (USA)* vol.3 no.6 (1985) p.1631-6]

[3] M. Kubo, Y. Sasai, Y. Yoshioka, M. Ogura [*J. Cryst. Growth (Netherlands)* vol.73 (1985) p.439-44]

[4] A.F.S. Penna et al [*Solid State Commun. (USA)* vol.51 no.6 (1984) p.425-8]

[5] K.H. Goetz et al [*J. Appl. Phys. (USA)* vol.54 (1983) p.4543-52]

[6] A. Louati, C. Charreaux, A. Nouailhat, G. Guillot, P.N. Favennec, M. Salvi [*Solid State Commun. (USA)* vol.62 (1987) p.31-4]

[7] M.A. Herman, D. Bimberg, J. Christen [*J. Appl. Phys. (USA)* vol.70 (1991) p.R1-R52]

[8] H.B. Bebb, R.K. Williams [*Semicond. Semimet. (USA)* vol.8 (Academic Press, New York, 1972) ch.4 p.181-320]

[9] E.F. Schubert, E.O. Göbel, Y. Horikoshi, K. Ploog, H.J. Queisser [*Phys. Rev. B (USA)* vol.30 (1984) p.813-20]

[10] C. Charreaux, G. Guillot, A. Nouailhat [*J. Appl. Phys. (USA)* vol.60 (1986) p.768-72]

[11] J. Singh, K.K. Bajaj [*J. Appl. Phys. (USA)* vol.57 (1985) p.5433-7]

[12] J. Singh, K.K. Bajaj [*Appl. Phys. Lett. (USA)* vol.44 (1984) p.1075-7 and vol.48 (1986) p.1077-9]
[13] R.F. Schnabel, R. Zimmermann, D. Bimberg, H. Nickel, R. Lösch, W. Schlapp [*Phys. Rev. B (USA)* vol.46 (1992) p.9873-6]
[14] J. Christen, D. Bimberg [*Phys. Rev. B (USA)* vol.42 (1990) p.7213-9]
[15] P.W. Yu, D.C. Walters, W.C. Mitchel [*J. Appl. Phys. (USA)* vol.60 (1986) p.3864-7]
[16] E. Zielinski, H. Schweizer, K. Streubel [*J. Appl. Phys. (USA)* vol.59 (1986) p.2196-2204]
[17] I.C. Bassignana, C.J. Miner, N. Puetz [*J. Appl. Phys. (USA)* vol.65 (1989) p.4299-4305]
[18] K.H. Goetz [Ph.D thesis, Technical University Aachen, 1984 unpublished]
[19] R.E. Nahory, M.A. Pollack, W.D. Johnston Jr., R.L. Barns [*Appl. Phys. Lett. (USA)* vol.33 (1978) p.659-61]
[20] Y. Takeda, A. Sasaki, Y. Imamura, T. Takagi [*J. Electrochem. Soc. (USA)* vol.125 (1978) p.130]
[21] A.N. Pikhtin [*Sov. Phys. Semicond. (USA)* vol.11 (1977) p.245-64]
[22] T.Y. Wu, G.L. Pearson [*J. Phys. Chem. Solids (UK)* vol.33 (1972) p.409-15]
[23] R. People [*Appl. Phys. Lett. (USA)* vol.50 (1987) p.1604-6]
[24] C.P. Kuo, S.K. Vong, R.M. Cohen, G.B. Stringfellow [*J. Appl. Phys. (USA)* vol.57 (1985) p.5428-32]
[25] H. Asai, K. Oe [*J. Appl. Phys. (USA)* vol.54 (1983) p.2052-6]
[26] J.W. Matthews, A.E. Blakeslee [*J. Cryst. Growth (Netherlands)* vol.27 (1974) p.118-25]
[27] R. People, J.C. Bean [*Appl. Phys. Lett. (USA)* vol.47 (1985) p.322-4]
[28] H. Temkin, D.G. Gershoni, S.N.G. Chu, J.M. Vandenberg, R.A. Hamm, M.B. Panish [*Appl. Phys. Lett. (USA)* vol.55 (1989) p.1668-70]
[29] M. Gal, P.C. Taylor, B.F. Usher, P.J. Orders [*J. Appl. Phys. (USA)* vol.62 (1987) p.3898-3901]
[30] M.J. Ekenstedt, S.M. Wang, T.G. Andersson [*Appl. Phys. Lett. (USA)* vol.58 (1991) p.854-5]
[31] T.G. Andersson, Z.G. Chen, V.D. Kulakovskii, A. Uddin, J.T. Vallin [*Appl. Phys. Lett. (USA)* vol.51 (1987) p.752-4]
[32] M. Grundmann, U. Lienert, J. Christen, D. Bimberg, A. Fischer-Colbrie, J.N. Miller [*J. Vac. Sci. Technol. B (USA)* vol.8 (1990) p.751-7]
[33] I.J. Fritz, P.L. Gourley, L.R. Dawson [*Appl. Phys. Lett. (USA)* vol.51 (1987) p.1004-6]
[34] P.L. Gourley, I.J. Fritz, L.R. Dawson [*Appl. Phys. Lett. (USA)* vol.52 (1988) p.377-9]
[35] M. Grundmann, U. Lienert, D. Bimberg, A. Fischer-Colbrie, J.N. Miller [*Appl. Phys. Lett. (USA)* vol.55 (1989) p.1765-7]
[36] D. Morris, A.P. Roth, R.A. Masut, C. Lacelle, J.L. Brebner [*J. Appl. Phys. (USA)* vol.64 (1988) p.4135-40]
[37] E. Towe [*J. Appl. Phys. (USA)* vol.53 (1982) p.5136-9]
[38] T.P. Pearsall, L. Eaves, J.C. Portal [*J. Appl. Phys. (USA)* vol.54 (1983) p.1037-47]
[39] P.W. Yu, E. Kuphal [*Solid State Commun. (USA)* vol.49 (1984) p.907-10]
[40] Y.P. Varshni [*Physica (Netherlands)* vol.34 (1967) p.149-54]
[41] J.D. Lambkin, D.J. Dunstan [*Solid State Commun. (USA)* vol.67 (1988) p.827-9]
[42] M. Altarelli, F. Bassani [in *Handbook on Semiconductors* Ed. T.S. Moss (North-Holland, 1982) vol.1 Ed. W. Paul ch.5 p.269-322]
[43] Yu-Ssu Chen, O.K. Kim [*J.Appl. Phys. (USA)* vol.52 (1981) p.7392-6]

[44] A.F.S. Penna, J. Shah, A.E. DiGiovanni [*Solid State Commun. (USA)* vol.51 (1984) p.217-20]

[45] P.W. Yu, C.K. Peng, H. Morkoç [*J. Appl. Phys. (USA)* vol.65 (1989) p.2427-30]

[46] P.K. Bhattacharya, M.V. Rao, M-J. Tsai [*J. Appl. Phys. (USA)* vol.54 (1983) p.5096-5102]

[47] K.L. Fry, C.P. Kuo, R.M. Cohen, G.B. Stringfellow [*Appl. Phys. Lett. (USA)* vol.46 (1985) p.955-7]

[48] Z. Chen, W. Korb, R.K. Bauer, D. Bimberg [*Appl. Phys. Lett. (USA)* vol.55 (1989) p.645-7]

[49] M. Feng, J.D. Oberstar, T.H. Windhorn, L.W. Cook, G.E. Stillman, B.G. Storetman [*Appl. Phys. Lett. (USA)* vol.34 (1979) p.591-3]

[50] G. Guillot et al [*Semicond. Sci. Technol. (UK)* vol.5 (1990) p.391-4]

[51] S.G. Bishop [in *Deep Centers in Semiconductors* Ed. S.T. Pantelides (New York, 1986) ch.8 p.541-626]

[52] D. Bimberg, B. Srocka [Datareview in this book: *6.2 Impurity levels in InGaAs*]

[53] M.V. Rao, P.K. Bhattacharya [*J. Appl. Phys. (USA)* vol.57 (1985) p.333-7]

[54] T.H. Zhang, N.N. Ledentsov, K. Ploog [*Appl. Phys. A (Germany)* vol.53 (1991) p.437-41]

[55] C. Charreaux, A. Louati, A. Nouailhat, G. Guillot, J.L. Benchimol, S. Slempkes [*Semi-Insulating III-V Materials*, Ohmsha LTD. 1986, p.515-20]

[56] K.H. Goetz et al [*Appl. Phys. Lett. (USA)* vol.46 (1985) p.277-9]

[57] T. Yagi, Y. Fujiwara, T. Nishino [*Jpn. J. Appl. Phys. (Japan)* vol.22 (1983) p.L467-9]

[58] A. Salokatve, M. Hovinen [*J. Appl. Phys. (USA)* vol.67 (1990) p.3379-81]

[59] E.H. Reihlen, D. Birkedal, T.Y. Wang, G.B. Stringfellow [*J. Appl. Phys. (USA)* vol.68 (1990) p.1750-6]

[60] H. Beneking, P. Narozny, N. Emeis [*Appl. Phys. Lett. (USA)* vol.47 (1985) p.828-30]

[61] K. Yamada, K. Wada [*Inst. Phys. Conf. Ser. (USA)* vol.106 (1989) ch.4 p.153-8]

[62] J.P. Laurenti, P. Roentgen, K. Wolter, K. Seibert, H. Kurz, J. Camassel [*J. de Physique (France)* coll.C4 suppl.9 vol.49 (1988) p.C4-693-6]

[63] W. Walukiewicz [*Appl. Phys. Lett. (USA)* vol.54 (1989) p.2009-11]

[64] A.R. Peaker [in *Properties of Gallium Arsenide, 2nd Edition,* EMIS Datareviews Series no.2 (INSPEC, IEE, 1990)]

[65] B. Elman, E.S. Koteles, P. Melman, K. Ostreicher, C. Sung [*J. Appl. Phys. (USA)* vol.70 (1991) p.2634-40]

[66] A.P. Roth et al [*Can. J. Phys. (Canada)* vol.67 (1989) p.330]

[67] P.N. Uppal, R.P. Leavitt, S.P. Svensson, J.S. Ahearn, R. Herring [*Inst. Phys. Conf. Ser. (UK)* vol.96 (1989) p.191-7]

[68] K. Oe, H. Takeuchi [*Jpn. J. Appl. Phys. (Japan)* vol.26 (1987) p.L120-2]

[69] W.A. Bonner, B.J. Skromme, E. Berry, H.L. Gilchrist, R.E. Nahory [*Inst. Phys. Conf. Ser. (UK)* vol.96 (1988) ch.5 p.337-42]

[70] T. Bryskiewizs, P. Edelman, Z. Wasilewski, D. Coulas, J. Noad [*J. Appl. Phys. (USA)* vol.68 (1990) p.3018-20]

[71] A.P. Roth et al [*J. Cryst. Growth (Netherlands)* vol.113 (1991) p.379-84]

6.2 Impurity levels in InGaAs

D. Bimberg and B. Srocka

November 1992

A INTRODUCTION

Energy levels introduced by impurities in InGaAs are reviewed in this Datareview. Since the ternary alloy has not been as widely investigated as binary compounds, only a few experimental data are available. These data mostly exist for the lattice matched composition.

B LATTICE MATCHED MATERIAL - $In_{0.53}Ga_{0.47}As/InP$

B1 Shallow Levels

TABLE 1 lists values of binding energies for impurities acting as shallow acceptors. The ionisation energy values are derived from the wavelength position of the donor-acceptor-pair and the conduction band-acceptor transition lines with respect to the excitonic lines. Since microscopic fluctuations of the alloy composition lead to a variation of the energy gap, the PL-lines are much broader than those in binary material [1]. A slight shift to higher energies often results from the convolution of the theoretical line shape with a Gaussian broadening function [2]. This has not always been taken into account in the literature [3-5]. Thus determination of the exact ionisation energies is not straightforward. Additional problems arising from potential fluctuations are discussed by Yu and co-workers [6], with more details given in [1,2,6].

TABLE 1 Shallow acceptor levels in InGaAs.

Dopant	Ionisation energy (meV)	Refs
C	13 ± 1	[2,3,11]
Mg	16 ± 2	[4]
Cd	19	[5]
	23.4 .. 6.1 for 7×10^{15} .. 5×10^{17} cm^{-3} doping concentration	[7] (Hall meas.)
Zn	22 ± 1	[2,11]
	28 ± 1	[3]
Be	about 27	[12]
Si	25 ± 1	[2]
	33 ± 2	[3]
Hg	68 ± 5	[4]

With increasing doping concentration, the higher excited states merge, forming a tail in the density of states. The activation energy is thus lowered and may even vanish for the degenerate case. This dependence was investigated by Wehmann et al [7].

The binding energy of the shallow donors (Sn, S, Te, Si, Se in [8]) is calculated with good precision from a simple hydrogenic model [2] as 3 meV. Direct determination by photoluminescence is impossible because of the strong alloy broadening. Some investigators [9,10] have claimed larger values (between 7 and 10 meV) for the donor binding energy, but the interpretation of their PL-spectra seems to be inconsistent.

B2 Deep Levels

Deep level transient spectroscopy (DLTS) - the most common measurement method for deep states - is also affected by alloy broadening [13,14]. Thus, as in the case of photoluminescence, the exact determination of activation energies is complicated. This holds for optically assisted methods (DLOS, photocapacitance). On the other hand, temperature dependent Hall measurements are less precise, since ambipolar conduction together with poorly known mobility causes serious problems [15]. This may explain some of the differences between results of different authors. TABLE 2 gives a summary of known transition metal-related deep levels and of hydrogen.

TABLE 2 Impurity related deep levels in InGaAs ('+' above valence band edge, '-' below conduction band edge, a - acceptor, d - donor).

Impurity	Energy level (meV)	Type	Ref	Measurement method
Cr	-24	d	[17]	PL
	-30	d	[17]	Hall
Mn	$+52.3 \pm 1.8$	a	[24]	Hall
	$+52.5 \pm 2.5$	a	[25]	Hall
	$+53 \pm 3$	a	[26]	PL
H	about -100	d	[23]	Hall
Fe	-340	a $Fe^{2+/3+}$	[16]	DLTS, DLOS, PL
	-370	a	[18]	DLTS
	+350	a	[15]	Hall
	+460	a	[17]	Hall
	± 150	Fe-related	[20,21]	PL
Ti	-370	d $Ti^{3+/4+}$	[5,27]	DLTS

The best investigated deep level so far is that at E_c - 0.34 eV belonging to the $Fe^{2+/3+}$ acceptor transition [15-18]. Guillot et al [16] combined several experimental techniques and obtained a consistent picture of this dopant. The apparent electron capture cross section is $3 \pm 2 \times 10^{-16}$ cm^{-2} for $T \rightarrow \infty$ [16]. The trap, investigated by Loualiche et al [19], and suggested to be oxygen-related, seems to be also caused by the Fe acceptor. Sometimes a level with an activation energy of about 150 meV is also ascribed to an Fe-related complex [20,21]. But the real nature of this level is not yet clear, since Forrest et al [22] found a trap at 160 meV in undoped material.

The location of the Cr-level should be considered with caution, since it is not clear if the detected level is an isolated impurity, a complex or an accidentally introduced different dopant [17]. It can hardly be the isolated Cr donor level, since then a strong contradiction to the predictions of the internal reference rule (see below) would exist.

The value for the H-related level is approximate [23] and the passivation mechanism of hydrogen is not fully understood.

Since the theory of Ledebo and Ridley was put forth that deep levels are linked to the vacuum level and are independent of the particular III-V semiconductor [28], a broad discussion has emerged. The now widely accepted result is that indeed transition metal (TM) impurities produce deep levels pinned with respect to an internal reference level unaffected by the individual host crystal within one class of semiconductors (III-V or II-VI) e.g. [29-32]. This model, originally called VRBE (Vacuum Referred Binding Energy) was recently re-named the Internal Reference Rule, since the reference level has been shown [31] to be the same as for the heterointerface alignment and not simply the vacuum. This rule may be used to extrapolate from levels known in one compound to positions in another one, if the conduction and valence band discontinuities are known. TABLE 3 summarises such values for lattice matched InGaAs expected from the known data for GaAs and InP. The band discontinuity data are taken from a comparison of the Ti level position in each compound [27]. Keeping in mind the uncertainties of the band edge discontinuities and level positions, one can see from TABLES 2 and 3, that the experimental values for Fe match the prediction while those for Cr and Mn do not.

Data for the electric field dependence of the activation energy are, to our knowledge, not reported so far.

TABLE 3 Expected energy levels for 3d TMs in InGaAs from internal reference rule [27,32,33]. For notation see TABLE 2.

Impurity	Expected energy level (meV)	Type
Ti	-370	d $Ti^{3+/4+}$
V	E_c + 35	a $V^{2+/3+}$
	E_v - 160	d $V^{3+/4+}$
Cr	+570	a $Cr^{2+/3+}$
	+150 .. +240	d $Cr^{3+/4+}$
Mn	E_v - 60 .. 170	a
Fe	+400	a $Fe^{2+/3+}$
Co	E_v - 40	a $Co^{2+/3+}$
Ni	+30 .. +110	a $Ni^{2+/3+}$
	-50 .. E_c + 100	d $Ni^{3+/4+}$

C COMPOSITION DEPENDENCE

To obtain the variation of binding energies with alloy composition, one has to interpolate between the data from GaAs, InGaAs and InAs [2,32-34]. The validity of this procedure arises simply from continuity considerations. For the Mn-acceptor, Phatak et al give an experimental value of 77 meV for an intermediate alloy with x_{Ga} = 79% [35].

For the TMs, the internal reference rule states that the band edges shift with composition while the deep levels remain fixed. In lattice matched systems (AlGaAs/GaAs and InGaAsP/InP) this prediction has been found to be valid [18,31,36].

D STRAIN EFFECTS

Theoretical considerations and experiments show that acceptor binding energies are strongly affected by strain [37-40]. Depending on the actual acceptor binding energy, the degenerate acceptor levels split and shift in energy. The acceptor deformation potentials are reduced with respect to those of the valence band.

E CONCLUSION

Exact determination of impurity introduced energy levels is somewhat complicated by the alloy broadening effects. Further work is expected to broaden our knowledge of impurity levels in InGaAs.

REFERENCES

[1] D. Bimberg, B. Srocka [Datareview in this book: *6.1 Photoluminescence of pure InGaAs alloys*]

[2] K.H. Goetz et al [*J.Appl. Phys. (USA)* vol.54 (1983) p.4543-52]

[3] P.K. Bhattacharya, M.V. Rao, M-J. Tsai [*J.Appl.Phys. (USA)* vol.54 (1983) p.5096-102]

[4] A. Louati, C. Charreaux, A. Nouailhat, G. Guillot, P.N. Favennec, M. Salvi [*Solid State Commun. (USA)* vol.62 (1987) p.31-4]

[5] Z. Chen, W. Korb, R.K. Bauer, D. Bimberg [*Appl. Phys. Lett. (USA)* vol.55 (1989) p.645-7]

[6] P.W. Yu, C.K. Peng, H. Morkoç [*J. Appl. Phys. (USA)* vol.65 (1989) p.2427-30]

[7] H.H. Wehmann, F. Fiedler, A. Schlachetzki [*Electron. Lett. (UK)* vol.22 (1986) p.1338-40]

[8] F. Fiedler, H.H. Wehmann, A. Schlachetzki [*J. Cryst. Growth (Netherlands)* vol.74 (1986) p.27-38]

[9] Yu-Ssu Chen, O.K. Kim [*J. Appl. Phys. (USA)* vol.52 (1981) p.7392-6]

[10] A.F.S. Penna, J. Shah, A.E. DiGiovanni [*Solid State Commun. (USA)* vol.51 (1984) p.217-20]

[11] K.L. Fry, C.P. Kuo, R.M. Cohen, G.B. Stringfellow [*Appl. Phys. Lett. (USA)* vol.46 (1985) p.955-7]

[12] M. Feng, J.D. Oberstar, T.H. Windhorn, L.W. Cook, G.E. Stillman, B.G. Storetman [*Appl. Phys. Lett. (USA)* vol.34 (1979) p.591-3]

[13] P. Omling, L. Samuelson, H.G. Grimmeiss [*J. Appl. Phys. (USA)* vol.54 (1983) p.5117-22]

[14] J. Criado, A. Gomez, E. Munoz, E. Calleja [*Appl. Phys. Lett. (USA)* vol.49 (1986) p.1790-2]

[15] B. Tell, U. Koren, B.I. Miller [*J. Appl. Phys. (USA)* vol.61 (1987) p.1172-5]

[16] G. Guillot et al [*Semicond. Sci. Technol. (UK)* vol.5 (1990) p.391-4]

[17] M.V. Rao, P.K. Bhattacharya [*J. Appl. Phys. (USA)* vol.57 (1985) p.333-7]

[18] M. Sugawara, M. Kondo, T. Takanohashi, K. Nakajima [*Appl. Phys. Lett. (USA)* vol.51 (1987) p.834-6]

[19] S. Loualiche, A. Gauneau, A. Le Corre, D. Lecrosnier, H.L. Haridon [*Appl. Phys. Lett. (USA)* vol.51 (1987) p.1361-3]

[20] K.H. Goetz et al [*Appl. Phys. Lett. (USA)* vol.46 (1985) p.277-9]
[21] C. Charreaux, A. Louati, A. Nouailhat, G. Guillot, J.L. Benchimol, S. Slempkes [*Semi-Insulating III-V Materials*, Ohmsha LTD., 1986, p.515-20]
[22] S.R. Forrest, O.K. Kim [*J.Appl. Phys. (USA)* vol.53 (1982) p.5738-45]
[23] E.M. Omeljanovsky, A.V. Pakhomov, A.Y. Polyakov [*Phys. Lett. A (Netherlands)* vol.141 (1989) p.75-7]
[24] P.S. Whitney, C.G. Fonstad [*J. Appl. Phys. (USA)* vol.57 (1985) p.4663-7]
[25] N. Chand, P.A. Houston, P.N. Robson [*Electron. Lett. (UK)* vol.17 (1981) p.726-7]
[26] T.H. Zhang, N.N. Ledentsov, K. Ploog [*Appl. Phys. A (Germany)* vol.53 (1991) p.437-41]
[27] N. Baber, H. Scheffler, A. Ostmann, T. Wolf, D. Bimberg [*Phys. Rev. B (USA)* vol.45 (1992) p.4043-7]
[28] L.A. Ledebo, B.K. Ridley [*J. Phys. C (UK)* vol.15 (1982) p.L961-4]
[29] J.M. Langer, H. Heinrich [*Phys. Rev. Lett. (USA)* vol.55 (1985) p.1414-7]
[30] M. Hamera, W. Walukiewicz, D.D. Nolte, E.E. Haller [*Phys. Rev. B (USA)* vol.39 (1989) p.10114-9]
[31] J.M. Langer, C. Delerue, M. Lanoo, H. Heinrich [*Phys. Rev. B (USA)* vol.38 (1988) p.7723]
[32] A.M. Hennel [*Semicond. Semimet. (USA)* (Academic Press, New York, 1992) ch.6, in print]
[33] A.M. Hennel [*Acta Phys. Pol. A (Poland)* vol.79 (1991) p.15-29]
[34] Landoldt-Börnstein [*Numerical Data and Functional Relationships in Science and Technology*, New Series, vol.III/17a, Ed. O. Madelung (Springer, 1982)]
[35] S.B. Phatak, S.M. Bedair, S. Fujita [*Solid-State Electron. (UK)* vol.23 (1980) p.839-44]
[36] T. Tanahashi, S. Okuda, M. Kondo, M. Sugawara, S. Isozumi, K. Nakajima [Mat. Res. Soc. - Fall Meeting 1987]
[37] J.P. Loehr, Y.C. Chen, J. Rhee, P.K. Bhattacharya, J. Singh [*J. Vac. Sci. Technol. (UK)* vol.10 (1992) p.797-9]
[38] J.P. Loehr, J. Singh, Y.C. Chen, P.K. Bhattacharya, D. Biswas [*Inst. Phys. Conf. Ser. (UK)* vol.106 (1990) p.441-6]
[39] Yia-Chung Chang [*Physica B&C (Netherlands)* vol.146 no.1-2 (1987) p.137-49]
[40] L.C. Andreani, S. Fraizzoli [*Phys. Rev. B (USA)* vol.42 (1990) p.7641-2]

6.3 Deep levels in InGaAs/GaAs and InGaAs/InP

P.K. Bhattacharya

February 1993

A INTRODUCTION

In evaluating the quality of semiconducting materials for device applications, it has been found that centres with deep energy levels in the forbidden energy gap of semiconductors play an important role. Deep levels essentially act as carrier recombination or trapping centres and adversely affect device performance. In addition to the energy position of the deep levels in the forbidden energy gap, information is needed about their capture and emission rates for electrons and holes, concentrations, capture cross-sections and spatial distributions. The capture cross-section of a trap distinguishes electron traps, hole traps and recombination centres. The thermal emission rate of carriers from trap levels to the respective band minima is given by

$$e_{n,p} = \sigma_{n,p} v_{th} N_{c,v} \exp(-E_T/kT) \tag{1}$$

where

$$\sigma_{n,p} = \sigma_{\infty} \exp(-\Delta E_B/kT) \tag{2}$$

is the thermally activated capture cross-section with a barrier energy ΔE_B. v_{th} is the thermal velocity and $N_{c,v}$ are the effective density of states in the conduction and valence bands. The capture rate of electrons at a deep level is given by

$$c_n = \sigma_n v_{th} n \tag{3}$$

where n is the concentration of electrons in the conduction band. A similar equation is true for the capture of holes. The technique most widely used to identify and characterise traps is the capacitance transient measurement (and its variation, Deep Level Transient Spectroscopy) on Schottky barriers or p-n junction diodes. Other commonly used techniques are admittance spectroscopy and conductance spectroscopy.

The characteristics of traps and deep impurity levels in $In_xGa_{1-x}As$ and $In_{1-x}Ga_xAs_yP_{1-y}$ are reported here. Very little work has been done to identify traps in $In_xGa_{1-x}As$ on GaAs, which are lattice-mismatched. More work has been done to identify traps in $In_{0.53}Ga_{0.47}As$ lattice-matched to InP. This ternary compound constitutes the end-point alloy of the quaternary system $In_{1-x}Ga_xAs_yP_{1-y}$ ($y = 2.16x$) lattice matched to InP and is technologically very important for optoelectronic device applications.

B TRAPS IN $In_xGa_{1-x}As/GaAs$

The first report of deep levels in $In_xGa_{1-x}As$ ($0 < x < 0.21$) grown by vapour phase epitaxy on n^+ GaAs substrates (3° off (100)) has been made by Mircea et al [1]. Thick layers were grown, with some initial grading, so that the dislocation density was in the 10^5 cm^{-2} range. A dominant electron trap was observed in the alloys with $E_T = 0.83$ eV for x = 0 (GaAs) and decreasing with x. The energy position of the trap with respect to the conduction band edge in the alloys, with the valence band edge as a reference, is shown in FIGURE 1. In the composition range from x = 0 to x = 0.2 the bandgap decreases by 18% while the trap energy varies by 9%. The main electron trap in GaAs is the well known EL2 level, and it appears that the centres identified in the ternary alloys are related to it.

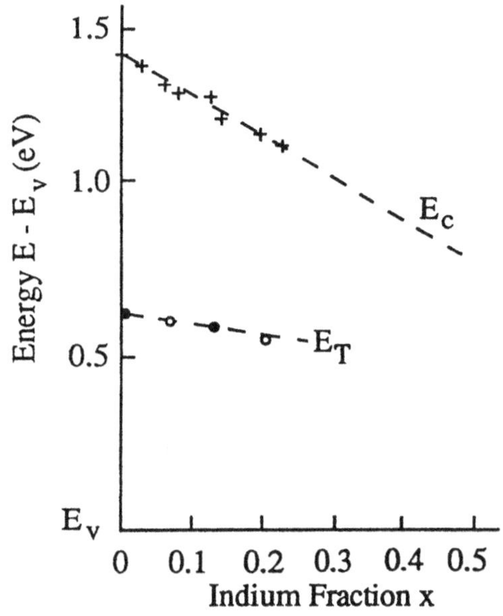

FIGURE 1 Dependence on indium fraction x of the bottom of the conduction band E_c and of the deep level E_T for deep electrons in $In_xGa_{1-x}As$. Reference is top of the valence band.

Traps in undoped and Cr-doped bulk liquid encapsulated Czochralski (LEC) $In_xGa_{1-x}As$ ($0.01 < x < 0.12$) have been identified and characterised by Bao et al [2]. These crystals were grown along the [111] direction with GaAs or InGaAs seeds. The activation energy of these traps decreases as x increases. The amount of change in E_T is about one-third of the bandgap shrinkage due to increased In concentration. The activation energies of the traps are listed in TABLE 1.

TABLE 1 Traps in bulk LEC $In_xGa_{1-x}As$ [2].

Material	E_T (eV)
Undoped $In_{0.06}Ga_{0.94}As$	0.17 0.34 0.52
Cr-doped $In_{0.055}Ga_{0.945}As$	0.23 0.39 0.52

C TRAPS IN LATTICE MATCHED $In_{0.53}Ga_{0.47}As$ AND $In_{1-x}Ga_xAs_yP_{1-y}/InP$

The first identification of traps in $In_{0.53}Ga_{0.47}As$ was made by Bhattacharya et al [3] and in $In_{0.85}Ga_{0.15}As_{0.39}P_{0.61}$ by Sasai et al [4]. Their results are summarised in TABLES 2 and 3, respectively. Since then, electron and hole traps in these alloys and in InP have been identified and characterised by several groups [5-30]. The data are summarised in TABLES 4 and 5 and FIGURE 2. The physico-chemical origins of these centres are largely unknown. In addition, a dominant trap level at (E_c - (0.16 ± 0.01)) eV or (E_c - 0.10) eV has been identified in lattice-matched $In_{0.53}Ga_{0.47}As/InP$ or mismatched InGaAs/InP [31-35].

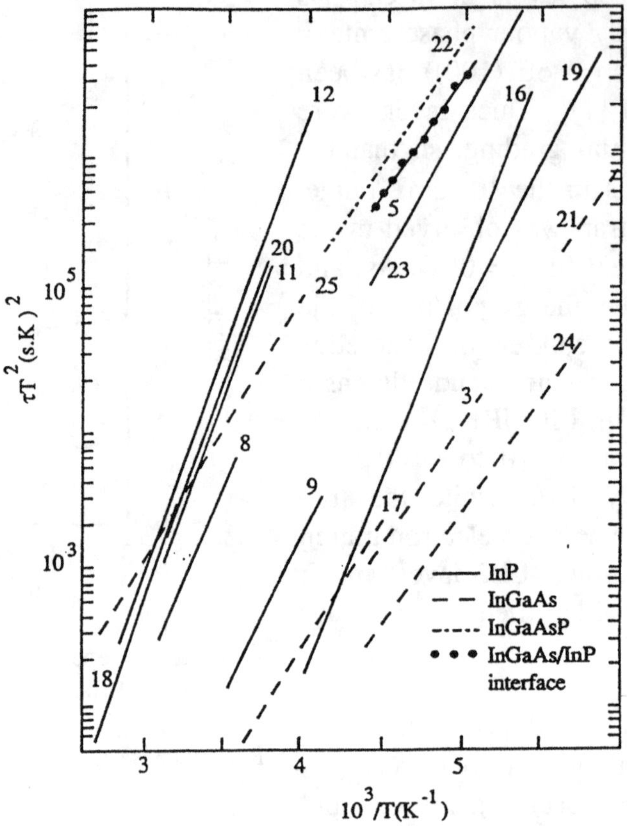

FIGURE 2 Arrhenius plots of electron traps identified in InP and lattice matched InGaAs and InGaAsP.

TABLE 2 Parameters of trap levels identified in n- and p-type LPE $In_{1-x}Ga_xAs_yP_{1-y}$ [3].

Material	E_g at 300 K (eV)	($E_T + \Delta E_B$) (eV)	σ^e_∞ (cm^2)	N_T (cm^{-3})
p-$In_{0.81}Ga_{0.19}As_{0.45}P_{0.55}$	1	0.59 ± 0.03 0.22 ± 0.02	1.5 ± 0.7 x 10^{-18} 4.3 ± 0.2 x 10^{-21}	1.2 ± 0.1 x 10^{15} 5.0 ± 0.5 x 10^{13}
n-$In_{0.71}Ga_{0.29}As_{0.64}P_{0.36}$	0.95	0.81 ± 0.03	2.3 ± 0.1 x 10^{-13}	1.0 ± 0.4 x 10^{13}
n-$In_{0.73}Ga_{0.27}As_{0.58}P_{0.42}$	0.98	0.81 eV trap may be present		
p-$In_{0.53}Ga_{0.47}As$	0.74	0.35 ± 0.02	9.4 ± 0.5 x 10^{-16}	6.8 ± 1.4 x 10^{13}
p-InP	1.35	0.37 ± 0.03	1.1 ± 0.05 x 10^{-17}	3.0 ± 0.5 x 10^{15}

TABLE 3 Parameters of electron traps observed in $In_{0.85}Ga_{0.15}As_{0.39}P_{0.61}$ [4].

	E_T (eV)	σ_n (cm^2)	N_T (cm^{-3})
E_1	0.27	1.0 x 10^{-14}	9.2 x 10^{13}
E_2	0.30	4.6 x 10^{-16}	1.1 x 10^{14}
E_3	0.51	3.0 x 10^{-14}	1.5 x 10^{14}

TABLE 4 A summary of traps identified in $In_{1-x}Ga_xAs_yP_{1-y}$ and their parameters [11].

Material and epitaxy methods	Electron Traps		Hole Traps		Ref
	E_T (eV)	σ_n (10^{-16} cm^2)	E_T (eV)	σ_p (10^{-16} cm^2)	
n-$Ga_{0.1}In_{0.9}As_{0.21}P_{0.79}$ (VPE)	0.81 0.35	590 9×10^{-3}	0.25	5×10^{-5}	[5]
n-$Ga_{0.2}In_{0.8}As_{0.45}P_{0.55}$ (LPE)	0.59 0.22	0.015 4.3×10^{-5}			[3,5]
n-$Ga_{0.3}In_{0.7}As_{0.64}P_{0.36}$ (LPE)	0.83	230	0.27	2.5×10^{-5}	[3,5]
n-$Ga_{0.2}In_{0.8}As_{0.44}P_{0.56}$ (LPE)	0.60 0.67	2.5×10^{6} 1.6×10^{5}			[6]
n-$Ga_{0.18}In_{0.82}As_{0.39}P_{0.61}$ (LPE)	0.27 0.30 0.51	100 4.6 300			[4]
p-$Ga_{0.28}In_{0.72}As_{0.6}P_{0.4}$ (LPE)			0.25 0.40 0.69	0.5 0.04 60	[11]
p-$Ga_{0.47}In_{0.53}As$ (LPE)	0.35	9.4			[3,5]
p-$Ga_{0.47}In_{0.53}As$ (LPE)			0.15		[10]
n-$Ga_{0.47}In_{0.53}As$ (LPE)	0.57				[8]
n-$Ga_{0.47}In_{0.53}As$ (LPE)	0.16	0.3			[7]
n-$Ga_{0.47}In_{0.53}As$ (VPE)	0.42 0.35	0.06 9	0.23	3	[9]
p-$Ga_{0.47}In_{0.53}As$ (LPE)			0.08	6×10^{-4}	[11]
n-$Ga_{0.47}In_{0.53}As$ (MBE)	0.33	1×10^{-16}			[12]
n-$Ga_{0.47}In_{0.53}As$ (MBE)	0.45 (attributed to Fe)				[13]

TABLE 5 Parameters of hole traps in InP [11].

Materials (growth techniques and type)	Hole traps		Ref
	E_T (eV)	σ_p (10^{-16} cm^2)	
InP (VPE) p-type	0.09 0.22 0.29 0.41 0.50	0.06 0.04 4 4 3	[26]
InP (VPE) n-type	0.27 - 0.28	10^3 - 10^4	[27]
InP (VPE) n-type	18	5	[28]
InP (LEC) p-type	0.52		[29]
InP (LPE) p-type	0.22	0.1	[11]

D DEEP LEVELS IN $In_{0.53}Ga_{0.47}As$:Fe

For a number of applications, like passivation of active layers, current confinement, or for photodetectors, it is desirable to have InGaAs epitaxial layers of near-intrinsic nature exhibiting high resistivity. Liquid phase epitaxial growth of semi-insulating $In_{0.53}Ga_{0.47}As$ doped with Fe with $N_D - N_A$ as low as 10^{12} cm^{-3} was first demonstrated by Rao and Bhattacharya [14]. From analysis of temperature-dependent Hall data on these samples the Fe acceptor ionisation energy was found to be 0.46 eV. Subsequently deep levels in Fe-doped $In_{0.53}Ga_{0.47}As$ have been characterised by several groups [10,13,15-17]. From DLTS measurements, deep levels attributed to Fe have been identified at 0.41 eV [15], 0.33 eV [16], 0.45 eV [16] and 0.37 eV [17] above the valence band. The Fe-related deep level is believed to be an electron trap. Sugawara et al [17] have experimentally determined the energy position of the deep level for the entire lattice matched $In_{1-x}Ga_xAs_yP_{1-y}$ composition range. From infrared excitation spectroscopy results Dansas et al [13] identified a 0.32 eV electron trap in $In_{0.53}Ga_{0.47}As$ and attributed its presence to Fe impurities. In addition to the deep level, two shallow acceptor levels at 110 meV and 150 meV above the valence band were also identified by these authors. It is not clear whether these levels contribute to the high resistivity behaviour or not.

E CONCLUSION

A review of dominant deep levels in $In_xGa_{1-x}As$/GaAs, $In_{0.53}Ga_{0.47}As$/InP and $In_{1-x}Ga_xAs_yP_{1-y}$/InP has been made. Some of these may be native defects, or are related to specific impurities such as Fe. The physico-chemical origin of most of them is largely unknown.

REFERENCES

[1] A. Mircea, A. Mitonneau, J. Hallais [*Phys. Rev. B (USA)* vol.16 (1977) p.3665]

[2] X.J. Bao et al [*J. Electron. Mater. (USA)* vol.20 (1991) p.207]

[3] P.K. Bhattacharya, J.W. Ku, S.J.T. Owen, S.H. Chiao, R. Yeats [*Electron. Lett. (UK)* vol.15 (1979) p.753]

[4] Y. Sasai, Y. Yamazoe, M. Okuyama, T. Nishino, Y. Hamakawa [*Jpn. J. Appl Phys. (Japan)* vol.18 (1979) p.1415]

[5] P.K. Bhattacharya, J.W. Ku, S.J.T. Owen, G.H. Olsen, S.-H.Chiao [*IEEE J. Quantum Electron. (USA)* vol.17 (1981) p.150]

[6] A. Tamura, K. Oka, M. Inoue, J. Shirafuji, Y. Inuishi [*Jpn. J. Appl. Phys. (Japan)* vol.19 Suppl.19.1 (1980) p.479]

[7] S.R. Forrest, O.K. Kim [*J. Appl. Phys. (USA)* vol.53 (1982) p.5738]

[8] R. Trommer, H. Albrecht [*Jpn. J. Appl. Phys. (Japan)* vol.22 (1983) p.L364]

[9] J.N. Patillon, S. Makaram-Ebeid, J.P. Chane, B.G. Martin, G.M. Martin [in *Proc. 11th Int. Conf. on Gallium Arsenide and Related Compounds*, Biarritz, 1984, Inst. Phys. Conf. Ser. vol.74 (1985) p.475]

[10] K.-H. Goetz et al [*Appl. Phys. Lett. (USA)* vol.46 (1985) p.277]

[11] J.L. Pelloie, G. Guillot, A. Nouailhat, A.G. Antolini [*J. Appl. Phys. (USA)* vol.59 (1986) p.1536]

[12] S. Loualiche, A. Gauneau, A. LeCorre, D. Lecrosnier, H. L'Haridon [*Appl. Phys. Lett. (USA)* vol.51 (1987) p.1361]
[13] P. Dansas, D. Pascal, C. Bru, S. Laval, L. Giraudet, M. Allovon [*J. Appl. Phys. (USA)* vol.67 (1990) p.1384]
[14] M.V. Rao, P.K. Bhattacharya [*J. Appl. Phys (USA)* vol.57 (1985) p.333]
[15] G. Guillot et al [*Semicond. Sci. Technol. (UK)* vol.5 (1990) p.391]
[16] Z. Chen, T. Wolf, W. Korb, D. Bimberg [*J. Appl. Phys. (USA)* vol.64 (1988) p.4574]
[17] M. Sugawara, M. Kondo, T. Takanohashi, K. Nakajima [*Appl. Phys. Lett. (USA)* vol.51 (1987) p.834]
[18] U.K. Mishra, J.F. Jensen, A.S. Brown, M.A. Thompson, L.M. Jelloian, R.S. Baubein [*IEEE Electron Device Lett. (USA)* vol.9 (1988) p.482]
[19] O. Wada, A. Majerfeld, A.N.M.M. Choudhury [*J. Appl. Phys. (USA)* vol.51 (1980) p.423]
[20] G. Bremond, A. Nouailhat, G. Guillot, B. Cockayne [*Electron. Lett. (UK)* vol.17 (1981) p.55]
[21] M.J. Caldas, A. Fazzio, A. Zunger [*Appl. Phys. Lett. (USA)* vol.45 (1984) p.671]
[22] M.C. Joncour, R. Mellet, M.N. Charasse, J. Burgeat [*J. Cryst. Growth (Netherlands)* vol.75 (1986) p.295]
[23] J.-C. Renaud, N.L. Nguyen, M. Allovon, P. Blanconnier, S. Vuye, A. Scavennec [*J. Lightwave Technol. (USA)* vol.6 (1988) p.1507]
[24] L. Giraudet [Doctoral thesis (Paris XI University Orsay, 1988)]
[25] S. Krishnamurthy, A. Sher [*J. Appl. Phys. (USA)* vol.61 (1987) p.1475]
[26] M. Inuishi, B.W. Wessels [*Electron. Lett. (UK)* vol.17 (1981) p.685]
[27] A.N.M.M. Choudhury, P.N. Robson [*Electron. Lett. (UK)* vol.15 (1979) p.247]
[28] A.T. Macrander, W.D. Johnston Jr. [*J. Appl. Phys. (USA)* vol.54 (1983) p.806]
[29] S.S. Li, W.L. Wang, E.H. Shaban [*Solid State Commun. (USA)* vol.51 (1984) p.15]
[30] G.P. Watson et al [*J. Appl. Phys. (USA)* vol.71 (1992) p.3399]
[31] M. Ogura, M. Mizuta, K. Onaka, H. Kukimoto [*Jpn. J. Appl. Phys (Japan)* vol.22 (1983) p.1502]
[32] Y. Takeda, S. Takigawa, M. Kasu, A. Sasaki [*Proc. 12th Int. Symp. on Gallium Arsenide and Related Compounds*, Karuizawa, Japan, Inst. Phys. Conf. Ser. no.79 (1985) p.301]
[33] D. Pogany, F. Ducroquet, S. Ababou, G. Bremond [*J. Electrochem. Soc. (USA)* vol.140 (1993) p.560]
[34] P.S. Whitney, C.G. Fonstad [*J. Cryst. Growth (Netherlands)* vol.83 (1987) p.219]
[35] C.D. Lee, S.R. Forrest [*J. Appl. Phys. (USA)* vol.69 (1991) p.342]

6.4 Auger recombination coefficients in $In_{0.53}Ga_{0.47}As$

J. Shah

February 1993

A INTRODUCTION

Recombination of carriers plays an important role in semiconductor physics as well as in the operation of devices such as semiconductor lasers. For example, the relative importance of radiative and non-radiative processes determines the quantum efficiency and threshold of lasers. Most of the non-radiative processes are extrinsic in that they involve impurities or defects in the semiconductor. There is, however, an intrinsic non-radiative process, the Auger recombination process, which was first considered by Beattie and Landsberg [1]. In the band-to-band radiative process, an electron in the conduction band recombines with a hole in the valence band and emits a photon. In the band-to-band Auger process, the electron-hole recombination is not accompanied by the emission of a photon but by the excitation of an adjacent carrier to a higher energy state determined by energy and momentum conservation. This process is mediated by the Coulomb interaction between the carriers. The Auger recombination coefficients in InGaAs grown lattice-matched to InP and their quantum wells are discussed in this Datareview. For a general discussion of Auger recombination in semiconductors, the reader is referred to several reviews and recent articles [2-10].

B THE AUGER RECOMBINATION PROCESSES

There are several different Auger processes depending on which initial and final states are involved, and whether or not a phonon is involved. FIGURE 1(a) illustrates one of the simplest Auger processes in a direct-gap zinc-blende semiconductor such as InGaAs or GaAs in which interband recombination is accompanied by the excitation of an electron to a higher energy state. This process is classified as CCCH, and its efficiency is determined by the Coulomb interaction between the two initial electrons and the band structure. FIGURE 1(b) illustrates the CCCHP process which is similar to the CCCH process but additionally involves a phonon (P). Other possible processes include CHHL, CHHS, CHHLP and CHHSP, where L and S denote light and spin-split-off hole respectively.

Different considerations apply if quantum wells are involved. In this case, the final state can be a bound state within a quantum well or an unbound state above the barrier [8, 11-13]. Since there are many different electron and hole subbands, a number of different final states, both bound in a well and unbound in the barrier, are possible. An additional complication is the non-parabolic dispersion curves for valence bands in quantum wells. These complications make it extremely difficult to calculate the Auger coefficients in quantum wells. If band structure parameters are the same, no significant difference in the Auger coefficient is expected for bulk and quantum wells [8].

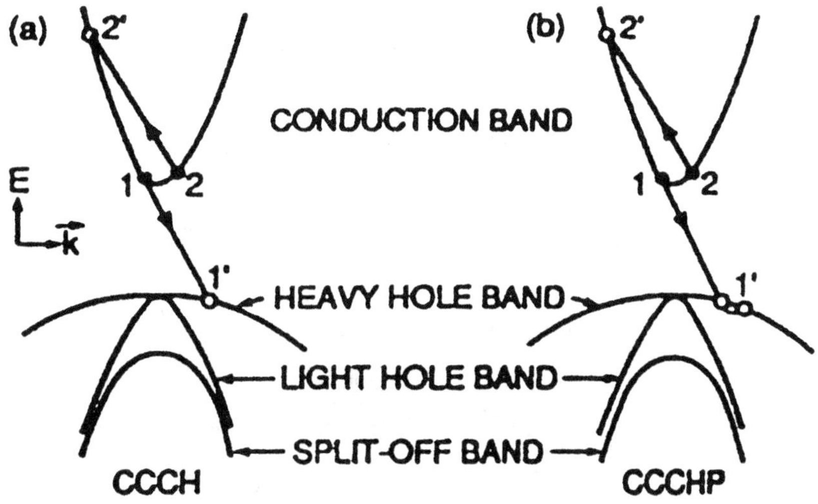

FIGURE 1 Band-to-band Auger recombination with the excitation of an electron to a higher energy state; (a) without (CCCH), and (b) with (CCCHP) phonon assistance.

The calculation of Auger recombination coefficient C is carried out using Fermi's golden rule [1] and a detailed discussion is given in a recent book [5]. The Auger recombination rate R_A and the Auger recombination time τ_A are defined as

$$R_A = Cn^3 \tag{1}$$

$$\tau_A = 1/(Cn^2) \tag{2}$$

where n is the injected electron-hole pair density. The n^3 and the n^2 factors have to be suitably modified if the number of electrons and holes are different. The Auger recombination coefficient C for each of the Auger processes is different and depends on many factors, such as the bandgap energy, the temperature, the effective masses and other band structure parameters, the type of doping, doping density and degeneracy. C generally increases with a reduction in the bandgap energy and an increase in the temperature. Thus, Auger recombination is more important in the smaller bandgap semiconductor InGaAs compared to GaAs and InGaAsP (1.3 μm bandgap). At very high densities, screening of the Coulomb interaction reduces the Auger coefficient. The phonon-assisted Auger processes are not important for $In_{0.53}Ga_{0.47}As$ at room temperature but become important at low temperatures. Finally, a differential Auger recombination lifetime, for recovery of a small density perturbation near a steady-state density n_0, is frequently used in the laser literature and is 3 times shorter than given by Eqn (2) at $n = n_0$.

C CALCULATED VALUES OF AUGER COEFFICIENTS

TABLE 1 shows a compilation of values for bulk $In_{0.53}Ga_{0.47}As$ and for InGaAs quantum wells.

TABLE 1 Calculated Auger coefficients for various band-to-band Auger processes in $In_{0.53}Ga_{0.47}As$.

Auger Process Type	Auger Coefficient C ($cm^6 s^{-1}$)		Comments	
CCCH	1.3×10^{-27} 5×10^{-31}	(300 K) (200 K)	bulk; $n_o = 10^{18}$ cm^{-3}	[6]
CHHS	9.1×10^{-28} 1.5×10^{-30}	(300 K) (200 K)	bulk; $p_o = 10^{18}$ cm^{-3}	[6]
CHHL	1.3×10^{-28} $\sim 1.3 \times 10^{-32}$	(300 K) (200 K)	bulk; $p_o = 10^{18}$ cm^{-3}	[6]
CCCHP	1.4×10^{-29}	(300 K)	bulk; $n_o = 10^{18}$ cm^{-3}	[6]
CHHSP	1.7×10^{-29}	(300 K)	bulk; $p_o = 10^{18}$ cm^{-3}	[6]
CHHLP	5.6×10^{-30}	(300 K)	bulk; $p_o = 10^{18}$ cm^{-3}	[6]

D EXPERIMENTAL VALUES OF AUGER COEFFICIENTS

Experimental values of Auger coefficients have been extracted from measurements of the decay time constant as a function of the carrier density introduced by doping or photoexcitation. For doped samples, Auger coefficients are determined from the measurements of decay times and quantum efficiency as a function of doping density [14]. Time-resolved luminescence as well as pump-and-probe transmission techniques have been used for measuring Auger coefficients at high photoexcitation densities [15-18]. TABLE 2 presents a compilation of values from the literature for bulk and quantum well $In_{0.53}Ga_{0.47}As$.

TABLE 2 Experimentally determined Auger coefficients in $In_{0.53}Ga_{0.47}As$ at 300 K.

Experimental Parameters	Auger Coefficient C ($cm^6 s^{-1}$)	Comments
bulk, $n = p \simeq 3 \times 10^{19}$ cm^{-3}	9.8×10^{-29}	[17]
bulk, $p_o = 2 \times 10^{18}$ cm^{-3}	3.6×10^{-29}	[15]
bulk, $n_o = 2 \times 10^{18}$ cm^{-3}	5×10^{-30}	[15]
bulk, $n = p \simeq 2 \times 10^{18}$ cm^{-3}	$(2.5 \pm .5) \times 10^{-28}$	[14]; the value reduces to 1.8×10^{-28} if the carrier temperature is assumed to be 400 K
bulk, $n = p \simeq 1.9 \times 10^{18}$ cm^{-3}	3.2×10^{-28}	[16]; density from luminescence spectra; strong temperature dependence explained by CHHS + phonon-assisted processes
11 nm quantum well, $n = p \simeq 1.9 \times 10^{18}$ cm^{-3}	9×10^{-29}	[16]; no significant temperature dependence. C increases with well-width
bulk, $n = p \leq 2.5 \times 10^{18}$ cm^{-3}	7×10^{-29}	[18]; smaller C for larger density
10 nm quantum well, $10^{18} < n = p < 4 \times 10^{18}$ cm^{-3}	6×10^{-29}	[18]; smaller C for larger density
~9 nm quantum well, $10^{12} < n = p < 4 \times 10^{12}$ cm^{-2}	9×10^{-29}	[19]; C decreases for compressive strain and constant for tensile strain
10 nm quantum well	1×10^{-28}	[20]; C same for strained quantum wells
5 nm quantum well	5×10^{-30}	[21]; C increases with In concentration

E CONCLUSION

TABLE 1 shows that CCCH and CHHS processes dominate at room temperature in bulk InGaAs and combine to give $C \simeq 2 \times 10^{-27} cm^6/s$, (Auger lifetime: 125 ps at 2×10^{18} cm^{-3}). There is a wide variation in the experimental values reported in the literature and even the largest value (TABLE 2) is about a factor of 5 smaller than the theoretical value, leading to an Auger lifetime of ~ 780 ps. For quantum wells, the Auger recombination is a factor of three smaller and is independent of temperature, a result of different effective masses making CCCHP the dominant process in quantum wells [11].

REFERENCES

[1] A.R. Beattie, P.T. Landsberg [*Proc. R. Soc. Lond. A (UK)* vol.259 (1959) p.16]

[2] P.T. Landsberg [*Phys. Status Solidi (Germany)* vol.41 (1970) p.457]

[3] P.T. Landsberg [*Solid-State Electron. (UK)* vol.28 (1985) p.137]

[4] P.T. Landsberg [*Solid-State Electron. (UK)* vol.30 (1987) p.1107]

[5] G.P. Agrawal, N.K. Dutta [*Long-Wavelength Semiconductor Lasers* (Van Nostrand Reinhold Co., New York, 1986), ch.3 and 9]

[6] N.K. Dutta, R.J. Nelson [*J. Appl. Phys. (USA)* vol.53 (1982) p.74]

[7] A.Haug [*J. Phys. Chem. Solids (UK)* vol.49 (1988) p.599]

[8] R.I. Taylor, R.A. Abram, M.G. Burt, C. Smith [*Semicond. Sci. Technol. (UK)* vol.5 (1990) p.90-104]

[9] A. Sugimura [*IEEE J. Quantum Electron. (USA)* vol.17 (1981) p.627-35; ibid vol.19 (1983) p.930-2]

[10] D. Nichols et al [*IEEE J. Quantum Electron. (USA)* vol.28 (1992) p.1239]

[11] A. Haug [*Electron. Lett. (UK)* vol.26 (1990) p.1415-6]

[12] A. Haug [*Semicond. Sci. Technol (UK)* vol.7 (1992) p.1337]

[13] A. Haug [*Appl. Phys. A (Germany)* vol.51 (1990) p.354]

[14] M.E. Prise, M.R. Taghizadeh, B.H. Wherrett, S.D. Smith [*J. Lumin. (Netherlands)* vol.8 (1985) p.320]

[15] C.H. Henry, R.A. Logan, F.R. Merritt, C.G. Bethea [*Electron. Lett. (UK)* vol.20 (1984) p.358]

[16] S. Hauser, G. Fuchs, A. Hangleiter, K. Streubel, W.T. Tsang [*Appl. Phys. Lett. (USA)* vol.56 (1990) p.913-5]

[17] E. Wintner, E.P. Ippen [*Appl. Phys. Lett. (USA)* vol.44 (1984) p.999]

[18] B. Sermage, D.S. Chemla, D. Sivco, A.Y. Cho [*IEEE J. Quantum Electron. (USA)* vol.22 (1986) p.774]

[19] M.C. Wang, K. Kash, C.E. Zah, R. Bhat, S.L. Chuang [*Appl. Phys. Lett. (USA)* vol.62 (1993) p.166]

[20] G. Fuchs, C. Schiedel, A. Hangleiter, V. Harle, F. Scholz [*Appl. Phys. Lett. (USA)* vol.62 (1993) p.396]

[21] L. Davis, Y. Lam, D. Nichols, J. Singh, P.K. Bhattacharya [*IEEE Photonics Technol. Lett. (USA)* vol.5 (1993) p.120]

CHAPTER 7

OPTICAL AND ELECTRO-OPTIC PROPERTIES OF InGaAs

7.1 Dielectric functions and refractive index of $In_xGa_{1-x}As$
7.2 Raman spectra of $In_xGa_{1-x}As$
7.3 Optical properties of pseudomorphic InGaAs/GaAs heterostructures and quantum wells
7.4 Optical properties of InP-based lattice-matched InGaAs quantum wells
7.5 Linear and quadratic electro-optic coefficients in InGaAsP

7.1 Dielectric functions and refractive index of $In_xGa_{1-x}As$

S.A. Alterovitz

A INTRODUCTION

We review the ellipsometrically measured refractive indices, extinction coefficients and dielectric functions of $In_xGa_{1-x}As$, in the range 1.5 - 6 eV, for several values of the composition x. Two types of material are considered: bulk and thin films. The results for the dielectric functions are further described by the dependence of the critical points on the composition. An approximate algorithm to calculate the dielectric functions for an arbitrary value of the composition is recommended. Strain effects in thin $In_xGa_{1-x}As$ films grown on GaAs are also discussed.

B MATERIALS

Most of the applications of $In_xGa_{1-x}As$ are in the electronics and optoelectronics areas. Thus, samples are usually thin films, grown on a binary III-V semiconductor, mostly GaAs, InP and InAs. In this Datareview, we will denote material with lattice matched compositions to the above mentioned binaries as the 'bulk' material. At these compositions (x values of 0, 0.53 and 1.0) very good quality, thick films or crystals are available. Extreme care in the surface preparation yielded very high quality ellipsometric results for these compositions: x = 0 in [1], x = 0.53 in [2] and x = 1.0 in [3]. We will use the results of these three materials to find most of the quantitative data given below.

Thin films of $In_xGa_{1-x}As$ with x different from lattice matched values were also measured by ellipsometry [4-6]. Two types of film were measured: thin pseudomorphic and 'thick'. The 'thick' films were 1 - 2 μm thick, while the thin films were assumed to be pseudomorphic, i.e. thinner than the critical thickness [7]. The pseudomorphic films were assumed to be fully strained, and the 1 - 2 μm thick films on GaAs were treated as fully relaxed [4-6]. However, in a recent publication [8] it is shown that a considerable amount of strain is maintained even in thick $In_xGa_{1-x}As$ films on GaAs. For example, the 1 μm $In_xGa_{1-x}As$ film on GaAs used in [6] is 28 times thicker than the critical thickness [7]. However, according to [8], at this thickness the film is only 55% relaxed. Even the 2 μm $In_xGa_{1-x}As$ film used in [4], which is roughly 200 times thicker than the critical thickness, still may have 30% residual strain. It must be noted that the results of [8] are, probably, not universal, as they would depend on the deposition method and substrate temperature. However, the problem of the remnant strain in the films is further complicated by the possibility of propagation of misfit dislocations to the surface. Experimental ellipsometric work on 0.3 - 1.0 μm thick $In_xGa_{1-x}As$ on InP showed strong effects of misfit dislocations on the results [9]. We believe that our former results of the dielectric function of $In_xGa_{1-x}As$ on InP [4] have also been affected by these effects.

C REFRACTIVE INDEX

In this section we discuss the complex refractive index that consists of a real part, i.e. the regular refractive index n, and an imaginary part, i.e. the extinction coefficient k. The absorption coefficient α is given by $\alpha = 4\pi k/\lambda$, where λ is the wavelength. Results from [1-4] for n and k vs. the wavelength λ are shown in FIGURES 1 and 2 for x = 0, 0.3, 0.53 and 1.0. There is a clear systematic change in both n and k versus the indium concentration x. For example, the value of the n peak goes down while its location moves to higher wavelengths, both with increasing x. Films with higher x values have higher extinction coefficients above 4500 Å, approaching the bandgap.

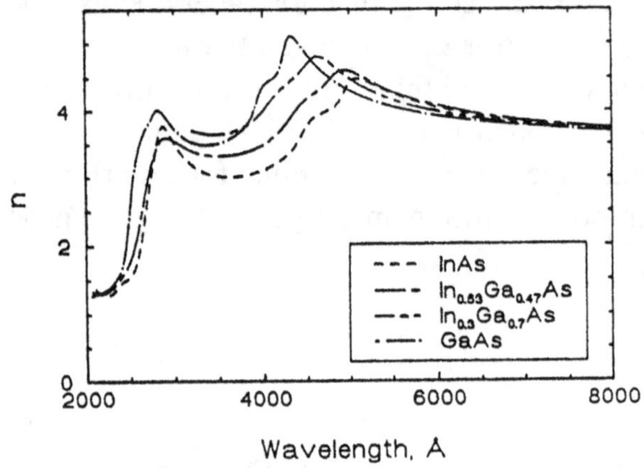

FIGURE 1 Refractive index n vs. wavelength, for several $In_xGa_{1-x}As$ compounds.

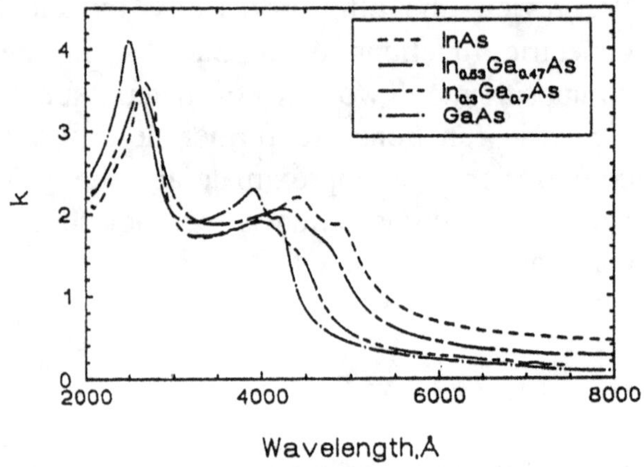

FIGURE 2 Extinction coefficient k vs. wavelength, for several $In_xGa_{1-x}As$.

D DIELECTRIC FUNCTIONS

The dielectric function consists of a real part ε_1 ($\varepsilon_1 = n^2 - k^2$) and an imaginary part ε_2 ($\varepsilon_2 = 2nk$). These functions can be easily calculated from the results given in FIGURES 1 and 2. For completeness, ε_1 and ε_2 functions vs. the energy are displayed in FIGURES 3 and 4 respectively, for x = 0, 0.3, 0.53 and 1.0. Again, we see clear trends in these functions vs. x.

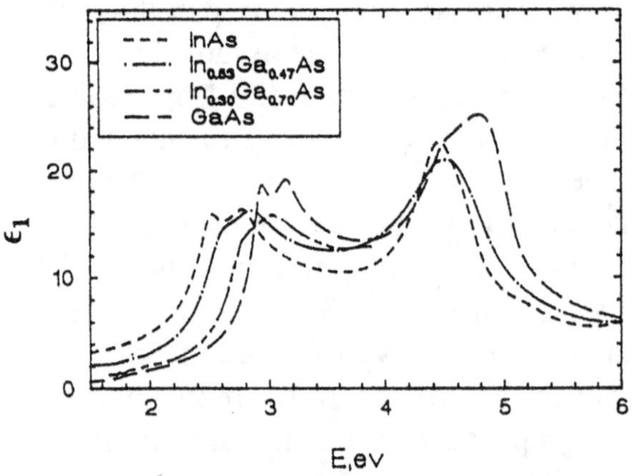

FIGURE 3 Real part of the dielectric function, ε_1, vs. energy, for several $In_xGa_{1-x}As$ compounds.

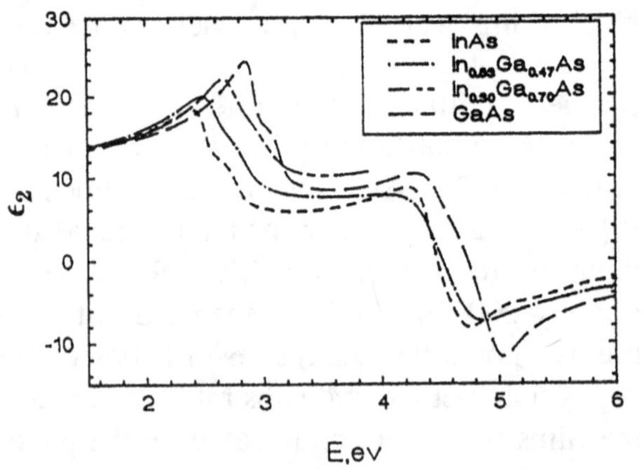

FIGURE 4 Imaginary part of the dielectric function, ε_2, vs. energy, for several $In_xGa_{1-x}As$ compounds.

However, one of the striking features of all four figures is the fact that the x = 0.3 function does not have the fine structure of the other functions, with the peaks and shoulders being less sharp than for the 'bulk' samples. The same observation was also found for the $In_xGa_{1-x}As$ dielectric functions given in [6].

The complex dielectric function ε vs. the energy E (E = ħω) has structure which is attributed to interband critical points [10]. These critical points are analysed using the standard analytical line shapes

$$\varepsilon(\omega) = C - Ae^{i\phi}(\omega - E_c + i\Gamma)^n \tag{1}$$

where E_c is the critical point energy threshold, Γ is the broadening, A its amplitude and φ is an excitonic phase angle. The value of n gives the nature of the transition, with most common values n = -1 for a Lorentzian shape due to discrete excitons and n = 0 (i.e. logarithmic) for two dimensional transitions. C is a constant. We performed a Marquardt least square fit of the second derivative of the dielectric function to obtain E_c, Γ, A and φ for each critical point. There are four easily observable critical points in the experimental range, namely E_1, $E_1+\Delta_1$, E_o' and E_2 (see [10] for their description). In FIGURE 4, we can see clearly these four critical points of the GaAs ε_2 function. They are: E_1 - the peak near 2.9 eV, $(E_1+\Delta_1)$ - the shoulder around 3.1 eV, E_o' - the peak at 4.4 eV and E_2 - the minimum around 5 eV. We have fitted all four critical points for all ellipsometrically measured samples reported in the literature [1-4,6]. As the quality of the dielectric functions of 'bulk' material is so much superior, we will only present these results. The fits were done using n = -1 for E_1 and $E_1+\Delta_1$ and n = 0 for E_o' and E_2, in order to be consistent with the GaAs results [10]. Results are given in TABLE 1. The obtained critical point values are very similar to those measured by electroreflectance [11]. Quadratic polynomials describing the results, in eV, are as follows:

$$E_1(x) = 2.91 - 0.88x + 0.47x^2 \tag{2}$$

$$(E_1+\Delta_1)(x) = 3.15 - 1.00x + 0.63x^2 \tag{3}$$

$$E_o'(x) = 4.45 - 0.29x + 0.22x^2 \tag{4}$$

$$E_2(x) = 4.97 - 0.75x + 0.52x^2 \tag{5}$$

TABLE 1 Critical points of $In_xGa_{1-x}As$ determined from ellipsometry.

x	E_1, eV	$E_1+\Delta_1$, eV	E_o', eV	E_2, eV
0.0	2.914	3.155	4.448	4.968
0.53	2.578	2.803	4.356	4.718
1.0	2.503	2.787	4.381	4.741

Calculated values of the critical points using Eqns (2)-(5) were compared to experimental values for presumably relaxed thin film results for x = 0.1, 0.2, 0.24 and 0.30. Except for one point (E_o' of x = 0.1), the calculated and experimental values agree to better than 0.05 eV, and, in most cases, to within 0.02 eV. Discrepancies in the 0.02 - 0.05 eV range are, probably, due to residual strain effects (see below) or imperfect surface quality.

The effect of strain was studied experimentally in [5,6]. Quantitative results of the stress effect on E_1, $E_1+\Delta_1$, are given in [6] using the theory for biaxial strain developed previously [12]. The experimental and theoretical results show that the E_1 point moves up in energy less than the $E_1+\Delta_1$ point, i.e., the value of Δ_1 increases with strain. Using a linear approximation, for $x \leq 0.24$, the result of [6] can be summarised as follows: (E_1 in eV)

$$\text{relaxed} \quad E_1 = 2.91 - 0.79x \tag{6}$$

$$\text{strained} \quad E_1 = 2.91 - 0.58x \tag{7}$$

This result means that a strained film with an actual composition x will have the E_1 point at the same energy that corresponds to a composition value x_{eff} of the relaxed material, with $x_{eff} = 0.73x$. The approximate result $x_{eff} = 0.7x$ was also obtained in [5], when the value of the critical point E_1 was used for the interpolations.

E INTERPOLATIONS

Two main ways to interpolate the dielectric function of a ternary III-V semiconductor were proposed [13,14]. The first one [13] requires estimates of 21 parameters for each calibration dielectric function. Due to its complexity, it will not be considered here. We will dwell on the second method [14], which is a numerical approach, based on known calibration dielectric functions at several concentrations x_i and the composition dependence of the critical points. An interpolated dielectric function for a composition x, with the nearest composition calibration functions ε_a and ε_b at x_a and x_b respectively, is calculated by shifting the ε_a function in energy so that $x_a = x$ and shifting ε_b in energy the other way until $x_b = x$. Then, interpolate the amplitude of the ε_a and ε_b functions with the weights $(x - x_a)/(x_b - x_a)$ and $(x_b - x)/(x_b - x_a)$. For energies between the critical points, a linear approach is used [14]. The interpolation algorithm requires a critical point with energy below the experimental range. The bandgap critical point E_o was chosen. E_o was measured by photoreflectance [15] on 1 μm thick $In_xGa_{1-x}As$ films on GaAs, presumably relaxed. The result, in eV at room temperature, is:

$$E_o(x) = 1.42 - 1.53x + 0.45x^2 \tag{8}$$

An interpolation algorithm using the critical points E_o, E_1 and E_2 was implemented [5] using relaxed material dielectric functions. The algorithm was successful in obtaining correct compositions for relaxed films, but needed a correction (see Section D above) for strained material. For improved application of the algorithm, additional, accurate calibration dielectric functions are required, both for relaxed and strained material.

F CONCLUSION

A review of the ellipsometrically determined dielectric functions of $In_xGa_{1-x}As$ was given. The review includes a graphic display of the functions, and a table of the values of four of the critical points for several compositions x, for strain-relaxed material. Two approximations were used for the dependence of the critical points on the composition parameter x: (1) A quadratic equation for strain-relaxed material. (2) A linear equation for material coherently strained on GaAs, for $x \leq 0.24$. An empirical way to interpolate the dielectric functions for an arbitrary composition x within the experimental range was also described.

ACKNOWLEDGEMENT

The author would like to thank D.E. Aspnes and C. Pickering for providing diskettes of their experimental dielectric functions, and R.M. Sieg for technical assistance.

REFERENCES

[1] D.E. Aspnes, S.M. Kelso, R.A. Logan, R. Bhat [*J. Appl. Phys. (USA)* vol.60 (1986) p.756-67]

[2] D.E. Aspnes, H.J. Stocker [*J. Vac. Sci. Technol. (USA)* vol.21 (1982) p.413-16]

[3] E.D. Palik, R.T. Holm [in *Handbook of Optical Constants of Solids*, Ed E.D. Palik (Academic Press, Orlando, 1985) p.479-89]

[4] S.A. Alterovitz et al [*Proc. 2nd Int. Conf. on Electronic Materials*, Eds T. Sugano, R.P.H. Chang, H. Kamimura, I. Hayashi, T. Kamiya (MRS Press, Pittsburgh, 1990) p.187-92]

[5] S.A. Alterovitz, P.A. Sekula-Moise, R.M. Sieg, M.N. Drotos, N.A. Bogner [*Thin Solid Films (Switzerland)* vol.220 (1992) p.241-6]

[6] C. Pickering, R.T. Carline, M.T. Emeny, N.S. Garawal, L.K. Howard [*Appl. Phys. Lett. (USA)* vol.60 (1992) p.2412-14]

[7] T.G. Andersson, Z.G. Chen, V.D. Kulakovskii, A. Uddin, J.T. Vallin [*Appl. Phys. Lett. (USA)* vol.51 (1987) p.752-4]

[8] V. Krishnamoorthy, Y.W. Lin, L. Calhoun, H.L. Liu, R.M. Park [*Appl. Phys. Lett. (USA)* vol.61 (1992) p.2680-82]

[9] B.R. Bennett, J.A. del Alamo [*Appl. Phys. Lett. (USA)* vol.58 (1991) p.2978-80]

[10] P. Lautenschlager, M. Garriga, S. Logothetidis, M. Cardona [*Phys. Rev. B (USA)* vol.35 (1987) p.9174-89]

[11] A.G. Thompson, J.C. Woolley [*Can. J. Phys. (Canada)* vol.45 (1967) p.2597-606]

[12] F.H. Pollak [*Semicond. Semimet. (USA)* Ed T.P. Pearsall, vol.32 (Academic Press, Boston, 1990) p.17-53]

[13] M. Erman, J.B. Theeten, P. Chambon, S.M. Kelso, D.E. Aspnes [*J. Appl. Phys. (USA)* vol.56 (1984) p.2664-71]

[14] P.G. Snyder, J.A. Woollam, S.A. Alterovitz, B. Johs [*J. Appl. Phys. (USA)* vol.68 (1990) p.5925-6]

[15] Z. Hang, D. Yang, F.H. Pollak, G.D. Pettit, J.M. Woodall [*Phys. Rev. B (USA)* vol.44 (1991) p.10546-50]

7.2 Raman spectra of $In_xGa_{1-x}As$

R. Merlin

January 1993

A INTRODUCTION

As for other III-V compounds and alloys, Raman spectra of $In_xGa_{1-x}As$ [1-36] are dominated by transverse (TO) and longitudinal (LO) optical phonons of wavevector $k \approx 0$. The $In_xGa_{1-x}As$ system exhibits two-mode behaviour, i.e. there are two sets of modes (a total of four) that can be ascribed to InAs-like and GaAs-like vibrations [1,3,4,37]. The relative intensity of these features depends primarily on the scattering geometry [1,3,4,36]. In backscattering from (001) faces, TO scattering is forbidden; Raman traces show the InAs-like LO phonon and the much stronger GaAs-like LO mode. Instead, the InAs-like TO mode is the strongest peak in (110) backscattering where LO phonons are not allowed [36]. Because their frequency varies strongly with molar fraction and stress, the optical modes are very useful markers for characterising samples. Additional though harder to quantify information on the layer quality is provided by measurements of the phonon widths [6-7].

Other than $k \approx 0$ phonons, Raman data show disorder-activated (DA) first-order and weaker second-order scattering [1,3,36]. The most prominent DA-features are due to longitudinal-acoustic (referred to as DALA) and TO phonons (DATO) with $k \neq 0$ [1,3].

Most of the Raman work on $In_xGa_{1-x}As$ focuses on conventional phonon scattering in undoped or lightly-doped samples; the pressure behaviour of the GaAs-like optical modes is reported in [35]. Time-resolved Raman studies dealing with nonequilibrium phonon effects and cooling of hot carriers are reported, respectively, in [15] and [26]. The latter work discusses electronic single-particle scattering in photoexcited plasmas. Raman spectra due to coupled LO phonon-plasmon excitations for p-type $In_xGa_{1-x}As$ are considered in [24].

B THICK LAYERS AND LATTICE-MATCHED $In_xGa_{1-x}As$

In this section, we consider samples of thickness $L >> l_c$ where l_c is the critical thickness. Raman studies have been reported for films grown on InP [2-4,19], GaAs [1,3,29] and Ge [25] substrates using the techniques of molecular-beam (MBE) [6,19,29,36], liquid-phase (LPE) [1-6] and vapour-phase (VPE) epitaxy [6,9]. The discussion also applies to the lattice-matched system $In_{0.53}Ga_{0.47}As/InP$ [5,6,9,36].

FIGURE 1 shows room-temperature spectra of $\geq$ 12 μm-thick $In_xGa_{1-x}As$ layers at various compositions. Here, the geometry is (001)-backscattering allowing only LO modes. A close inspection of the spectra reveals DATO bands [1-3] in the range of the $k \approx 0$ phonons; the presence of these bands is possibly the reason why early work [4] showed an apparent violation of Raman selection rules. In quantitative studies, it is particularly important to distinguish the DA-features from the allowed modes [3,36]. This can be achieved by analysing the spectral lineshape [36] or by separating the different symmetry components [3].

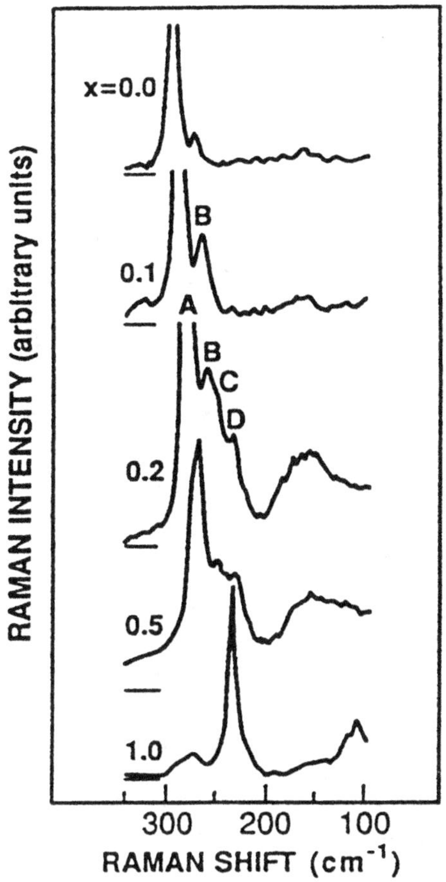

FIGURE 1 Raman spectra of $In_xGa_{1-x}As$ grown on GaAs [3]. The GaAs-like LO and TO phonons correspond to peaks A and B. C and D are the InAs-like TO and LO modes. Other features are due to disorder-activated or second-order scattering. The so-called DALA-band [1,3] is at $\approx 150\ cm^{-1}$ for x = 0.2. Data were obtained using the 514.5-nm line of an Ar^+ laser.

Values of the TO and LO frequencies for the compounds at T = 300 K, as well as best second-order fits to the frequency of the four $k \approx 0$ modes of the alloy [19] are given in TABLE 1. The latter are plotted in FIGURE 2 together with representative data from Raman [4,19] and infrared [37] experiments.

TABLE 1 Frequencies of optical phonons (in cm^{-1}) at T = 300 K (GaAs [38] and InAs [39]). Alloy expressions [19] are recommended for $0.2 \leq x \leq 0.8$.

Phonon	GaAs	InAs	$In_xGa_{1-x}As$
GaAs-like ω_{LO}	291.9(2)		$-32.4x^2 - 18.6x + 290.0$
GaAs-like ω_{TO}	268.6(2)		$-29.0x^2 - 5.3x + 265.1$
InAs-like ω_{LO}		238.6(2)	$8.92x^2 - 7.7x + 234.9$
InAs-like ω_{TO}		217.3(2)	$-16.5x + 233.7$

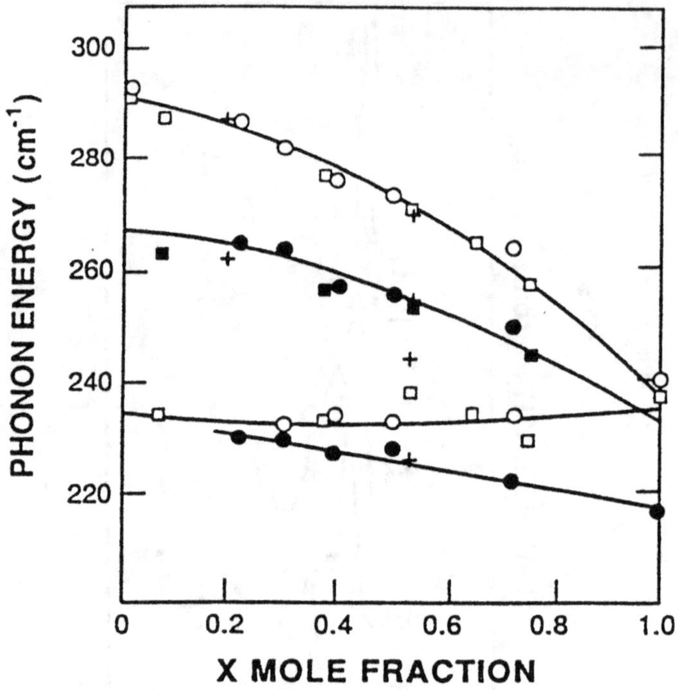

FIGURE 2 Dependence of optical phonon-frequencies on composition [19]. From top to bottom, the traces correspond to the GaAs-like LO, TO and InAs-like LO and TO phonons. Solid lines were calculated using the expressions given in TABLE 1. Box and filled box are from [19], while + are results from [4]; circle and filled circle are from infrared data [37].

C STRAINED $In_xGa_{1-x}As$

Thin layers ($L << l_c$) of $In_xGa_{1-x}As$ deposited on, say, GaAs adopt the lattice constant of the substrate in the plane orthogonal to the growth axis. This so-called pseudomorphic growth is accompanied by the development of strain manifesting itself in shifts and/or splitting of the optical phonons of the alloy. A general discussion of strain-effects on phonons for III-V compounds is given in [43,44]. For our purposes, it is sufficient to consider strain effects on LO modes for (001)-grown $In_xGa_{1-x}As$ and on TO modes for (011) substrates; as mentioned earlier, the TO modes in (001)- and the LO modes in (011)-backscattering are Raman forbidden. Other than superlattices, Raman data have been reported for strained layers grown by MBE [14] and metalorganic chemical vapour deposition (MOCVD) [28,31] on (001) GaAs.

For (011) faces, TO modes are allowed in backscattering. The strain lifts the degeneracy of this mode and leads to frequency shifts given by [43,44]

$$\Delta\omega_{TO} = \left[\frac{q_{TO}S_{12}}{S_{11}+S_{12}} + 2\,(p_{TO}+q_{TO}) \pm \frac{r_{TO}S_{44}}{3(S_{11}+S_{12})}\right](\varepsilon/\omega_{TO}) \qquad (1)$$

Here, $\varepsilon = (a_S/a_{InGaAs}-1)$ and S_{11}, S_{12} and S_{44} are elastic compliance constants. The lattice parameters of the substrate and of $In_xGa_{1-x}As$ are, respectively, a_S and a_{InGaAs}; p, q and r are phenomenological parameters measuring changes in spring constants due to an applied stress

[43,44]. In the case of (001) growth, LO phonons are the relevant Raman modes; their frequency shifts according to

$$\Delta\omega_{LO} = \left[\frac{p_{LO}S_{12}}{S_{11}+S_{12}} + q_{LO}\right](\varepsilon/\omega_{LO}) \tag{2}$$

Values of the parameters appearing in the above equations are reproduced in TABLE 2. Notice that constants p, q and r differ slightly for TO and LO phonons in GaAs [43] reflecting the fact that the TO-LO splitting depends very weakly on the stress [43,44]. The difference has not as yet been determined for InAs [44]; quoted values are the same for TO and LO modes.

TABLE 2 Room temperature parameters used to determine strain-induced frequency shifts in pseudomorphic $In_xGa_{1-x}As$. S_{11}, S_{12} and S_{44} are elastic compliance constants [40] while a is the lattice parameter (GaAs [41] and InAs [42]). Experimental values of p, q and r are from [43,44]. Note that the expressions for $In_xGa_{1-x}As$ are linear interpolations; for p, q and r, they do not distinguish between GaAs- and InAs-like modes.

Parameter	GaAs	InAs	$In_xGa_{1-x}As$
p_{LO}/ω^2_{LO}	-1.75 ± 0.08	-0.94 ± 0.42	0.81x - 1.75
q_{LO}/ω^2_{LO}	-2.45 ± 0.07	-2.08 ± 0.34	0. 37x - 2.45
p_{TO}/ω^2_{TO}	-2.26 ± 0.08	-0.94 ± 0.42	1.32x - 2.26
q_{TO}/ω^2_{TO}	-2.56 ± 0.07	-2.08 ± 0.34	0.48x - 2.56
r_{TO}/ω^2_{TO}	-0.88 ± 0.03	-0.76 ± 0.15	0.12x - 0.88
S_{11} (GPa^{-1})	1.175×10^{-2}	1.945×10^{-2}	$(0.77x + 1.175) \times 10^{-2}$
S_{12} (GPa^{-1})	-0.365×10^{-2}	-0.685×10^{-2}	$-(0.32x + 0.365) \times 10^{-2}$
S_{44} (GPa^{-1})	1.69×10^{-2}	2.53×10^{-2}	$(0.84x + 1.69) \times 10^{-2}$
a (Å)	5.65325(2)	6.0583(1)	0.405x + 5.6533

D SUPERLATTICES BASED ON $In_xGa_{1-x}As$

The available literature focuses on strained $GaAs/In_xGa_{1-x}As$ structures grown by MBE [8,16,35] and VPE [20], as well as lattice-matched $InP/In_{0.53}Ga_{0.47}As$ superlattices grown by MBE [11], MOCVD [12], metalorganic vapour phase (MOVPE) [18] and migration-enhanced epitaxy [34]. Raman studies of MBE-grown $In_xGa_{1-x}As/Al_{1-y}In_yAs$ are reported in [17]. Superlattice effects were observed in [8], [12], and [16-18]. Evidence of scattering by interface modes [45] is given in [12]. This work, using resonant excitation, reveals unexpected red-shifts of LO phonons and a strong power-dependence of the spectra which are poorly understood. Possibly related results on the $GaAs/Al_xGa_{1-x}As$ system are described in [46]. Scattering by the so-called folded acoustic phonons [45] is reported in [16-18] for periodic and in [16] for non-periodic structures. Such studies probe the modulation of the photoelastic coefficient and give a measure of the superlattice period [45].

E RESONANT SCATTERING

Resonant Raman scattering refers to situations where the laser energy is in the vicinity of an optical critical point of the material, often the bandgap [47]. Typically, resonances lead to scattering cross sections which are larger by several orders of magnitude, and also to an enhanced sensitivity to probe defects (associated with forbidden scattering). The resonant studies reported for ≈ 2 μm-thick $In_xGa_{1-x}As$ layers (x = 0.19, 0.24, 0.53) grown on InP [9] and lattice-matched $InP/In_{0.53}Ga_{0.47}As$ quantum-well structures [12] involve the E_1-gap. Interestingly, the E_1-resonance has not been investigated much in the case of $GaAs/Al_xGa_{1-x}As$ structures (for GaAs, see [47]). There remains to be seen whether the differences in the resonant behaviour of the two systems stem from gap or alloying effects [9].

F CONCLUSION

Raman scattering by optical modes in undoped and by coupled LO-plasmons in doped $In_xGa_{1-x}As$ has been shown to be a very useful tool to probe the alloy composition and the doping level of the layers and, in strained material, to characterise the strain. The main features of the spectra are fairly well understood except for very thin layers where efforts to explore interface and quantum confinement effects [45] have been very limited.

REFERENCES

[1] R. Carles, N. Saint-Cricq, A. Zwick, M.A. Renucci, J.B. Renucci [*J. Phys. C (France)* vol.42 C-6 (1981) p.105-7]; N. Saint-Cricq, R. Carles, J.B. Renucci, A. Zwick, M.A. Renucci [*Solid State Commun. (UK)* vol.39 (1981) p.1137-41]

[2] G.P. Schwartz, J.E. Griffiths, G.J. Gualtieri [*Thin Solid Films (Switzerland)* vol.94 (1982) p.213-22]

[3] K. Kamimoto, T. Katoda [*Appl. Phys. Lett. (USA)* vol.40 (1982) p.826-8]

[4] T.P. Pearsall, R. Carles, J.C. Portal [*Appl. Phys. Lett. (USA)* vol.42 (1983) p.436-8]

[5] K. Kakimoto, T. Katoda [*Appl. Phys. Lett. (USA)* vol.42 (1983) p.811-3]

[6] P. Parayantal, F.H. Pollak [*Phys. Rev. Lett. (USA)* vol.52 (1984) p.1822-5]

[7] M.W. Muller [*Phys. Rev. B (USA)* vol.30 (1984) p.6196-8]

[8] M. Nakayama, K. Kubota, H. Kato, N. Sano [*Solid State Commun. (UK)* vol.51 (1984) p.343-5]

[9] K.P. Jain, R.K. Soni, S.C. Abbi, M. Balkanski [*Phys. Rev. B (USA)* vol.32 (1985) p.1005-8]

[10] M. Kubo, Y. Sasai, Y. Yoshioka, M. Ogura [*J. Cryst. Growth (Netherlands)* vol.73 (1985) p.439-44]

[11] M. Watt, C.M. Sotomayor-Torres, P.D. Hatton, H. Vass, P.A. Claxton, J.S. Roberts [*Superlattices Microstruct. (UK)* vol.3 (1987) p.75-8]

[12] J.A.C. Bland, W. Hayes, M.S. Skolnick, D.J. Mowbray, S.J. Bass [*Superlattices Microstruct. (UK)* vol.3 (1987) p.83-7]; D.J. Mowbray, W. Hayes, L.L. Taylor, S.J. Bass [*Semicond. Sci. Technol. (UK)* vol.5 (1990) p.83-9]

[13] D. Kirillov, M. Vichr, R.A. Powell [*Appl. Phys. Lett. (USA)* vol.50 (1987) p.262-4]

[14] G. Burns, C.R. Wie, F.H. Dacol, G.D. Pettit, J.M. Woodall [*Appl. Phys. Lett. (USA)* vol.51 (1987) p.1919-21]

[15] J.A. Kash, S.S. Jha, J.C. Tsang [*Phys. Rev. Lett. (USA)* vol.58 (1987) p.1869-72]
[16] M.W.C. Dharma-wardana, A.H. McDonald, G.C. Aers, D.J. Lockwood, W.T. Moore, R.L.S. Devine [*Superlattices Microstruct. (UK)* vol.3 (1987) p.665-71]; D.J. Lockwood, M.W.C. Dharma-wardana, W.T. Moore, R.L.S. Devine [*Appl. Phys. Lett. (USA)* vol.51 (1987) p.361-3]
[17] S.T. Davey, E.G. Scott, B. Wakefield, G.J. Davies [*Semicond. Sci. Technol. (UK)* vol.2 (1987) p.683-6]
[18] S.T. Davey, P.C. Spurdens, B. Wakefield, A.W. Nelson [*Appl. Phys. Lett. (USA)* vol.51 (1987) p.758-60]
[19] S. Emura, S. Gonda, Y. Matsui, H. Hayashi [*Phys. Rev. B. (USA)* vol.38 (1988) p.3280-6]
[20] F. Iikawa et al [*Phys. Rev. B (USA)* vol.38 (1988) p.8473-6]
[21] R.K. Soni [*J. Cryst. Growth (Netherlands)* vol.94 (1989) p.347-50]
[22] Y.K. Su, M.C. Wu, B.S. Chiu [*J. Cryst. Growth (Netherlands)* vol.96 (1989) p.47-51]
[23] S.J. Yu, H. Asahi, H. Sumida, S. Emura, S. Gonda, H. Tanoue [*Inst. Phys. Conf. Ser. (UK)* vol.106, Gallium Arsenide and Related Compounds, 1989 (Institute of Physics, 1990) p.527-32]
[24] T. Akatsuka, R. Miyake, S. Nozaki, T. Yamada, M. Konagai, K. Takahashi [*Jpn. J. Appl. Phys. (Japan)* vol.29 (1990) p.L537-9]
[25] T. Kanata, H. Takakura, M. Matsunaga, Y. Hamakawa [*Appl. Phys. Lett. (USA)* vol.56 (1990) p.1752-4]
[26] D. Kim, P.Y. Yu [*Appl. Phys. Lett. (USA)* vol.56 (1990) p.2210-2]
[27] Z. Sobiesierski, D.I. Westwood, R.H. Williams [*Mater. Sci. Eng. B, Solid-State Mater. Adv. Technol. (Switzerland)* vol.5 (1990) p.265-8]
[28] G. Ambrazevichius, S. Marcinkevichius, R. Bendorius, T. Kideikis, K. Naudzhius [*Lith. Phys. J. (USSR)* vol.31 (1991) p.210-4]
[29] W.C. Tang, H.J. Rosen [*Appl. Phys. Lett. (USA)* vol.58 (1991) p.1644-6]
[30] K. Yano, T. Katoda [*J. Appl. Phys. (USA)* vol.70 (1991) p.7036-41]
[31] O. Brafman, D. Fekete, R. Sarfati [*Appl. Phys. Lett. (USA)* vol.58 (1991) p.400-2]
[32] J.P. Estrera, W.M. Duncan, Y.C. Kao, H.Y. Liu, E.A. Beam III [*J. Electron. Mater. (USA)* vol.20 (1991) p.983-7]
[33] J. Finders et al [*J. Cryst. Growth (Netherlands)* vol.107 (1991) p.151-5]
[34] H. Asahi et al [*Appl. Surf. Sci. (Netherlands)* vol.60/61 (1992) p.625-30]
[35] V. Lemos, T. Ritter, B.A. Weinstein [*Appl. Phys. Lett. (USA)* vol.61 (1992) p.1417-9]
[36] J.P. Estrera et al [*Appl. Phys. Lett. (USA)* vol.61 (1992) p.1927-9]; J.P. Estrera et al [*Proc. SPIE Int. Soc. Opt. Eng. (USA)* vol.1678 Spectrosc. Charact. Tech. Semicond. Technol. IV (1992) p.120-30]
[37] M.H. Brodsky, G. Lucovsky [*Phys. Rev. Lett. (USA)* vol.21 (1968) p.990-3]; G. Lucovsky, M.F.Chen [*Solid State Commun. (UK)* vol.8 (1970) p.1397-401]
[38] A. Mooradian, G.B. Wright [*Solid State Commun. (UK)* vol.4 (1966) p.431-4]
[39] R. Carles, N. Saint-Cricq, J.B. Renucci, M.A. Renucci, A. Zwick [*Phys. Rev. B (USA)* vol.22 (1980) p.4804-15]
[40] R.M. Martin [*Phys. Rev. B (USA)* vol.1 (1970) p.4005-11]
[41] J.B. Mullin, B.W. Straughan [*Crit. Rev. Solid State Sci. (USA)* vol.5 (1975) p.441-56]
[42] J.V. Ozolin'sh, G.K. Averkieva, A.F. Ilvin'sh, N.A. Goryunova [*Kristollografiya (USSR)* vol.7 (1962) p.850-3; English Translation in *Sov. Phys.-Crystallogr. (USA)* vol.7 (1963) p.691-3]
[43] P. Wickbold, E. Anastassakis, R. Sauer, M. Cardona [*Phys. Rev. B (USA)* vol.35 (1987) p.1362-8]

[44] F. Cerdeira, C.J. Buchenauer, F.H. Pollak, M. Cardona [*Phys. Rev. B (USA)* vol.5 (1972) p.580-93]

[45] B. Jusserand, M. Cardona [in *Light Scattering in Solids V*, Eds M. Cardona, G. Güntherodt (Springer, Berlin, 1989) ch.3 p.49-152]

[46] G. Gammon, L. Shi, R. Merlin, G. Ambrazevicius, K. Ploog, H. Morkoc [*Superlattices Microstruct. (UK)* vol.4 (1988) p.405-7]

[47] M. Cardona [in *Light Scattering in Solids II*, Eds M. Cardona, G. Güntherodt (Springer, Berlin, 1982) ch.2 p.19-178]

7.3 Optical properties of pseudomorphic InGaAs/GaAs heterostructures and quantum wells

M. Dutta and J. Pamulapati

January 1993

A INTRODUCTION

Good quality layers are possible in InGaAs/GaAs heterostructures when the 7% lattice mismatch between GaAs and InAs is totally accommodated by the elastic strain. In this elastic limit these heterostructures on GaAs substrates have only the InGaAs layers strained and the strain amplitude is independent of the thickness below the critical limit.

To date most of the photoluminescence, absorption and reflectance studies have been done to characterise the layer quality and to determine the band offsets in strained InGaAs/GaAs systems. Effective mass theories are followed in order to fit the transition energies with in most cases the GaAs deformation potentials being used for the strain calculations. In other cases, linear interpolations of the values between GaAs and InAs are used. These are for GaAs, a = -8.4 eV, b = -2 eV [1] and for InAs, a = -6 eV, b = -1.8 eV [2]. Although in some of the initial calculations, the effect of strain was added in the bandgap, the later works have the strain terms included in the Hamiltonian in order to obtain the subband energies and to compare to the experimentally measured transition energies. Without the inclusion of strain, the bandgap as a function of indium concentration is taken in most of these studies to be

$$E_g = 1.430 - 1.53x + 0.45x^2 \text{ at } 300 \text{ K [3]}$$

$$E_g = 1.508 - 1.47x + 0.375x^2 \text{ at } 77 \text{ K [4]}$$

$$E_g = 1.519 - 1.5837x + 0.475x^2 \text{ at } 2 \text{ K [5]}$$

The elastic stiffness constants used by most of these studies are again linear interpolations of the values for GaAs and InAs. These are $C_{11} = 11.88 \times 10^{11}$ dyn/cm^2, $C_{12} = 5.38 \times 10^{11}$ dyn/cm^2 for GaAs while for InAs the values in units of 10^{11} dyn/cm^2 are $C_{11} = 8.33$ and $C_{12} = 4.53$ (ref [3]).

The moderately strained InGaAs/GaAs systems with indium concentration of up to 30% will be covered, while the results of the more heavily strained binary InAs/GaAs structures will not be included. The results of the band offsets as measured by photoluminescence, absorption/transmission and reflectance/modulated spectroscopy are in Sections B, C and D, tabulated in TABLES 1, 2 and 3. Since the conclusions of these experiments are not yet definitive, a representative number of references have been summarised. Only undoped quantum wells with GaAs as the barrier materials will be discussed and the results with samples where the higher gap ternary AlGaAs is used are left out of this Datareview. Results of magneto-optical studies and hydrostatic pressure studies are summarised in TABLES 4 and 5, in Sections E and F, and the external uniaxial pressure work will be summarised in Section G.

The data from miscellaneous optical experiments will be grouped in Section H, with the conclusions in Section I.

B PHOTOLUMINESCENCE

The results of the photoluminescence experiments given in the following table include photoluminescence excitation or were in some cases done with an additional technique to identify the excited states for a better set of transition energies to fit to the calculations where the band offset was used as an adjustable parameter.

TABLE 1.

In Mole Fraction	Samples InGaAs/GaAs	Temp. of Expt. (K)	Band Offset	Comment
13%	110/190 Å	2 & 300 K	Interpolation of ref [6] for InAs/GaAs offset	Ref [7]. Light hole type II while heavy hole is type I.
7-30%	100-180 Å well widths	2, 77 & 300 K	$Q_c = 0.8$	Ref [8]. Independent of indium concentration.
4 & 12%	200/200 Å 144/450 Å 160/480 Å	77 K	Q_c In concn dependent	Ref [9]. High intensity pulsed laser used to determine excited states. Polarisation dependence for identifying heavy and light hole states.
18 & 25%	80/1000 Å 120/1000 Å	1.6 K	$Q_c = 0.6$	Ref [10]. Electronic Raman scattering, TEM and CSA were used to further characterise the samples.
5%	70, 100, 140 & 180 Å well widths	9 K	$Q_c = 0.57$	Ref [11]. The 180 Å quantum well sample was chemically released which then allowed unambiguous identification of heavy and light hole states.
13-30%	85 & 120 Å	8 K	$Q_c = 0.65$	Ref [12]. When the misfit strain exceeds 2% there is no longer a layer by layer growth, but island formation.
10%	2, 4, 6, 10 & 17 monolayer well widths	2 K	Upper limit of $Q_v = 0.4$	Ref [13]. No conclusions were drawn for the 17 monolayer sample since the value of x could not be established.

C ABSORPTION/TRANSMISSION

TABLE 2.

In Mole Fraction	Samples InGaAs/GaAs	Temp. of Expt.	Band Offset	Comment
15%	50/200 Å, 100/200 Å & 120/200 Å	77 K	Q_v= 0.3	Ref [14]. Polarisation selection rules used to identify the hole states.
13-19%	85, 106, 158, 213 Å well with 200 Å barriers	2 K, 4 K	Q_v = 0.3	Ref [15]. The offset was determined to be independent of In mole fraction.
12-25%	120/120 Å	Room	Q_c = 0.65	Ref [16]. Offset determined to be independent of In conc.

D REFLECTANCE/MODULATED SPECTROSCOPY

TABLE 3.

In %	Samples	Temperature	Band Offset	Comment
12%	30/100 Å, 50/100 Å	77 K & 300 K	Q_v = 0.3	Ref [17]. First observation of unconfined states as well as miniband dispersion.
11% 19%	100 Å well 107 Å well	300 K	Band offset varies with conc. Good agreement with ref [9].	Ref [18]. Photoreflectance with the polarisation dependence of the total internal reflection mode, which allows the unambiguous identification of the heavy and light hole transitions.
10% & 20%	49, 68, 104 & 272 Å	2 K	Q_v = 0.34	Ref [19]. External uniaxial stress was also applied to assist in the assignment of the spectral features.

E MEASUREMENTS OF EXCITON BINDING ENERGY AND EFFECTIVE MASSES

TABLE 4.

In %	Samples InGaAs/GaAs	Technique & Temp.	Binding Energy	Comments
15%	48/200 Å 25/400 Å 14/400 Å	PLE at 4 K	9.2 meV 8.7 meV 6.5 meV	Ref [20]. Binding energy from separation of the sharp heavy hole and the continuum edge. In-plane heavy hole mass of 0.16 m_0 and 0.21 m_0 used with better fit obtained with the latter.
7% 11% 12% 13% 13%	80/150 Å 80/150 Å 80/150 Å 80/150 Å 100 Å well 65 Å well	PLE in magnetic field up to 6 T and at 10 K	7 meV 7.5 meV 7.9 meV 8.1 meV 8 meV 8.5 meV	Ref [21]. RHEED and double crystal X-ray used for In calibration. In-plane heavy hole mass of 0.18 m_0 and 0.22 m_0 used with better fit obtained with the latter value.
12%	25/100 Å 50/100 Å 100/100 Å 200/100 Å	Reflectivity at 4 K in magnetic field up to 16 T	4.1 meV 6.9 meV 8.0 meV 6.5 meV	Ref [22]. The heavy hole effective masses were found to be 0.36, 0.25, 0.23 and 0.19 m_0 for the 25, 50, 100 and 200 Å well widths respectively. Reduction in binding energy due to coupling through the barriers.
10%	2, 4, 6, 10 & 17 monolayer single quantum wells	PL at 2 K in magnetic field up to 3.6 T		Ref [23]. The diamagnetic shift increases with decreasing well size, tending to the bulk GaAs value. An effective heavy-hole g-value of 1.1 obtained from the spin splitting.
18%	30/500 Å 40/500 Å 60/500 Å 110/500 Å	PL & PLE in fields up to 6 T	8.8 meV 11.3 meV 8.0 meV 7.2 meV	Ref [24]. The experimental values are compared with a variational-perturbation calculation, but theory underestimates the magnitude of the experimental peak.

F HYDROSTATIC PRESSURE DEPENDENCE

TABLE 5.

In %	Samples	Technique	Temp.	Comment
16 & 25%	20, 40, 65, 90 & 120 Å well with 150 Å barrier	PL with pressure 0 - 45 kbar	77 K	Ref [25]. The pressure coefficient varied between 9.94 to 10.51 meV/kbar and increased for decreasing well widths.
9-22 % 18%	100 Å well 30, 40, 60, 120 Å wells	PL with 0 - 40 kbar pressure	20 K & 80 K	Ref [26]. The pressure coefficient is low and does not follow a linear interpolation or the value predicted from a third order elasticity theory, but a value of 10.7 - 6.0x meV/kbar. Well width dependence is similar to ref [25].
10-20%	80 & 150 Å wells	Modulated transmission; 0 - 45 kbar	300 K	Ref [27]. Pressure coefficient decreases for increasing In concentration and is lower than linearly interpolated values.
19%	30, 70, 110 & 160 Å with 200 Å barriers	PL with 0 - 65 kbar	77 K	Ref [28]. Pressure coefficient decreases for increasing well width. The increase of the barrier height with pressure is believed to be the major cause of change in the pressure coefficient.

G EXTERNAL UNIAXIAL PRESSURE STUDIES

In [29] external uniaxial stress of up to 18 kbar in the [110] direction shows several symmetry forbidden transitions in the photoreflectance spectra. These features are not observed when a stress of up to 8 kbar is applied in the [100] direction and are due to an electric field along the [001] direction induced by the piezoelectric coupling. At 17.8 kbar this field is 1.85 x 10^5 V/cm. In [30] the reflectivity is measured with the application of external uniaxial stress. The data show significant stress induced mixing between the first light hole and second heavy hole minibands.

H OTHER MEASUREMENTS

There have been a couple of studies on the measurements of binding energy of acceptors in the strained InGaAs systems. These values are affected by the fact that the biaxial strain has a substantial effect on the valence band structure. Binding energies of carbon and magnesium acceptors in a 12% InGaAs/GaAs quantum well have been measured by low temperature PL [31]. The binding energies are reduced from the values of the thick unstrained layers of 27 meV for magnesium and 24 meV for carbon, and are larger only for acceptors at the centre of the very narrow well. For the magnesium acceptors the measured values varied from 18 - 28 meV as a function of well width. The values for the carbon acceptors in the strained wells are not quoted. Beryllium doped InGaAs quantum wells with different amounts of lattice mismatch were studied in [32]. The binding energy for the Be acceptor was seen to vary between 30 meV at no lattice mismatch and 18 meV at a lattice mismatch of 1.4% corresponding to an In concentration of 20%.

A number of studies have used photoluminescence as a method for estimating the critical thickness in these strained layer systems. These studies have not been included here since information obtained from the intensity of luminescence has so many factors involved that the information obtained may be very qualitative.

I CONCLUSION

As can be seen from the above tables there is as yet no strong consensus on a number of the important material parameters. In the studies where numbers are extracted, input parameters such as the deformation potentials and elastic stiffness constants are introduced with linear interpolations done between the values of the binaries GaAs and InAs, as there are no measured values available for the ternary alloy at the various indium concentrations. There is clearly room for more systematic measurements of many of these parameters. Furthermore errors are introduced in the numbers due to the strain not being completely uniform, or completely elastic, or in the case of thick strain-released samples, residual strains affect the results. However the above tables show that although there is no exact agreement in the values the trends are clear.

REFERENCES

[1] M. Chandrasekhar, F.H. Pollak [*Phys. Rev. B (USA)* vol.15 (1977) p.2127]

[2] J.F. Tsay, S.S. Mitra, B. Bendow [*Phys. Rev. B (USA)* vol.10 (1974) p.1476]

[3] S. Adachi [*J. Appl. Phys. (USA)* vol.53 (1982) p.8775-92]

[4] Y.T. Leu, F.A. Theil, H. Scheiber, B.I. Miller, J. Bachman [*J. Electron. Mater. (USA)* vol.8 (1979) p.663]

[5] K.H. Goetz et al [*J. Appl. Phys. (USA)* vol.54 (1983) p.4543]

[6] W.A. Harrison [*J. Vac. Sci. Technol. (USA)* vol.14 (1977) p.1016]

[7] J.Y. Marzin, E.V.K. Rao [*Appl. Phys. Lett. (USA)* vol.43 (1983) p.560-2]

[8] T.G. Andersson, Z.G. Chen, V.D. Kulakovskii, A. Uddin, J.T. Vallin [*Phys. Rev. B (USA)* vol.37 (1988) p.4032-38]

[9] M.J. Joyce, M.L. Johnson, M. Gal, B.F. Usher [*Phys. Rev. B (USA)* vol.38 (1988) p.10978-80]

[10] J.-P. Reithmaier, R. Hoger, H. Riechert, A. Heberle, G. Abstreiter, G. Weimann [*Appl. Phys. Lett. (USA)* vol.56 (1990) p.536-8]

[11] M.J. Joyce, Z.X. Xu, M. Gal [*Phys. Rev. B (USA)* vol.44 (1991) p.3144-49]

[12] J. Pamulapati, P. Bhattacharya, R.L. Tober, J.P. Loehr, J. Singh [*J. Appl. Phys. (USA)* vol.71 (1992) p.4487-91]

[13] D.C. Reynolds, K.R. Evans, C.E. Stutz, B. Jogai, C.R. Wie, P.W. Yu [*Phys. Rev. B (USA)* vol.45 (1992) p.11156-60]

[14] J.-Y Marzin, M.N. Charasse, B. Sermage [*Phys. Rev. B (USA)* vol.31 (1985) p.8298-301]

[15] G. Ji, D. Huang, U.K. Reddy, T.S Henderson, R. Houdre, H. Morkoc [*J. Appl. Phys. (USA)* vol.62 (1987) p.3366-73]

[16] S. Niki, C.L. Lin, W.S.C. Chang, H.H. Wieder [*Appl. Phys. Lett. (USA)* vol.55 (1989) p.1339-41]

[17] S.H. Pan et al [*Phys. Rev. B (USA)* vol.38 (1988) p.3375-82]

[18] A. Kzendov, H. Shen, F.H. Pollak, D.P. Bour [*Solid State Commun. (USA)* vol.73 (1990) p.11-14]

[19] G. Arnaud, J. Allegre, P. Lefebvre, H. Mathieu, L.K. Howard, D.J. Dunstan [*Phys. Rev. B (USA)* vol.46 (1992) p.15290-301]

[20] K.J. Moore, G. Duggan, K. Woodbridge, C. Roberts [*Phys. Rev. B (USA)* vol.41 (1990) p.1090-94]

[21] H.Q. Hou, Y. Segawa, Y. Aoyagi, S. Namba, J.M. Zhou [*Phys. Rev. B (USA)* vol.42 (1990) p.1284-9]

[22] N.J. Pulsford et al [*Phys. Rev. B (USA)* vol.43 (1991) p.2246-54]

[23] D.C. Reynolds, K.R. Evans, C.E. Stutz, P.W. Yu [*Phys. Rev. B (USA)* vol.43 (1991) p.4244-8]

[24] M.J.L.S. Haines et al [*Phys. Rev. B (USA)* vol.43 (1991) p.11944-49]

[25] H.Q. Hou, L.J. Wang, R.M. Tang, J.M. Zhou [*Phys. Rev. B (USA)* vol.42 (1990) p.2926-31]

[26] V.A. Wilkinson et al [*Phys. Rev. B (USA)* vol.42 (1990) p.3113-9]

[27] W. Shan et al [*Phys. Rev. B (USA)* vol.43 (1991) p.14615-20]

[28] G.-H. Li, B.-Z. Zheng, H.-X. Han, Z.-P Wang, T.G. Andersson, Z.G. Chen [*Phys. Rev. B (USA)* vol.45 (1992) p.3489]

[29] H. Qiang, F.H. Pollak, C. Mailhiot, G.D. Petit, J.M. Woodall [*Phys. Rev. B (USA)* vol.44 (1991) p.9126-8]

[30] P. Boring, J.-M. Jancu, B. Gil, D. Bertho, C. Jouanin, K.J. Moore [*Phys. Rev. B (USA)* vol.46 (1992) p.4764-8]

[31] A.P. Roth, D. Morris, R.A. Masut, C. Lacelle, J.A. Jackman [*Phys. Rev. B (USA)* vol.38 (1988) p.7877-80]

[32] J.P. Loehr, Y.C. Chen, D. Biswas, P. Bhattacharya, J. Singh [*Appl. Phys. Lett. (USA)* vol.57 (1990) p.180-82]

7.4 Optical properties of InP-based lattice-matched InGaAs quantum wells

J. Pamulapati and M. Dutta

January 1993

A INTRODUCTION

The InGaAs system lattice matched to InP substrates is attracting considerable attention for applications in high speed electronics and photonics. This material system is appealing due to its optical (energy gap at ~ 0.73 eV) and transport (high room temperature mobility > 10^4 cm^2/V s) properties which are capitalised upon in a myriad of applications such as high-frequency millimeter-microwave devices, detectors, and lasers in the 1.3 - 1.6 μm spectral region, with the latter finding a market in fibre-optic communication systems.

With the incorporation of InGaAs in a quantum well, i.e. surrounded by a higher energy gap material, there is an added dimension of tailorability of bandgap/wavelength, refractive index, exciton lifetime, etc. A number of these properties are altered due to the difference in the density of states which changes from a $\rho(E) \sim E^{1/2}$ to a step-like dependence. We discuss here several of the optical properties - photoluminescence (PL) in Section B, reflectance in Section C, and absorption/transmission in Section D - related to InGaAs quantum wells lattice matched to InP.

For proper lattice matching to InP, $In_xGa_{1-x}As$ must have a composition of x = 53.2% to match the lattice constant of InP(5.8688) [1]. This is determined from a linear interpolation between the lattice parameters of InAs(6.0584) and GaAs(5.6533). The energy gap of $In_{0.53}Ga_{0.47}As$ has been determined by a number of authors and techniques with a summary of the results listed in TABLE 1.

TABLE 1 Fundamental energy gap of $In_{0.53}Ga_{0.47}As$ lattice matched to InP.

Energy Gap	Experimental Technique	Comment
0.790	PL and photoreflectance	This work fits the energy gap with measured values for In mole fraction less than 15% yielding a slightly quadratic dependence of energy gap on In mole fraction at 300 K [2].
0.858	Photoresponse from an electrolyte-semiconductor junction	Work measures the gap of InGaAs material up to a 50% In mole fraction at 300 K [3].
0.728	Photoreflectance	Measures lattice matched material as a function of temperature and determines a fit to the temperature dependence [4].
0.739	Wavelength modulated spectroscopy	300 K measurements [5].

In addition, more recent work fits the temperature dependence of the bandgap for lattice matched InGaAs to a Varshni equation shown in Eqn (1), where $E_0(0)$ is 0.8029 eV, α is 4.0×10^{-4} eV/K and β is 226 K [4].

$$E_0(T) = E_0(O) - \frac{\alpha T}{T + \beta} \tag{1}$$

Another dominant factor which determines the optical properties of lattice matched InGaAs quantum wells is the band offset. The conduction band offset has been determined to be approximately 0.51 ± 0.04 eV for the $In_{0.53}Ga_{0.47}As/In_{0.52}Al_{0.48}As$ system [6]. This represents 70% of the total bandgap discontinuity as measured by current-voltage measurements on n^+ InGaAs/InAlAs/n^- InGaAs capacitors. For the InGaAs/InP quantum well system, the conduction band discontinuity has more recently been determined to be 42% of the bandgap discontinuity [7]. This fraction was determined by the use of admittance spectroscopy on both p^+n and n^+p junctions. The results agree with earlier results of band offset using capacitance-voltage measurements.

B PHOTOLUMINESCENCE

Photoluminescence has been one of the prime characterisation tools used to study the quality of InGaAs quantum wells. Using photoluminescence, the energy of the lowest level transition is delineated. In addition, luminescence can be used to indirectly extract the surface roughness between the well and barrier layers including the broadening of the excitonic linewidth due to alloy disorder [8].

B1 $In_{0.53}Ga_{0.47}As/In_{0.52}Al_{0.48}As$

This particular system is the easiest system to grow with solid source molecular beam epitaxy (MBE) and has been found to be particularly effective for use in microwave and millimetre wave integrated circuits. The usefulness of this system is due in large part to its large conduction band discontinuity (~ 0.5 eV). Detailed in TABLE 2 is a representative sampling of the luminescence quality of InGaAs/InAlAs quantum well systems. From this table one is able to discern that the linewidth of the transition generally increases with a decrease in the quantum well width. It should also be noted that the linewidths for this system are considerably larger than those observed in the GaAs/AlGaAs system. One reason for this increase in linewidth is due to the alloy disorder in the barrier and well region and background impurity concentration in the well layer. The linewidth due to alloy broadening is given by [8]:

$$\sigma_a = \sqrt{\sigma_1^2 + \sigma_2^2} \tag{2}$$

where

$$\sigma_1 = \sigma_{B1}^{a} \frac{\Delta_{QW}}{\Delta_{B1}} \frac{1}{\sqrt{P_0^{ex}}} \tag{3}$$

TABLE 2 Photoluminescence linewidth (FWHM) as a function of InGaAs well width.

Well Width (Å)	Linewidth (FWHM) (meV)	Comments
15 30 50 100	10.8 7.4 6.7 5.7	Grown using pulsed molecular beam method. The barrier used for this sample was $In_{0.52}(Ga_{0.5}Al_{0.5})_{0.48}As$. Indicates the ability to limit the interface roughness, alloy disorder and background impurity concentration [9].
6 12 24 41 88	21 32 33 22 16	Grown by organometallic vapour phase epitaxy (OMVPE). Linewidths for the bulk InAlAs and InGaAs were 12 meV and 2.2 meV, respectively. Combined with the linewidths for the thinner wells, this indicates that the effective fluctuation of the well thickness was less than one monolayer [10].
15 21 29 73 100 103 132	82 49 59 22.9 20 34.7 42.3	Grown by molecular beam epitaxy (MBE). This work found that the emission intensity increases as the quantum well width decreases to approximately 27 Å and then decreases for thinner wells [11-12]. In addition, the linewidth broadening has been correlated with the presence of free carriers within the quantum well for $L_z > 50$ Å, and with variations in quantum well thickness for $L_z < 50$ Å.
100 120	5.8 12	Grown by MBE. These works correlate the linewidth to broadening mechanisms such as alloy broadening, interface roughness, and impurity incorporation [13,14]. The work also finds that the linewidth of the quantum well transition decreases with increasing growth interrupt times with a commensurate decrease in luminescence intensity.

$$\sigma_2 = \sigma^a_{B2} \frac{\Delta_{QW}}{\Delta_{B2}} \frac{1}{\sqrt{1 - P_0^{ex}}} \qquad (4)$$

σ^a_{B1} and σ^a_{B2} represent the PL linewidth in the InAlAs and InGaAs, respectively. Δ_{QW} represents the variation of the electron-hole subband energy difference with variation in the alloy composition, and Δ_B's represent the variation in the bandgap of the material with alloy composition. P_0^{ex} is the exciton fraction in the barrier. In addition to the linewidth due to alloy broadening, interface roughness also affects the linewidth. The interface roughness can be described by

$$\sigma_{IR} = \delta_1 \frac{\partial E^{eh}}{\partial W} \mid W_0 \qquad (5)$$

where δ_1 is the height of the interface roughness, E^{eh} is the separation between the electron and hole subbands, W is the well size, and W_0 is the mean well size of the quantum well [8].

B2 $In_{0.53}Ga_{0.47}As/In_{1-x}Ga_xAs_yP_{1-y}$

For proper lattice matching of $In_{1-x}Ga_xAs_yP_{1-y}$ to InP, the relation between the compositions x and y is expressed as [1]

$$x \simeq \frac{0.1894y}{0.4184 - 0.013y} \quad (0 \leq y \leq 1) \tag{6}$$

A representative sampling of the material quality available through several growth techniques is given in TABLE 3. What is noticeable in TABLE 3 is that the material quality as determined through luminescence linewidths is significantly smaller than for the InGaAs/InAlAs quantum well system. An advantage of the InGaAs/InGaAsP system is with the use of InP barriers (y = 0), when the broadening due to alloy clustering in the barrier is eliminated.

TABLE 3 Photoluminescence linewidth for the InGaAs/InGaAsP quantum well system.

Well Width (Å)	Linewidth (FWHM) (meV)	Comments
10 53 91	12.5 10.7 4.4	Grown using OMVPE. The barrier used for this sample was $In_{0.77}Ga_{0.23}As_{0.5}P_{0.5}$, indicating good control of both the interface and barrier and well quality [15].
6 15 24 40 60 100 160	10 6.9 5.0 3.7 2.8 2.9 2.4	Grown by OMVPE [16]. This work represents the best luminescence linewidths reported for the InGaAs/InP quantum well system. Earlier works [17-20] report similar findings, although not able to achieve interface and material quality control to the same degree.
10 20 40 60 80 110 150	8.5 7 5 3.5 3 3 3	Grown by chemical beam epitaxy (CBE) [21]. The interface roughness was held to < 2.93 Å. The linewidths for the narrower wells are indicative of this as they are narrower than calculations using Eqn (5) dictate. The authors also find a discrepancy of the band-offset parameter for well widths less than 50 Å. For smaller well widths a 20% conduction band discontinuity is estimated.
200	8.3	Grown by metallorganic chemical vapour deposition (MOCVD) [22]. Work indicates that the conduction band discontinuity is closer to 20-30% of the band gap discontinuity. Linewidths indicate control of interface fluctuations greater than 2.93 Å.

C REFLECTANCE

Reflectance measurements have been used to determine the refractive indices of InGaAs/InAlAs multiple quantum wells as a function of photon energy and barrier width [23]. The functional relation derived from reflection measurements is

$$n^2 = 1 + \left(\frac{E_0 E_d}{\left(E_0^2 - (h\nu)^2 \right)} \right) \tag{7}$$

with

$$E_0 = 2.21\bar{x} + 2.02$$

$$E_d = 13.2\bar{x} + 20.9 \quad (8)$$

$$\bar{x} = \frac{0.48L_b}{(L_b + L_z)}$$

The authors noted a discrepancy between their measured values and those obtained from a linear extrapolation between the binaries. The linear extrapolation is noted to be a good fit to vapour phase epitaxial layers and initiated a study of differences between MBE, metallorganic MBE and liquid phase epitaxy (LPE). The conclusion was a noticeable difference in the optical properties of the material due to growth temperature differences.

In the only reported measurements of photoreflectance and electroreflectance on lattice matched InGaAs quantum wells, the quantum wells show well resolved ground state exciton and higher confined transitions [24,25]. The quantum wells were grown using MBE and had InP barriers. More detailed investigation of the spectra indicated that the InP barrier contained approximately 10% arsenic. This observation is confirmed by X-ray diffraction.

D ABSORPTION

An optical absorption study of InGaAs/InP quantum wells has been performed both theoretically and experimentally [26]. The integrated intensity was experimentally determined to be 1.1×10^{-4} eV which is 30% smaller than for GaAs/AlGaAs. This is due to a larger two-dimensional exciton radius caused by the larger static dielectric constant and the smaller in-plane reduced effective mass. The optical absorbance of an exciton resonance in the quantum well is given by

$$\alpha_{ex} L_{QW} = SB(\hbar\omega - E_{ex}) \quad (9)$$

where S is the integrated intensity of the exciton spectrum based on the effective mass approximation, $B(\hbar\omega - E_{ex})$ is the normalised broadening function and L_{QW} is the period in a multiple quantum well structure. This work also suggests that to clearly resolve an exciton at room temperature for this particular system, an inhomogeneous broadening needs to be less than 10 meV. Similar measurements have found that the temperature dependence of the half-width at half maximum (HWHM) is described by

$$\Gamma = \Gamma_0 + \frac{\Gamma_{ph}}{\exp(\hbar\omega_{LO}/k_B T) - 1} \quad (10)$$

where Γ_0 is the inhomogeneous broadening width, Γ_{ph} is the coupling constant with excitons, and $\hbar\omega_{LO}$ is the LO-phonon energy [27].

For the InGaAs/InAlAs system, Eqns (9) and (10) are equally valid. Absorption measurements in this system have shown that Γ_0 is ~ 4 meV and Γ_{ph} is 9 meV, almost a factor of 2 greater than values measured in the InGaAs/InP quantum well system [28,14]. In addition, the carrier lifetime in the InGaAs/InAlAs quantum well has been determined to be 860 ps as determined by photoluminescence correlation spectroscopy. This lifetime is close to the value typically obtained for the GaAs/AlGaAs and the InGaAs/InP quantum well systems.

E CONCLUSION

The InGaAs quantum well system, lattice matched to InP, has indeed found many applications from monolithic microwave integrated circuits to optical fibre communication systems. But unfortunately, there appears to be little data available on several key parameters for the InGaAs system such as the band offset. While there seems to be general consensus on the band offset parameter for the InGaAs/InAlAs system, this is not the case with the InGaAs/InGaAsP system. In addition, there is little work on the relevant parameters such as the index of refraction for these material systems. Clearly, there is room for further study in the InGaAs quantum well system.

REFERENCES

[1] S. Adachi [*J. Appl. Phys. (USA)* vol.53 (1982) p.8775-92]

[2] R.E. Nahory, M.A. Pollack, J.C. DeWinter [*J. Appl. Phys. (USA)* vol.46 (1975) p.775-82]

[3] B.J. Baliga, R. Bhat, S.K. Ghandhi [*J. Appl. Phys. (USA)* vol.46 (1975) p.4608]

[4] D.K. Gaskill, N. Bottka, L. Aina, M. Mattingly [*Appl. Phys. Lett. (USA)* vol.56 (1990) p.1269-71]

[5] T.W. Nee, A.K. Green [*J. Appl. Phys. (USA)* vol.68 (1990) p.5314-17]

[6] C.K. Peng, A. Ketterson, H. Morkoç, P.M. Solomon [*J. Appl. Phys. (USA)* vol.60 (1986) p.1709-12]

[7] D.V. Lang et al [*Appl. Phys. Lett. (USA)* vol.50 (1987) p.736-8]

[8] J. Singh, K.K. Bajaj [*J. Appl. Phys. (USA)* vol.57 (1985) p.5433-7]

[9] T. Fujii, Y. Nakata, S. Muto, S. Hiyamizu [*Jpn. J. Appl. Phys. (Japan)* vol.25 (1986) p.L598-L600]

[10] H. Kamei, K. Hashizume, M. Murata, N. Kuwata, K. Ono, K. Yoshida [*J. Cryst. Growth (Netherlands)* vol.93 (1988) p.329-35]

[11] D.F. Welch, G.W. Wicks, L.F. Eastman [*Appl. Phys. Lett. (USA)* vol.46 (1985) p.991-3]

[12] D.F. Welch, G.W. Wicks, L.F. Eastman [*Appl. Phys. Lett. (USA)* vol.43 (1983) p.762-4]

[13] F-Y. Juang, P.K. Bhattacharya, J. Singh [*Appl. Phys. Lett. (USA)* vol.48 (1986) p.290-2]

[14] S. Gupta, P.K. Bhattacharya, J. Pamulapati, G. Mourou [*J. Appl. Phys. (USA)* vol.69 (1991) p.3219-25]

[15] P.J.A. Thijs et al [*J. Cryst. Growth (Netherlands)* vol.93 (1988) p.863-9]

[16] H. Kamei, H. Hayashi [*Proc. 3rd Int. Conf. on Indium Phosphide and Related Materials*, Cardiff, UK, 8-11 April 1991 (IEEE, New York, 1991) p.389-94]

[17] K.W. Carey, R. Hull, J.E. Fouquet, F.G. Kellert, G.R. Trott [*Appl. Phys. Lett. (USA)* vol.51 (1987) p.910-12]

[18] D. Grützmacher, L. Wolter, H. Jürgensen, P. Balk, C.W.T. Bulle Lieuwma [*Appl. Phys. Lett. (USA)* vol.52 (1988) p.872-3]

[19] D. Moroni, J.N. Patillon, E.P. Menu, P. Gentric, J.P. André [*J. Phys. (France)* vol.48 (1987) p.143-6]

[20] B.J. Miller, E.F. Schubert, U. Koren, A. Ourmazd, A.H. Dayem, R.J. Capik [*Appl. Phys. Lett. (USA)* vol.49 (1986) p.1384-86]

[21] W.T. Tsang, E.F. Schubert [*Appl. Phys. Lett. (USA)* vol.49 (1986) p.220-22]

[22] M. Razeghi, J.P. Hirtz, U.O. Ziemelis, C. Delelande, B. Etienne, M. Voos [*Appl. Phys. Lett. (USA)* vol.43 (1983) p.585-7]

[23] S. Nojima, H. Asahi [*J. Appl. Phys. (USA)* vol.63 (1988) p.479-83]

[24] V.M. Airaksinen, H.K. Lipsanen, P. Ravila, T. Tuomi, P.A. Claxton [*Proc. SPIE Conf. on Modulation Spectroscopy*, San Diego, California, USA, 19-21 March 1990, p.238-43]

[25] H.K. Lipsanen, V.M. Airaksinen, T. Tuomi, P.A. Claxton [*Proc. SPIE Conf. on Modulation Spectroscopy*, San Diego, California, USA, 19-21 March 1990, p.300-5]

[26] M. Sugawara, T. Fujii, S. Yamazaki, K. Nakajima [*Phys. Rev. B. (USA)* vol.42 (1990) p.9587-97]

[27] Y. Kawaguchi, H. Asahi [*Appl. Phys. Lett. (USA)* vol.50 (1987) p.1243-5]

[28] J.S. Weiner, D.S. Chemla, D.A.B. Miller, T.H. Wood, D. Sivco, A.Y. Cho [*Appl. Phys. Lett. (USA)* vol.46 (1985) p.619-21]

7.5 Linear and quadratic electro-optic coefficients in InGaAsP

J.E. Zucker

February 1993

A INTRODUCTION

Electro-optic effects are essential to the operation of many bulk and integrated optic devices including amplitude and phase modulators, tunable wavelength filters, and spatial switches. With the growth of optical fibre communications systems operating near 1.3 and 1.55 μm wavelength, InGaAsP and related compounds have gained in technological importance. We discuss here the properties of these materials at near- and below-bandgap wavelengths that are relevant to the design of electrorefractive devices.

B DEFINITIONS

When an electric field E is applied, the change in the refractive index n is given by

$$\Delta\left[\frac{1}{n^2}\right]_{ij} = r_{ijk}E_k + R_{ijkl}E_kE_l$$

where the indices i, j, k, l give the components of the incident optical polarisation and applied field along the crystal principal axes. Permutation symmetry makes $r_{ijk} = r_{jik}$ and $R_{ijkl} = R_{jikl} = R_{ijlk}$ while the crystal symmetry for InGaAsP (class $\bar{4}3\,m$) ensures that some of the electro-optic coefficients will be zero and that other components of the tensor will be related [1]. With the convention of contracted indices such that ij = (11) ⇔ 1, (22) ⇔ 2, (33) ⇔ 3, (32) = (23) ⇔ 4, (13) = (31) ⇔ 5, and (12) = (21) ⇔ 6, the electro-optic coefficients can be summarised as:

$$r = \begin{bmatrix} 0 & 0 & 0 \\ 0 & 0 & 0 \\ 0 & 0 & 0 \\ r_{41} & 0 & 0 \\ 0 & r_{52} & 0 \\ 0 & 0 & r_{63} \end{bmatrix} \text{ with } r_{41} = r_{52} = r_{63} \text{ and } R = \begin{bmatrix} R_{11} & R_{12} & R_{13} & 0 & 0 & 0 \\ R_{21} & R_{22} & R_{23} & 0 & 0 & 0 \\ R_{31} & R_{32} & R_{33} & 0 & 0 & 0 \\ 0 & 0 & 0 & R_{44} & 0 & 0 \\ 0 & 0 & 0 & 0 & R_{55} & 0 \\ 0 & 0 & 0 & 0 & 0 & R_{66} \end{bmatrix}$$

with $R_{12} = R_{13} = R_{21} = R_{23} = R_{31} = R_{32}$, $R_{11} = R_{22} = R_{33}$ and $R_{44} = R_{55} = R_{66}$. In general, experimental measurements do not necessarily have the incident optical polarisation or electric field vectors along the above axes. Hence coordinate transformation [2] may be needed to obtain the refractive index change in some configurations. For example, crystals grown on (001) substrate orientation have as cleavage planes (110) and $(1\bar{1}0)$. FIGURE 1 shows this common configuration, with field E along [001] and optical propagation vector k along [110] or $[1\bar{1}0]$.

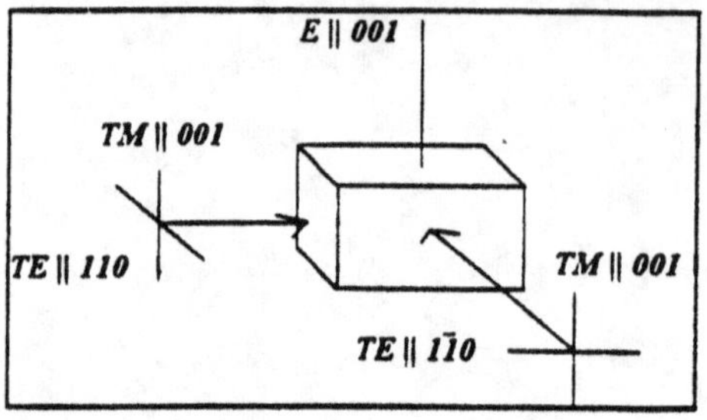

FIGURE 1 Cleavage planes of (001)-oriented crystal.

Using

$$\Delta\left[\frac{1}{n^2}\right] \approx \frac{-2\Delta n}{n^3}$$

where n is the zero-field refractive index one obtains

$$\Delta n_{TE} = -\frac{1}{2}n^3\left[R_{12}E_3^2 - r_{41}E_3\right]$$

for k|| [110] and

$$\Delta n_{TE} = -\frac{1}{2}n^3\left[R_{12}E_3^2 + r_{41}E_3\right]$$

for k|| $[1\bar{1}0]$. In both cases,

$$\Delta n_{TM} = -\frac{1}{2}n^3\left[R_{11}E_3^2\right]$$

Hence, depending on the direction of propagation, the quadratic contribution can either add or subtract from the linear electro-optic effect.

C POCKELS AND FRANZ-KELDYSH CONTRIBUTIONS TO ELECTRO-OPTIC EFFECTS

The linear electro-optic (or Pockels effect) has contributions from several physical phenomena. When measured at frequencies well above the acoustic resonances of the sample, r_{41}^S, the clamped (or constant-strain) value of r, arises from the direct modulation of electronic properties by the applied field as well as via the polarisability of the ions that drag the electron cloud. In III-V semiconductors, the lattice and electronic contributions have opposite sign and comparable magnitude [3], which accounts for the relatively small values of r_{41}. A low-frequency applied field produces a strain (inverse piezoelectric effect) that gives an additional elasto-optic contribution to the refractive index change. The unclamped (or constant stress) r^T_{41} is related to the clamped value by $r^S_{41} = r^T_{41} - r^P_{41}$ where $r^P_{41} \approx +2 \times 10^{-13}$ m/V is the small piezoelectric contribution [4]. The wavelength- and composition-dependence of r_{41} is weak for incident photon energies less than the bandgap.

As the operating wavelength λ approaches the bandgap wavelength λ_g of $In_{1-x}Ga_xAs_{1-y}P_y$ the changes in refractive index associated with the Franz-Keldysh (FK) effect begin to dominate over the linear electro-optic effect. The FK effect is a field-induced extension of the low-energy absorption tail [5], with the corresponding change in refractive index Δn_{FK} at below-gap wavelengths positive, and nonlinear with applied field [6,7]. In addition, Δn_{FK} has strong dispersion near the band edge [8,21,24], so the energy separation $\Delta\omega$ between λ and λ_g is an important parameter. For moderate applied fields, Δn_{FK} is quadratic with applied field. The quadratic (or Kerr) coefficient R scales as $e^{(-\Delta\omega)}$ as shown in FIGURE 2, where some measured values of $|R|$ in GaAs are plotted versus $\Delta\omega$. While GaAs is relatively well-characterised [9-13], measurements of electro-optic coefficients for other compositions of InGaAsP are scarce. Hence it may be useful to consult the theoretical estimates by Adachi and Oe [4,14].

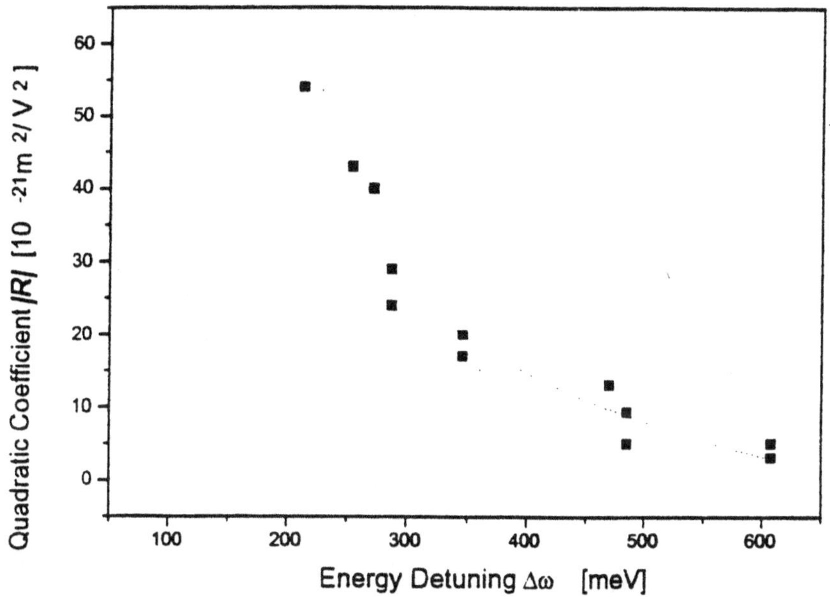

FIGURE 2 Magnitude of quadratic electro-optic coefficients R_{11} and R_{12} for GaAs (from TABLE 2) as a function of energy detuning from the bandgap.

D MEASUREMENTS

TABLES 1 and 2 give some measured values for electro-optic coefficients for operating wavelengths 1.0 μm > λ > 2.0 μm and compositions x,y of $In_{1-x}Ga_xAs_{1-y}P_y$. Often InGaAsP is characterised not by mole fraction but rather by bandgap wavelength λ_g so when only λ_g is given in a reference we have converted to x,y using the formula of Flemish et al [15] with the assumption that the material is lattice-matched to InP substrate. In addition, since $\Delta n \propto n^3$, the values for the zero field refractive index n are also provided. If not given in the original reference, n is taken from [16] or [17]. Since R is highly dispersive, the energy separation between $\Delta\omega$ and λ_g is also given.

TABLE 1 $In_{1-x}Ga_xAs_{1-y}P_y$ linear electro-optic coefficients.

r_{41} coefficient (10^{-12} m/V)	Zero-field refractive index	Wavelength (μm)	x,y	Ref	Notes
-1.43	3.36	1.25	0.15, 0.33	[18]	
-1.44	3.33	1.32	0.15, 0.33	[18]	
-1.34	3.29	1.32	0.10, 0.22	[18]	
-0.141	3.5	1.54	0.24, 0.51	[19]	
-0.107	3.45	1.65	0.24, 0.51	[19]	
-1.32	3.29	1.064	0,1	[20]	T
-1.34	3.29	1.064	0,1	[20]	S
-1.49	3.23	1.208	0,1	[20]	T
-1.54	3.23	1.208	0,1	[20]	S
-1.53	3.20	1.306	0,1	[20]	T
-1.59	3.20	1.306	0,1	[20]	S
-1.63	3.17	1.50	0,1	[20]	T
-1.68	3.17	1.50	0,1	[20]	S
-1.20	3.49	1.0	1,0	[10]	T
-1.8	3.48	1.06	1,0	[13]	
-1.17	3.48	1.064	1,0	[20]	T
-1.33	3.48	1.064	1,0	[20]	S
-1.72	3.468	1.09	1,0	[12]	
-1.65	3.42	1.09, 1.15	1,0	[11]	
-1.17	3.46	1.1	1,0	[10]	T
-1.68	3.447	1.15	1,0	[12]	
-1.43	3.447	1.15	1,0	[23]	T
-1.25	3.43	1.2	1,0	[10]	T
-1.32	3.413	1.3	1,0	[10]	T
-1.6	3.409	1.3	1,0	[13]	
-1.54	3.409	1.32	1,0	[9]	
-1.3	3.40	1.4	1,0	[10]	T
-1.33	3.38	1.5	1,0	[10]	T
-1.50	3.374	1.52	1,0	[9]	
-1.36	3.37	1.6	1,0	[10]	T
-1.32	3.36	1.7	1,0	[10]	T
-1.41	3.355	1.8	1,0	[10]	T
-1.44	3.347	1.9	1,0	[10]	T
-1.51	3.341	2.0	1,0	[10]	T
-1.1	3.09	1.153	1,1	[22]	S

T = constant stress, unclamped value. S = constant strain, clamped value.

TABLE 2 $In_{1-x}Ga_xAs_{1-y}P_y$ quadratic electro-optic coefficients.

Coefficient	(10^{-21} m^2/V^2)	Wavelength (μm)	x,y	Ref	Δω (meV)
R_{11}	-29	1.09	1,0	[12]	287
R_{11}	-20	1.15	1,0	[12]	346
R_{12}	-24	1.09	1,0	[12]	287
R_{12}	-17	1.15	1,0	[12]	346
$R_{11} \cong R_{12}$	-43	1.06	1,0	[13]	254
$R_{11} \cong R_{12}$	-13	1.3	1,0	[13]	470
R_{11}	-9.3	1.32	1,0	[9]	485
R_{11}	-3.2	1.52	1,0	[9]	608
R_{12}	-5.1	1.32	1,0	[9]	485
R_{12}	-5.1	1.52	1,0	[9]	608
$R_{11} \cong R_{12}$	-54	1.09	1,0	[11]	213
$R_{11} \cong R_{12}$	-40	1.15	1,0	[11]	272
$R_{11} - R_{12}$	-58	1.25	0.15, 0.33	[18]	130
$R_{11} - R_{12}$	-31	1.32	0.15, 0.33	[18]	181
$R_{11} - R_{12}$	-8	1.32	0.10, 0.22	[18]	252

E CONCLUSION

The InGaAsP alloys are important for fibre-optic communication. Electro-optic devices form an integral part of such a communication system. The electro-optic properties of InGaAsP are presented in this Datareview.

REFERENCES

[1] J.F. Nye [*Physical Properties of Crystals* (Oxford, Clarendon Press, 1964)]

[2] S. Namba [*J. Opt. Soc. Am. (USA)* vol.41 (1961) p.76]

[3] S.H. Wemple, M. DiDomenico [*Appl. Solid State Sci. (USA)* vol.3 (1972) p.263]

[4] S. Adachi, K. Oe [*J. Appl. Phys. (USA)* vol.56 (1984) p.74]

[5] R.H. Kingston [*Appl. Phys. Lett. (USA)* vol.34 (1979) p.744]

[6] A. Alping, L.A. Coldren [*J. Appl. Phys. (USA)* vol.61 (1987) p.2430]

[7] B.R. Bennett, R.A. Soref [*Proc. SPIE* vol.836 (1988) p.158]

[8] D.E. Aspnes [*Phys. Rev. (USA)* vol.153 (1967) p.972]

[9] C.A. Berseth, C. Wuetrich, F.K. Reinhart [*J. Appl. Phys. (USA)* vol.71 (1992) p.2821]

[10] T.E. Walsh [*RCA Rev. (USA)* (1966) p.323]

[11] J. Faist, F.K. Reinhart, D. Martin, E. Tuncel [*Appl. Phys. Lett. (USA)* vol.50 (1987) p.68]

[12] J. Faist, F.K. Reinhart [*J. Appl. Phys. (USA)* vol.67 (1990) p.7006]

[13] S.S. Lee, R.V. Ramaswamy, V.S. Sundaram [*IEEE J. Quantum Electron. (USA)* vol.27 (1991) p.726]

[14] S. Adachi, K. Oe [*J. Appl. Phys. (USA)* vol.56 (1984) p.1499]

[15] J.R. Flemish, H. Shen, K.A. Jones, M. Dutta, V.S. Ban [*J. Appl. Phys. (USA)* vol.70 (1991) p.2152]

[16] E.D. Palik (Ed.) [*Handbook of Optical Constants of Solids* (Academic, New York, 1985) p.439]

[17] T.P. Pearsall [*GaInAsP Alloy Semiconductors* (John Wiley & Sons, New York, 1982) p.362]

[18] H.G. Bach, J. Krauser, H.P. Nolting, R.A. Logan, F.K. Reinhart [*Appl. Phys. Lett. (USA)* vol.42 (1983) p.692]

[19] J.M. Hammer [NASA Contract Report 3521, April 1982]

[20] N. Suzuki, K. Tada [*Jpn. J. Appl. Phys. (Japan)* vol.23 (1984) p.291]

[21] J.G. Mendoza-Alvarez et al [*J. Lightwave Technol. (USA)* vol.6 (1988) p.793]

[22] D.F. Nelson, E.H. Turner [*J. Appl. Phys. (USA)* vol.39 (1968) p.3337]

[23] M. Sugie, K. Tada [*Jpn. J. Appl. Phys. (Japan)* vol.15 (1976) p.421]

[24] T.E. Van Eck, L.M. Walpita, W.S.C. Chang, H.H. Wieder [*Appl. Phys. Lett. (USA)* vol.48 (1986) p.451]

CHAPTER 8

InGaAs TECHNOLOGY

8.1 Epitaxy of InGaAs
8.2 n- and p-type doping in InGaAs
8.3 Selective etching of InGaAs and its heterostructures
8.4 X-ray characterisation of InGaAs

8.1 Epitaxy of InGaAs

N. Chand and R.F. Karlicek Jr.

January 1993

A INTRODUCTION

Lattice-matched $In_{0.53}Ga_{0.47}As$ on InP can be grown by any major epitaxial technique such as liquid phase epitaxy (LPE), molecular beam epitaxy (MBE), chemical beam epitaxy (CBE), vapour phase epitaxy (VPE), and organometallic VPE (OMVPE). Ternary alloy layers alone on InP have limited device applications and normally a variety of binary, ternary or quaternary (e.g. InGaAsP) heterostructures on InP or GaAs are grown. Today, an impressive array of electronic and optoelectronic device structures based on these heterostructures can be fabricated, in part, because of the tremendous advances made in the past twenty years in the business of III-V compound semiconductor epitaxy. Each of the available epitaxial growth techniques offers certain advantages of performance, cost, safety, flexibility, and ease of use. Of course, some of these features are, to a certain extent, mutually exclusive, and a perfect, all-purpose epitaxial growth technique has yet to be found (in spite of the claims by some of the more zealous practitioners of certain techniques).

Historically, LPE was the first crystal growth technique used to grow device structures of any technological importance and it is still widely used today in the manufacturing of InGaAsP based lasers and LEDs. Chloride VPE (both trichloride and hydride source), developed primarily for the fabrication of GaAs microwave devices, has also been used to fabricate ternary and quaternary heterostructures. As a replacement for LPE, chloride VPE was an attractive technique because of the increased wafer size and improved growth morphology, but it could not be used for Al-containing compounds because of the reactivity of the hot-walled quartz reactor vessel in the presence of aluminium chloride. During the development of chloride VPE in the 1970s, the technological importance of Al-containing III-V compounds limited its deployment. Furthermore, most chloride VPE reactors were custom made and highly specialised. Recently, MBE, OMVPE and CBE have emerged as the most prominent epitaxial growth techniques for InGaAs/InP heterostructures. The utility of these growth techniques can be inferred from the commercial availability of sophisticated MBE, OMVPE and CBE reactor systems from a variety of vendors and from the number of technical meetings which are now dedicated to the subject of epitaxy by one or the other of these techniques. These methods offer tremendous flexibility for growth both in terms of alloy combinations and the type of heterostructures which can be grown. OMVPE is also called metalorganic VPE (MOVPE) as well as metalorganic chemical vapour deposition (MOCVD). CBE also includes its variants such as gas source MBE (GSMBE) and metalorganic MBE (MOMBE). In TABLE 1 we compare the capabilities of these epitaxial techniques.

TABLE 1 Comparison of epitaxial techniques.

Capability	LPE	Trichloride VPE	Hydride VPE	MBE	CBE	OMVPE
Cost	low	low	moderate	high	high	moderate
Materials	any	no Al	no Al	no P	any	any
Purity	moderate	very high	moderate	high	high	high
Growth uniformity	poor	fair	fair	good	good	fair
Throughput	low	moderate	moderate	moderate	moderate	high
Interface control	poor	fair	fair	excellent	excellent	good
Reproducibility	fair	good	good	excellent	excellent	very good
In-situ monitoring	none	difficult	difficult	extensive	extensive	some
Safety issues	few	few	many	few	some	many

B SUBSTRATE PREPARATION

The growth of high quality films in terms of purity, morphology and freedom from crystalline defects requires careful preparation of the growth face of the substrates, from polishing to in-situ cleaning stage. The surface must be free from any damage, contamination, and inhomogeneities as they tend to produce surface defects. Reproducibility demands that a rigid procedure for preparation of substrates needs to be carefully followed. Many vendors now offer 'epi-ready' wafers and often they are better than those prepared in house. Normal cleaning procedure involves sequential degreasing of the wafer in trichloroethylene, acetone and methanol followed by rinsing in water. Finally, the substrate is chemically etched followed by a thorough rinse in water. Normal etchants for GaAs are $4H_2SO_4$:$1H_2O_2$:$1H_2O$ or 10 NH_4OH:$2H_2O_2$:$1H_2O$. InP can be etched in 0.1 - 1% Br_3-CH_3OH or in 10 H_2SO_4:$1H_2O_2$:$1H_2O$. Normally, suitable cleaning techniques are empirically optimised and can vary considerably, depending on the growth technique and the device application.

C LPE GROWTH OF InGaAs

N. Chand

C1 The Technique

LPE is the simplest, most widely used technique for growing lattice-matched InP/InGaAs(P) heterostructures. LPE has produced lasers and LEDs of highest quantum efficiencies and lifetimes. Many references discuss the technique in great detail [1-62]. For growth of $In_{0.53}Ga_{0.47}As$, the technique involves the dissolution of a known quantity of high purity InAs and GaAs along with the desired dopant in molten In at an elevated temperature to form a saturated solution. Slowly cooling the solution below the liquidus (equilibrium) temperature induces supersaturation, and if a suitable single crystal InP substrate is placed in contact with the solution epitaxial growth occurs. Various cooling cycles are used such as isothermal step cooling where the solution is cooled suddenly by several degrees before growth is initiated. A variation of this is where cooling continues after growth has started. At the end of the growth, the solution and the substrate are separated mechanically. To grow a multilayer structure, the substrate is sequentially brought into contact with different growth solutions in a multi-well boat. The crystal growth occurs at a solid-liquid interface which usually results in films with lower dislocation density.

A problem with using InP as a substrate is the high vapour pressure of P over this binary which results in thermal decomposition before growth. Steps taken to minimise this effect include covering the substrate with InP during exposure to elevated temperatures and removing the damaged layer using an in-situ etch back (undersaturated solution). After growth no special protection is necessary for $In_{0.53}Ga_{0.47}As$ layers during the cooling period because the surface is free from P and the ternary has a negligible As vapour pressure up to the growth temperature. Protection is, however, necessary for an InP layer under the graphite slider following the termination of growth. The maximum quantity of the solid which can be grown is simply the amount which would establish the equilibrium (the supersaturation) and is thus fundamentally limited by the thermodynamics and the total size of the system, i.e. the melt volume, the cooling range and the area of the seed on which growth occurs. The growth rate is limited by mass transport, which occurs by some combination of convection and diffusion. Further, the growth of a crystal also depends on the density of unpaired bonds on the substrate surface which in turn is related to the crystallographic orientation of the surface and also to its chemical and physical condition. Hence, the morphology and the growth rate are influenced by the condition and orientation of the substrate as well as by the transport processes within the solution.

The first LPE systems were of 'tipping' and 'dipping' types. Tipping used by Nelson [1] in 1963 involves rolling the solution onto the substrate to achieve solution substrate contact, whereas in dipping, first described by Rupprecht in 1967 [59], the substrate is dipped into the solution. The most commonly used method is due to Panish et al [12], in which the layers are sequentially grown by sliding the substrate into contact with different solutions either in a horizontal or in a vertical furnace.

C2 Selection of Growth Temperature

To grow high quality layers of required thickness, it is essential to select the growth temperature carefully [31]. The solubility of solutes and their diffusion coefficients in the In solution increase with temperature resulting in higher growth rate. The distribution coefficients of impurities (both unwanted and intentionally added) however decrease with temperature resulting in purer layers [26]. Unfortunately, growth at high temperature has many disadvantages, for example:

(a) it may be far from equilibrium resulting in increased crystalline defects;
(b) loss of volatile group V atoms from the solutions and the substrate;
(c) some unwanted impurities may dissolve in the solutions which may otherwise remain undissolved at low temperature.

Better control of layer thickness, reduction in the compositional grading of InGaAs and the strain at the heterojunction during cooling as well as better protection of the substrate from thermal degradation can be obtained if the growth is carried out between 600°C and 650°C.

C3 Solution Composition

To grow high quality lattice-matched layers of InGaAs on InP substrates, it is essential to know the correct solution composition that produces a lattice-matched condition at the growth temperature. Smooth surfaces are only possible within a narrow range of compositions. Moving away from this range produces an ever-increasing roughening of the surface until finally the layer fails to grow [27,29]. The distribution coefficients of the constituent elements and the growth rate in the InGaAs system are reported to be strongly affected by the crystallographic orientation of the substrate [3,19,21,30]. Consequently, different solution compositions are required to grow lattice-matched layers on substrates of different orientations. The conditions for equilibrium LPE growth of lattice-matched $In_{0.53}Ga_{0.47}As$ layers on InP substrates of different orientations have been determined by several workers from phase diagrams and lattice constant measurements [3,21-23,28-30]. The lattice mismatch is insensitive to the level of supersaturation of the melt solution if it is less than 10°C.

A solution consisting of $X^l_{Ga} = 0.0244$, $X^l_{In} = 0.9161$ and $X^l_{As} = 0.0595$ (where X^l_i represents the atomic fraction of an element i in the solution) has been used to grow lattice-matched layers of $In_{0.53}Ga_{0.47}As$ on (100) InP. These data were determined by Nakajima et al [18] and the solution just saturates at 650°C. These atomic fractions of the constituents are equivalent to 0.035040 g of GaAs and 0.065852 g of InAs per gram of In. $In_{0.53}Ga_{0.47}As$ layers of excellent morphology and electrical characteristics were consistently grown from the solutions of this composition. For growth of $In_{0.53}Ga_{0.47}As$ on (111)A InP, Nakajima et al [10] reported a solution comprising $X^l_{Ga} = 0.0255$, $X^l_{As} = 0.0587$ and $X^l_{In} = 0.9158$ for a liquidus temperature of 650°C.

C4 Purity of Layers

To grow high purity material, it is essential to bake the growth solutions in flowing purified H_2 for a long time period before loading the substrate and starting the growth cycle. This has been a subject of extensive study in many laboratories and several baking schemes have been reported [26,31-38]. Without baking, the residual impurity of LPE InP and InGaAs can be as high as 10^{17} cm^{-3}. By adopting different baking cycles, some of the authors have achieved net electron concentration below 10^{15} cm^{-3} or even low 10^{14} cm^{-3} for $In_{0.53}Ga_{0.47}As$ and InP. The impurity concentration in layers grown on the (111)B InP substrate is reported to be twice as high as on the (100) substrates, irrespective of the baking period [36].

By SIMS analysis [36], Si, S and O were detected as the main background donor impurities. Their concentration can be minimised by baking the growth solutions. Most workers attribute the dominant source of impurities to Si from reduction of silica by H_2 and the high value of distribution coefficient of Si ($\simeq$ 30) [23]. During baking, the Si in the solution is believed to be oxidised to SiO_2 by virtue of traces of H_2O in the flowing H_2 or solution itself, and is no longer incorporated as a donor in the epilayer. Attempts [25,26] have been made to introduce H_2O vapour intentionally into the H_2 gas stream at ppm levels to improve the purity but this made the results worse. A wet H_2 atmosphere is principally undesirable in LPE, since the surface of the solution may be oxidised and since oxygen induces a deep trap in InP. The decomposition of the substrate is also enhanced under wet H_2. It is reported [32,38] that by baking the solution in stages, the Si concentration can initially be reduced at a faster rate at high temperatures and finally be reduced to a lower value at lower temperatures, thereby reducing the overall required baking time.

The purity of layers grown from solutions baked at 700°C has been found higher than when the solution is baked at 900°C [36]. Kuphal and Fritzsche [31] have reported considerable improvement in purity by first baking pure In alone in vacuum followed by another bake of the complete solution under dry H_2. Baking of InP growth solutions presents a difficulty due to the relatively high vapour pressure of P over an In-rich solution. This decreases the liquidus temperature significantly. Baking In alone is not very effective. Any attempt to resaturate the solution after baking with InP defeats the purpose. The initial solution is, therefore, prepared for a higher liquidus temperature (T^1) such that the solution has the desired T^1 after baking.

To prevent S from entering the LPE system, Aylett and Haigh [35] redesigned their system, which improved the purity of their layers. The O_2 content is reduced by keeping the system rigorously leak tight and purging it with H_2 for sufficient time between loading and switching the furnace on, until the water vapour content measured by the hygrometer drops down to less than 1 ppm. Evacuating the system after loading also helps to reduce the O_2 level [34]. Lee et al [60] found that preliminary high temperature (~ 1600°C) baking of the graphite boat was important to produce InGaAs layers with a carrier concentration of (2 - 3) x 10^{14} cm^{-3}.

By simple baking of In with InAs and GaAs source materials over the weekend at 670°C, the author grew several layers of $In_{0.53}Ga_{0.47}As$ with background impurity concentration (N_D - N_A) of 3 - 4 x 10^{15} cm^{-3} at 300 K and mobility of 33,000 $cm^2V^{-1}s^{-1}$ at 77 K. Most of the studies of high purity growth of $In_{0.53}Ga_{0.47}As$ were done more than a decade ago. Since then, the purity of available starting materials has improved and it should now be easier to grow high purity layers.

C5 Dopants

For highest doping levels, Te and Zn are used for n-type and p-type dopants, respectively. Sn and Cd are the easiest n-type and p-type species for critical dopant control at low levels. The use of Zn and Cd as p-type dopants to grow p-n junctions is difficult because of problems with high volatility nature in the growth solution and high diffusion coefficients, resulting in displaced p-n junctions with graded interfaces [42-48]. By covering the solutions to prevent cross-contamination and by growing at minimum temperatures for minimum time, these effects can be minimised [44,45]. Chand et al [62] and others [53,54] have used Mn as the suitable p-dopant with low diffusivity.

C6 Disadvantages of LPE

The major disadvantages of LPE growth are [7,58]: small size of wafers resulting in less throughput; higher growth rate, making it difficult to grow very thin layers such as QWs with good reproducibility; lack of abrupt heterointerfaces; cross-contamination of solutions due to poor wipe-off or higher vapour pressure of dopants; and difficulty in achieving flat and featureless surface morphology of the epitaxial layers due to In inclusions, poor wipe-off, island growth, terracing, and meniscus lines. Nevertheless, many companies the world over still use LPE to manufacture high performance long wavelength lasers and light emitting diodes (LEDs). Modern LPE reactors used in production are very sophisticated and are highly automated, and are able to handle several wafers at a time. However, because of its inherent limitations, LPE is slowly being displaced by MOCVD.

D MBE GROWTH OF $In_{0.53}Ga_{0.47}As$

N. Chand

MBE is an epitaxial growth process involving the reaction of one or more thermal beams of atoms or molecules with a crystalline heated surface under ultrahigh vacuum conditions. The beams are thermally generated in Knudson-type effusion cells which contain In, Al, Ga and As as the constituent elements of the epitaxial films, and elemental Si and Be for the n- and p-type dopants, respectively. The temperature of the cells is accurately controlled within ±0.25°C to give the thermal beams of appropriate intensity. The thermal beams escaping from orifices in the cells travel in rectilinear paths to the substrate where they condense and grow under kinetically controlled growth conditions. Extensive use of cryogenic panels in the chamber reduces the water vapour and hydrocarbon content of the vacuum. For speed of operation and vacuum cleanliness, an air-lock is used for rapid sample loading. Analysing techniques such as Auger and mass spectroscopy and electron diffraction enable surface characterisation to be carried out in-situ. The MBE technique lends itself to precise doping and compositional control and, because of the slow growth rate (typically ~1 $\mu m\ hr^{-1}$), extremely thin layers can be grown with high reproducibility. The technique has the ability to provide abrupt initiation or cessation of growth and smoothing of the surface of the growing crystal down to atomic steps. However, MBE has a problem of growing phosphorus-bearing compounds because of the high vapour pressure of P and the resulting difficulty with the pumping system. The growth of ternary $In_{0.53}Ga_{0.47}As$ by MBE poses no such problem. $In_{0.52}Al_{0.48}As$ is a wide bandgap substitute material for InP which is lattice-matched to InP and can be grown relatively easily. There are a number of excellent review articles and books which discuss various aspects of MBE in great detail [63-75].

Growth of $In_{0.53}Ga_{0.47}As$ and $In_{0.52}Ga_{0.48}As$ was pioneered by K.Y. Cheng, K. Alavi and A.Y. Cho [76-79]. A freshly etched InP substrate with a thin protective oxide is loaded into the MBE system through an air-lock where the substrate is first outgassed at 300 - 400°C for ~30 min. The cells are properly outgassed before commencing the growth. At the same time, the substrate is heated under the exposure of an As_4 beam in the growth chamber to desorb the surface oxide, which for InP occurs between 500 and 530°C and is monitored by in-situ reflection high-energy electron diffraction (RHEED). As discussed by Cheng and coworkers [76-78], arsenic has the lowest free energy of formation to form oxides compared with In and P at temperatures above 450°C. When As molecules are evaporated onto the heated InP substrate, the passivated surface oxides are gradually replaced by arsenic oxides which then vaporise at around 460°C. Under an As_4 vapour pressure of ~1.5×10^{-6} torr (1.24×10^{14} As_4/cm^2s), the InP substrate surface can be cleaned without thermal damage at temperatures as high as 530°C. As the surface becomes cleaner, the RHEED pattern shows streaks with increasing intensity and decreasing background scattering. When the surface is atomically flat and free of contamination, the 1/2 or 1/4 order reconstruction pattern as well as Kikuchi lines are observed. After the oxide desorption, the substrate is ready for growth which is started by opening the necessary cell shutters. The growth is also monitored by in-situ RHEED. The growth is carried out under excess arsenic flux and the growth rate is determined by the arrival rate of Ga and In atoms. The maximum growth temperature (T_S) of InGaAs is limited by the incongruent evaporation of Ga, In and As from the surface which occurs around 580°C. The T_S is normally between 500 and 530°C.

Two main challenges in the growth of InGaAs are: (1) different thermal stability of the binary end members (InAs and GaAs) of the InGaAs alloy; and (2) achieving lattice matching over a large substrate area due to (1) above and to spatial variations of the In and Ga arrival rates. At $T_S > 300°C$, the rate of loss of As from InAs is much higher than from GaAs. To prevent (1), relatively high $J_{As_4}/(J_{In} + J_{Ga})$ flux ratios (approximately 5/1) are used. To achieve lattice matching, a rotating sample holder with a rotation speed ≥ 3 rpm is used, and the In and Ga effusion cells are aligned with the centre of the substrate and placed adjacent to each other. The composition of the epilayer is then controlled by adjusting the In and Ga oven temperatures individually. At the onset of growth, lattice mismatch between the epitaxial layer and the InP substrate is monitored with RHEED. X-ray diffraction is used to determine the composition of the epilayers upon the completion of the growth.

If the lattice mismatch between the substrate and the epilayer is of the order of 10^{-4}, each streak in the original RHEED pattern changes into several sections upon starting the growth and indicates slight faceting in the epilayer [76]. As the growth proceeds for the first few minutes, these broken sections recover into perfect streak patterns with one-half or one-quarter order reconstruction. If the lattice mismatch ($\Delta a/a$) between the epilayer and the substrate is in the range of 10^{-3}, two distinct RHEED pattern variations are observed during the growth process. For epilayers with a smaller lattice constant than the substrates, the RHEED patterns change from streaks into dots instantly after the growth starts. These dots gradually elongate into broken sections as growth proceeds, but they never recover into perfect streaks. On the other hand, if the lattice constant of the epilayer is larger than the substrate, a spotty RHEED pattern is first observed at the beginning of the growth, and rapidly recovers into streaks with one-half or one-quarter order modulation. For epilayers with larger lattice constants than the substrate, a cross-hatch pattern always exists. For lattice mismatch in the range of $\pm 10^{-4}$, the final surface is featureless. If the epilayer has a smaller lattice constant than the substrate, the epilayer is under tension and shows a rough appearance.

If the substrate surface is not completely clean at the moment the growth is started, the morphology of the grown layer always shows an arrow-head-like structure extruding along the [110] direction. Under this circumstance, even in the lattice-matched case, the RHEED pattern is fuzzy and spotty during the initial stage of the growth. This pattern is gradually transformed into streaks if the electron beam is incident to the crystal surface along the direction of extrusions. Otherwise, the final RHEED pattern shows diffused streaks.

E MBE GROWTH OF STRAINED $In_xGa_{1-x}As$ ON GaAs SUBSTRATES

N. Chand

The pseudomorphic $In_xGa_{1-x}As$ (x = 0.15 - 0.25) layers on GaAs substrate have been used extensively as active regions of MODFETs and QW lasers. To fully utilise the advantages provided by strained layers, one must be able to deposit the pseudomorphic heterostructures with very high quality and atomically smooth interfaces. The central issues regarding the growth of strained $In_xGa_{1-x}As$ involve the understanding of critical layer thickness as a function of mismatch (or strain) and the optimisation of the growth process. High quality strained layers can be grown provided that their thicknesses are below the critical layer thicknesses above which the strain is relieved by formation of misfit dislocations, thereby degrading the quality of the epitaxial layers. Curves of the critical layer thicknesses of InGaAs versus In composition can be found elsewhere [80-89,100].

Growth of strained $In_xGa_{1-x}As$ layers on GaAs has been studied and optimised by RHEED oscillations in many laboratories [90,91]. These studies were done as a function of x, T_S, and As_4/Ga beam equivalent pressure ratio, R. Weng and coworkers [91] find that in contrast to GaAs, the growth of InGaAs on GaAs causes a sharp reduction in the overall RHEED intensity, although the oscillations remain discernible even after the critical thickness is reached. The RHEED intensity recovers slightly after both In and Ga shutters are closed. Nevertheless, the recovery is worse than that of GaAs, and is much more subdued at lower T_S (520°C) than at higher T_S (580°C). Subsequent growth of GaAs layers on this InGaAs surface greatly increases the RHEED intensity, with the most significant change occurring at lower T_S (520°C). From both InGaAs and GaAs RHEED oscillations, these workers also determine the In content of the InGaAs layer as a function of T_S and R. They find that the InAs growth rate (at a fixed In temperature) is independent of R and remains fairly constant for $T_S \leq 560°C$. However, it drops almost linearly with increasing $T_S \geq 560°C$, with a dropping rate of approximately 1% per 1°C, due to In desorption. The activation energy of In desorption is found to be independent of both arsenic-to-total-group III flux ratio and of composition. A value of 1.88 ± 0.15 eV is obtained for this energy.

Besides In desorption or re-evaporation [92-95], surface segregation of In also occurs as seen with AES by Radulescu and coworkers [92] at about $T_S = 590°C$. In atoms are on the surface of the as-grown InGaAs layer and not in the 150 Å QW, as is intended, due to surface segregation. These workers conclude that In re-evaporation and surface segregation are insignificant at $T_S = 480 - 540°C$. Furthermore, insignificant incorporation and segregation of In are observed for the smaller InAs contents (x = 0.15, 0.07). As a result, they observe that the magnitude of the In flux plays a big role in determining the segregation process of In during MBE growth of $In_xGa_{1-x}As$ on GaAs, with the arrival rate for x = 0.3 being significantly larger than the re-evaporation rate. It is also reported that for growth of $In_{0.25}Ga_{0.75}As$ at $T_S \leq 570°C$, a transition from two-dimensional to three-dimensional growth is observed at a certain thickness of InGaAs. The thickness is largest for QWs grown at lower T_S and decreases as the T_S increases. This transition from 2D to 3D growth is not observed at high temperatures (> 570°C) where an In-induced (4x2) surface reconstruction exists.

Chand and coworkers [95] report that in clean growth conditions, the optical properties of InGaAs QWs are insensitive to the As beam used (As_4 versus As_2) and the luminescence intensity is independent of T_S in the 570°C - 630°C range. The post growth thermal annealing also does not have any effect on the luminescence intensity of InGaAs QWs. Thus, the growth of InGaAs QWs at $T_S > 570°C$ does not improve the optical properties but significantly reduces the In sticking coefficient. To control the emission wavelength, one needs to precisely control the source fluxes and T_S. If the growth is carried out at a temperature where the In sticking coefficient < 1, uniform heating of the substrate is essential [96].

From the study of InGaAs QW lasers, several groups have found that the quality of InGaAs/AlGaAs heterostructures improves if a thin layer of GaAs is sandwiched between InGaAs and AlGaAs [96-99]. Besides reducing the growth temperature to 550°C - 570°C and keeping the width of InGaAs QWs below the critical thickness, MBE growth of strained InGaAs is similar to that of GaAs. For a Ga beam equivalent pressure (BEP) of 5 x 10^{-7} torr, In BEP should be roughly 2 x 10^{-7} torr for the growth of $In_{0.2}Ga_{0.8}As$ at a growth rate of 0.6 μm/hr. A higher V/III BEP ratio (about 20) is required for growing InGaAs than for growing GaAs, to increase the sticking coefficient of In.

F CBE AND GSMBE GROWTH OF InGaAs

N. Chand

F1 Introduction

CBE and GSMBE [101-105] can be viewed as evolutions from hydride VPE, OMVPE, and MBE. In gas source MBE (GSMBE), the elemental As and P sources are replaced by AsH_3 and PH_3, respectively. If the group III species are also delivered to the substrate using gas sources, organometallics group III species, the resulting technique is called chemical beam epitaxy (CBE), invented by P. Balk et al [101] and made into a major growth technique by W.T. Tsang [102]. The replacement of As and P elemental sources with the hydrides, AsH_3 and PH_3, first reported by Panish in 1980 [106], was deemed to be necessary because III-V compounds containing P, and particularly both As and P, are not among those for which MBE with elemental sources has shown significant accomplishments. Precise control of the intensity of a molecular beam originating from the evaporation of elemental P from an effusion oven has been difficult, presumably because of the presence of allotropic forms of phosphorus with different vapour pressures, and its exponential pressure dependence on temperature which is exacerbated by the low vaporisation temperature required (about 200°C). Furthermore, P_4 is the predominant vaporising species when solid P is evaporated. A very low accommodation coefficient for P_4 on the growing III-V surface may be inferred from the work of Asahi et al [107] and of Foxon et al [108]. Although P_2 molecules can readily be used for epitaxy, their generation by thermally cracking P_4 derived from the solid does not eliminate the underlying difficulties with control over rate of vaporisation cited above, and has added control problems because of the proximity of the relatively low temperature vaporisation region and the high temperature (about 900°C) thermal cracking region. In CBE and GSMBE, the As_2 and P_2 beams are generated by thermal decomposition of the hydrides. By replacing the group III elemental sources with simple organometallic compounds, Tsang et al [109] grew InGaAs with lower net carrier concentrations than were previously attainable by GSMBE. A major advantage of the use of the organic compounds is the elimination of high temperature effusion cells for beam generation and the accompanying problems in beam flux control.

F2 Thermal Cracker

The innovative and novel feature of CBE and GSMBE is the AsH_3 and PH_3 thermal cracker. These hydrides negligibly crack at the substrate surface and a thermal cracker is used to complete their decomposition to yield molecular As_2 and P_2 beams and to prevent degradation of the vacuum system due to accumulation of the liquid hydrides on cryopanels. This demands a cracking efficiency of 99.9+%. Two versions of thermal crackers have evolved [101-106], both designed to yield very efficient cracking of the hydrides in spite of the vacuum environment of MBE. The earliest [106] was a high pressure cracker that permitted the thermal decomposition to occur at pressures of hundreds of torr. The reaction products (group V tetramers and H_2) were then allowed to leak, while hot, into a low pressure region where further thermal cracking of the tetramers to dimers occurred. The second type of cracker, which was developed after Calawa [110] demonstrated that a heated Ta wire could be used to cause the decomposition, is simply a BN tube packed with crushed Ta foil topped with BN chips. The Ta apparently catalyses the decomposition [111,112] of the hydrides to the elements. The bed of crushed BN following the Ta is used to prevent direct exposure of the substrate surface to the heated metal. It is possible that the Ta catalyst may be omitted if sufficient inert surface area can be provided. Panish [104] shows that the use of beds with very large surface area is not necessary with Ta as a catalyst, and that the Ta is stable for long term use.

F3 Pumping Requirements

The pressure regimes for CBE and GSMBE are determined by the criterion that beam epitaxy occurs. If the occurrence of molecular flow is used to define beam epitaxy, then a pressure of about 10^{-3} torr is the maximum possible. The condition that the mean free path be greater than the source to substrate distance also leads to a maximum pressure of 10^{-3} torr. The flow conditions between the source and substrate are dictated only by the geometry of the sources themselves, and their arrangement in space relative to the substrate. When there is beam epitaxy, a boundary layer near the substrate is absent and the geometry of the surrounding growth chamber does not influence the vapour flows as in OMVPE. These differences from OMVPE imply much greater simplicity of system design for achieving uniform growth and for scale up. However, the loss of the boundary layer may also mean the loss of useful boundary layer reactions that limit carbon incorporation.

Some idea of the pertinent conditions during GSMBE of $In_{0.53}Ga_{0.47}As$ can be obtained [104] by considering typical pressure and flux conditions at the substrate. At a growth rate of 1 μm/hr, the impingement rate of In atoms, assuming no re-evaporation, is about 5.5 x 10^{14} atoms/cm^{2}s. The required impingement rate, assuming complete reaction, of the group V dimer is half that amount. Allowing for a factor of 2 safety margin to provide sufficient As_2 flux to prevent decomposition of the InGaAs gives an As_2/(In,Ga) flux ratio of 1. Further, for the 1 μm/hr growth rate, with a typical MBE source to substrate distance of 12.5 cm, about 4 x 10^{-7} mole H_2/s is generated by the hydride decomposition. To maintain a background H_2 pressure of 10^{-4} torr, a pumping speed of about 70 l/s is needed in the region of the source and the substrate. This translates into pumping speeds of from about 200 to 2000 l/s at the pump, depending upon the geometry of the growth and pumping systems. Under these conditions the impingement rate of H_2 on the substrate surface is about 1000 times that of In and Ga atoms or As_2 molecules. Of course, the addition of H_2 as a carrier gas for the organometallics, increased growth rates, and the high As_2 flux required to prevent surface decomposition resulting from too high a substrate temperature, can also contribute to the H_2 load. This means that the upper limit of the substrate temperature is reached when the pumping speed of the vacuum system is reached.

F4 Group III Alkyls

In CBE [101], the beam of group III alkyl molecules impinges directly (i.e. line-of-sight) onto the heated substrate surface as in a conventional MBE process. A blend of organometallic compounds is fed to a single gas source which is maintained at ~ 50°C. The organometallics are not decomposed until they are pyrolysed on the heated substrate surface. The temperature of the source need only be high enough to prevent condensation of the organometallics. These would otherwise condense on the walls because cryopanels cooled by LN_2 are positioned nearby. Since the organometallics are not pyrolysed before they impinge on the heated substrate, the chemical reactions at the growing surface in CBE are considerably more complex than in GSMBE, and since there is no boundary layer, the chemistry is different from that in MOCVD. There is no boundary layer in front of the substrate surface nor are there gas-phase molecular collisions on the path because of the long mean-free-path of the molecules at a pressure of $< 5 \times 10^{-4}$ torr. Thus, after a group III alkyl molecule strikes at the substrate surface, it can either acquire enough thermal energy from the heated substrate and dissociate all three alkyl radicals leaving the elemental group III atom on the surface, or re-evaporate undissociated or partially dissociated. The probability of whichever process occurs depends on the substrate temperature and arrival rate of the organometallics. Thus, at a high enough substrate temperature the growth rate is determined by the arrival rate of the group III alkyls, while at lower substrate temperatures the growth rate is limited by the rate of surface pyrolysis. In all growth processes the group V material is always supplied in excess.

Mass flow controllers are usually used in a manner similar to that employed in OMVPE. For the organometallics, the source is at a controlled temperature and a carrier gas (usually H_2) is bubbled through it at a fixed pressure. The mass flow control is of the combined gas stream. For the hydrides, the mass flow controller directly controls the flow of the pure gases into the thermal cracker. The amount of flow in GSMBE is typically several orders of magnitude lower than in OMVPE.

All the commonly used group III organometallics except trimethyl indium are liquids at room temperature with vapour pressures below ~150 torr and are conveniently fed to a CBE gas source with H_2 carrier gas. The quantity of H_2 carrier gas introduced into the vacuum system is small compared to the quantity generated from the decomposition of PH_3 and organometallics and does not significantly add to the pumping load. An organometallic compound is kept in a cylinder immersed in a thermal bath which holds the temperature stable to better than ± 0.05°C. In this way the partial pressure of the organometallic inside the cylinder is kept constant. Experimentally it is found that only when a newly filled cylinder is used or when a cylinder is almost empty does the group III flux vary.

In the case of $Ga_{0.47}In_{0.53}As$/InP heterostructures [104], the most natural sequence of molecular beam changes when going from the InP to the ternary layer requires that the InP layer be stopped by turning off the In beam before the P_2 beam, and then switching from P_2 to As_2 and starting the $Ga_{0.47}In_{0.53}As$ growth by turning on In and Ga beams. Thus, barring any interchange between P and As, the InP layer is terminated with a layer of P atoms. Similarly, the $Ga_{0.47}In_{0.53}As$ layer is expected to be terminated with As atoms. The result of this growth procedure is that the interfaces between the $Ga_{0.47}In_{0.53}As$ and the InP on each side of the quantum well are different, with the leading side having, ideally, one atomic bilayer strained positively and the trailing side with one atomic bilayer strained an equal amount negatively. In reality, however, in both OMVPE and CBE, As incorporates into the layer much more efficiently than P. Therefore, the leading side and trailing side interfaces will not necessarily have an equal amount of positive and negative stress.

F5 Purity

Tsang [102] has reported growth of very high quality epitaxial layers of InGaAs, InP and InGaAsP by CBE. $In_{0.53}Ga_{0.47}As$ was reproducibly grown by several laboratories having the (004) X-ray Bragg reflection linewidth as narrow as 17 arc sec for ~2 μm thick layers. Such extreme composition uniformity was also supported by results from Auger depth profiles and 2 K photoluminescence measurements. Very intense efficient luminescence peaks due to excitonic transitions with FWHM as narrow as 1.2 meV were obtained. This represents the narrowest linewidth ever reported for GaInAs grown. Hall measurements of 2 - 5 μm thick $In_{0.53}Ga_{0.47}As$ layers grown on InP substrates had electron mobilities of 10,000 - 12,000 and 40,000 - 67,000 cm^2/V s at 300 and 77 K respectively with n in the range 5×10^{14} - 5×10^{15} cm^{-3}.

F6 Doping Studies

Minority carrier devices such as the $Ga_{0.47}In_{0.53}As_xP_{1-x}$/InP lasers are particularly interesting from a materials point of view in that they present crystal growth challenges involving the need for high quality material and p-n junctions that are nearly ideal. One such structure that also requires layers that are very thin and very highly doped adjacent to low doped regions is the single-heterostructure bipolar transistor. What is unusual are the doping levels in the heavily n-type collector contact, and the heavily p-type base layers. It is extremely important that the base-collector and the base-emitter junctions be abrupt, and that the base-emitter junction be at the heterojunction between the $Ga_xIn_{1-x}As$ base and the InP emitter, and that the heavy doping in the collector contact layer does not carry over into the lightly n-type collector.

The commonly available dopants for MBE of III-V compounds are generally Sn and Si for n-type material and Be for p-type material. Extensive experience with Sn in GSMBE and MBE [104,113] suggests that although the upper doping level with GaAs, InP and $Ga_{0.47}In_{0.53}As$ is at least in the mid 10^{19} cm^{-3} range, sharp doping interfaces are not attainable with the In containing compounds. Some Sn appears to 'float' on the growing surface and cause unintentional doping of subsequent layers. The cap layer on the emitter can thus be doped with Sn. The Si and Be dopants are more interesting for other layers. At the GSMBE growth temperature of 500 - 525°C, the Be doping of $Ga_{0.47}In_{0.53}As$ saturates at about 3×10^{19} cm^{-3} with a fall off of mobility and carrier concentration, and degradation in surface morphology when more Be is added. All of this is symptomatic of having reached the solubility limit. However, reduction of the growth temperature by 100°C, while maintaining the same growth rate, permits the achievement of $p = 2.5 \times 10^{20}$ cm^{-3} while still growing high quality material [114]. It is doubtful that the solubility is increasing with decreasing temperature. More likely, the decreased mobility of the Be atoms on the surface provides a kinetic limitation to the growth of Be crystallites. The Be atoms must have sufficient mobility to reach group III lattice sites, but essentially be frozen into those sites as the crystal grows. It is most interesting that, under the conditions of these experiments, the crystal and p-n junction quality remain high until they very abruptly degrade at the upper doping limit. Most recently, Ritter and coworkers [115] found that the incorporation efficiency of Be in $In_{0.53}Ga_{0.47}As$ and InP during MOMBE was much lower than unity. The effect is attributed to the removal of Be atoms from the surface by the organic gases in the growth chamber. The upper limit for delta doping of $In_{0.53}Ga_{0.47}As$ before the onset of diffusion is 7×10^{12} cm^{-2}, and in InP it is $< 5 \times 10^{11}$ cm^{-2}.

Impurities doping from vapour sources is more advantageous over solid sources in CBE. Weyers and coworkers [116] have carried out a rather extensive study surveying the suitability of various vapour sources as n- and p-type dopants in GaAs and InP. The following dopant sources were studied by them: silane (SiH_4), disilane (Si_2H_6), triethylsilane (TESiH, $(C_2H_5)_3SiH$), diethyltelluride (DETe, $Te(C_2H_5)_2$) and tetraethyltin (TESn, $Sn(C_2H_5)_4$) as n-type dopants, biscyclopenta-dienylmagnesium (Cp_2Mg, $Mg(C_5H_5)_2$), diethylzinc (DEZn, $Zn(C_2H_5)_2$) and diethylberyllium (DEBe, $Be(C_2H_5)_2$) as p-type dopants. These authors show that the use of gaseous sources is a viable approach to doping during MOMBE growth although they experienced problems with almost all sources studied. The main criteria for the suitability of a dopant source are: the decomposition behaviour on the substrate surface; the vapour pressure of the dopant sources which these authors suggest should preferably be at least 1 hPa at room temperature; the purity of the source; and the behaviour of the dopant element itself after the complete decomposition of the source molecule, which includes surface segregation, diffusion and desorption from the hot substrate surface.

Carbon is becoming a very important p-type dopant in GaAs [102,105,117-122]. Its incorporation is particularly convenient in CBE and MOMBE growth [102,105]. It can be incorporated by simply using some TMGa flow and adjusting the growth conditions properly during growth of GaAs using TEGa. Konagai and coworkers reported [118] that carbon concentrations up to 10^{21} cm^{-3} have been obtained and they are electrically active. It is also confirmed that C does not segregate or diffuse. Lyon et al [120] have studied CBr_4 and $CHBr_3$ as C doping sources for GaAs by GSMBE with elemental Ga and thermally cracked AsH_3. Hole densities in excess of 1×10^{20} cm^{-3} were measured in both CBr_4- and $CHBr_3$-doped GaAs, with 100% electrical activity of the incorporated C. The efficiencies of C incorporation from CBr_4 and $CHBr_3$ were 750 and 25 times that of TMG, and were less sensitive to substrate growth temperature and V/III ratio. Chin et al [121] have used CCl_4 to grow highly C-doped p-type $Ga_{0.5}In_{0.5}As$ and $Ga_{0.5}In_{0.5}P$ by GSMBE. Growth temperatures slightly below conventional values were used to increase the C incorporation, and a short-duration post-growth anneal near the growth temperature was necessary to obtain the highest hole concentrations, which were $p = 3 \times 10^{19}$ cm^{-3} for $Ga_{0.5}In_{0.5}As$ and $p = 5 \times 10^{18}$ cm^{-3} for $Ga_{0.5}In_{0.5}P$. Reversible acceptor passivation from hydrogen species in the growth environment was the plausible explanation for the annealing behaviour. Similar observations were made by Stockman et al [122] during the growth of p-$In_xGa_{1-x}As$ $(0 < x \leq 0.53)$ by OMVPE using CCl_4 as the C source. They obtained p-type induction over the entire composition range with $p > 1 \times 10^{20}$ cm^{-3} for $x < 0.12$ and $\leq 1 \times 10^{19}$ cm^{-3} for $In_{0.53}Ga_{0.47}As$. Due to surface reactions (etching) involving Cl-containing compounds during growth, the alloy composition depended on CCl_4 partial pressure, V/III ratio, and growth temperature.

F7 Alternative Sources

The hazard posed by the group V hydrides, AsH_3 and PH_3, together with the desire for precursor molecules producing higher quality layers has motivated the development of new, metalorganic group V and even group III precursor molecules. Stringfellow [123] gives a very thorough survey. For group III alternative sources, triisobutyl-gallium and -aluminium (TIBGa and TIBAl), and alane ($AlH_3(CH_3)_3N$) are very promising sources for further reducing the carbon residual incorporation problem in the CBE and MOMBE growth of GaAs and AlGaAs. As for group V alternative sources, triisobutyl-arsine and -phosphine, and phenyl-arsine appear to be quite promising as replacements for the hydrides. When phenyl-arsine is used, no pre-cracking is needed. Excellent InGaAs/InP quantum well results were obtained by Ritter et al [124] using triisobutyl-arsine and -phosphine.

F8 Laser-Assisted CBE and ALE

Laser-assisted CBE and atomic layer epitaxy (ALE) has been studied employing a multi-line Ar^+ laser and RHEED technique by Doi et al [125]. Two major findings are: (1) the decomposition rate of TEGa is enhanced to a greater extent on an As surface in comparison with that on a Ga surface under Ar^+ laser irradiation, (2) the laser irradiation also enhances the surface migration of the absorbed surface molecules. Chiu et al [126] also conducted atomic layer epitaxy of GaAs employing CBE using TMGa. The one-monolayer self-limiting process is clearly observed using RHEED intensity oscillations.

G CHLORIDE VAPOUR PHASE EPITAXY TECHNIQUES

R.F. Karlicek Jr.

The growth of InGaAs(P) lattice-matched to InP can be accomplished by two types of chloride VPE: trichloride VPE and hydride VPE. Since both trichloride and hydride VPE involve transport of the group III species as the volatile monochlorides, these two techniques can generally be called chloride VPE, though that name is usually chosen to describe the trichloride technique. The techniques used depend on the device application required. For example, photodetector applications requiring high purity frequently use trichloride VPE, while devices requiring a range of alloy compositions in the InGaAsP system typically use hydride VPE.

Chloride VPE techniques have been in use for more than 25 years for the growth of GaAs, InP and various ternary and quaternary alloys and have been reviewed in recent literature [127,128]. While commercial chloride reactor systems have been available, these systems were usually constructed by the crystal grower for specific applications and a wide variety of custom designs have been explored [129]. Chloride VPE is not really useful for the growth of alloys containing aluminium because the AlCl used to transport aluminium reacts with the quartz reactor vessel. This limitation has probably played a role in reducing the importance of chloride VPE given the technological importance of Al containing III-V compounds.

Trichloride and hydride VPE involve essentially identical chemical processes for epitaxy and differ in the starting materials used. Growth results from the reaction of indium and gallium monochlorides (InCl, GaCl) and $As_4(P_4)$. The growth process is usually conducted at atmospheric pressure in a heated quartz reactor vessel consisting of three regions: growth region, mixing region and source region. In trichloride VPE, the source section consists of separated regions of polycrystalline InP, InAs and GaAs at a temperature higher than the growth region. This source material is transported to the growth region by HCl generated by the pyrolysis of $AsCl_3$ directly over the solid sources, forming the group III monochlorides and $As_4(P_4)$. These reactors can also include a separate region for pyrolysing $AsCl_3$ and condensing the arsenic in a trap outside the heated reactor in order to generate high purity HCl for use in controlling reactor growth (or in-situ etching) conditions.

In hydride VPE, the separate source regions contain only molten indium and gallium which react with HCl (from a compressed gas cylinder) to form the volatile monochlorides and a tube for introducing AsH_3 and PH_3 which pyrolyses to form the group V species. An alternative approach to generate the group III monochlorides involves the simultaneous introduction of HCl and organometallic indium and gallium compounds (as in OMVPE) which react in the heated reactor source region to form the metal chlorides and methane [130]. In hydride VPE, the pyrolysis of PH_3 is slow (relative to AsH_3) at the temperatures and flow rates typically used for growth so that, in addition to P_4, unpyrolysed PH_3 and P_2 can play an important role in the growth process [131].

Dopant sources for chloride VPE typically involve the use of Zn supplied from a heated elemental Zn or from a dilute source of dimethylzinc for p-type doping, and S, Si, Te or Se supplied as diluted hydrides for the n-type doping. In trichloride VPE, $SiCl_4$ is also used as the Si source. Semi-insulating InP has also been grown using Fe (supplied as the dichloride) as a dopant by both trichloride [132] and hydride [133] VPE, but because the volatile dichloride is unstable in the presence of H_2, InP:Fe growth by chloride VPE requires the use of either N_2 or an inert gas as the carrier gas.

Trichloride VPE is preferred for the routine growth of high purity InP, GaAs and InGaAs because of the ready availability of high purity $AsCl_3$ and PCl_3. While all of the growth techniques described here can be used to grow very high purity InP, GaAs and InGaAs, trichloride VPE grows the highest purity material most consistently and with the least amount of effort under manufacturing conditions. A background carrier concentration of $< 1 \times 10^{15}$ cm^{-3} and 77 K Hall mobility of 85,000 $cm^2/V\,s$ in InGaAs has been achieved for InP/InGaAs/InP structures used for detector applications [134]. Because the purity of material grown by trichloride VPE is consistently so high, this technique is primarily used for the growth of optoelectronic devices like PIN diodes and avalanche photodiodes (APDs). This technique is also selected when safety restrictions preclude the use of the much more toxic AsH_3 and PH_3. Because InP readily undergoes thermal decomposition at the temperatures typically used for growth (650°C to 720°C) and trichloride reactors have no source of phosphorus over pressure without the simultaneous production of HCl, the preservation of InP substrates prior to growth can be difficult. This problem can be addressed by the use of removable wafer covers or by growing a thin layer of polycrystalline InP which is etched using HCl before growth of the desired layer [135].

Hydride VPE is preferred over trichloride VPE for reactor systems used to grow quaternary alloys (InGaAsP) because the separate introduction of group III and group V species makes it easier to grow the entire range of InGaAs(P) alloy compositions lattice-matched to InP. For this reason, hydride VPE has been applied primarily to the growth of long wavelength LED and laser structures using quaternary alloys as the active (light emitting) layers [136-138]. Although the purity of layers grown by hydride VPE depends strongly on the purity of the source gases (HCl, AsH_3 and PH_3) and is not routinely capable of the purity levels of trichloride VPE, hydride VPE has also been used for the commercial production of InGaAs PIN diodes.

The large reactor volumes and source chemistry associated with chloride VPE makes the growth of quantum layer structures difficult, primarily because the monochlorides are soluble in the molten In and Ga sources and fast changes in the gas composition are difficult to achieve. These difficulties can, however, be circumvented with the use of the organometallic In(Ga)/HCl approach described above [139]. Numerous clever reactor designs involving the use of multiple redundant source regions and several growth regions have, however, been used to grow quantum well structures [140,141]. Because other growth techniques such as OMVPE and MBE and its variants are much better suited for these types of structures, chloride VPE is not typically used in such applications.

Chloride VPE, like LPE, is based on reversible chemical reactions producing epitaxial growth at near equilibrium conditions. Much effort has been spent on thermodynamic modelling of chloride VPE as a means to optimise the growth process and to guide the determination of the gas phase stoichiometry required to grow appropriate alloy compositions [142,143]. Most growth for device applications is performed on nominally <100> oriented substrates (occasionally misoriented by a few degrees to improve the growth morphology). Because chloride VPE is typically performed at temperatures below the regime where thermodynamic equilibrium alone determines the growth rate, the kinetics of the surface reactions involved in epitaxy produce a large growth rate anisotropy. The role of the surface in the epitaxial growth process also makes chloride VPE a superior technique for selective area epitaxy [144,145].

H OMVPE

R.F. Karlicek Jr.

OMVPE was developed in the late 1960s for the growth of a wide variety of compound semiconductors [146] and has been reviewed in recent monographs [147,148]. Like MBE and its variants, OMVPE is a non-equilibrium growth process and OMVPE reactors can grow essentially all III-V binary and alloy compositions. OMVPE reactors are commercially available from a number of vendors. Unlike most LPE and chloride VPE reactors, OMVPE reactors are available for growth on multiple wafers at one time for high throughput production of a wide variety of device structures.

OMVPE is performed in cold wall quartz or stainless steel reactors using a heated susceptor containing the substrates. Heating techniques usually involve radio frequency inductive heating or radiative heating with infrared tungsten-halogen lamps. Group III organometallic compounds such as trimethylindium or trimethylgallium are most commonly used along with AsH_3 and PH_3 for the growth of ternary and quaternary alloys lattice-matched to InP or GaAs. The metal alkyl compounds are supplied from small (< 1 litre) stainless steel vessels immersed in a temperature controlled bath which controls the vapour pressure of the source. The group V hydrides are obtained from compressed gas cylinders. These compounds undergo pyrolysis directly above the heated substrate resulting in epitaxial growth. Because pyrolysis of these compounds usually occurs at temperatures below those used for high quality epitaxial growth, OMVPE growth is usually performed in a temperature regime where the growth is mass transport limited and the growth rate is relatively insensitive to temperature and substrate orientation.

Dopant sources are most usually H_2S or SiH_4 for n-type and dimethyl- or diethylzinc for p-type. Semi-insulating InP (10^8 Ω-cm typically) can also be grown by OMVPE using Fe (from ferrocene) as a dopant. InP:Fe has many applications in the fabrication of optoelectronic devices where current blocking layers are required for good device performance (as in lasers, for example) and OMVPE is the best technique for the growth of semi-insulating InP. A wide variety of dopants for use in OMVPE have been studied and a detailed study of OMVPE doping issues has been published [149].

A considerable amount of work on alternate source chemistries for OMVPE has been performed, particularly to develop replacements for the toxic hydrides. In the growth of compounds containing phosphorus, the slow pyrolysis of PH_3 relative to AsH_3 and the organometallic compounds requires the use of very large V/III ratios in the feed gas composition for good surface morphology (values of 100:1 to 300:1 are typical of low pressure OMVPE). The resulting high PH_3 consumption rate is inconvenient particularly with regard to exhaust gas treatment. Tertiarybutylphosphine (TBP), a liquid source, is becoming more popular as a PH_3 replacement because it pyrolyses more rapidly at lower temperatures and is less toxic [150]. Since the thermal decomposition characteristics of TBP and AsH_3 are similar, this combination has been shown to produce epitaxial layers with greater compositional uniformity than is typically obtained with PH_3/AsH_3 mixtures [151].

Alternative chemistries have also been investigated for the organometallic compounds for a variety of reasons. Diethyl gallium chloride has been used in the growth of GaAs by OMVPE with improved interface properties and improved growth selectivity [152]. Since TMI is a solid at temperatures used for OMVPE, reproducible sublimation, crucial to reproducible growth characteristics, is not always easily obtained. Ethyldimethylindium has been considered as a possible liquid replacement but it is unstable with respect to the formation of different indium alkyl compounds [153]. Most work to improve the stability of indium delivery to the reactor has focused on using TMI in improved container designs [154].

As in the case of chloride VPE reactors, many different OMVPE custom reactor designs were studied before well-designed OMVPE reactors were commercially available. Because growth occurs on a heated substrate in a cold wall reactor, much work has been done on the difficult issue of modelling the gas flow parameters of various reactor designs to improve epitaxial uniformity and interface abruptness [155].

OMVPE can be performed at atmospheric pressure or low pressure (typically 50 to 100 mbar) but low pressure systems are usually preferred for the growth of thin layers with abrupt interfaces because, for practical input flow rates, the gas flow velocity above the substrate is faster at lower pressure. Most commercial systems also make use of specially designed valve blocks to permit fast gas switching so that abrupt interfaces can be grown. In the growth of quantum well devices, the precise switching sequence of the source gases can have an important effect on the interface properties [156]. Although a wide variety of quantum well device structures grown by OMVPE have been extensively described in the literature, it is probably fair to say that the control of the layer thicknesses and interface abruptness, while quite good, has not yet achieved that possible using MBE or MBE-like growth processes.

Selective area epitaxy is becoming an important technique for the fabrication of a wide variety of device structures including quantum wire (or dot) structures and integrated waveguide/laser (or detector) structures [157]. A wide variety of structures can be devised using OMVPE with special mask designs on planar or etched surfaces to control the local growth rate or film composition. Generally, OMVPE is not particularly well suited for many types of selective area epitaxy because of the tendency for polycrystalline deposits to form on the masked portion of the wafer surface. This problem is not usually severe when the masked region size is of the order of 10 to 20 μm or less. Selective area epitaxy is most extensively used for the growth of current confinement structures around etched features such as lasers. In most cases, reasonable regrowth morphology is obtained only if the mask extends beyond the edge of the etched feature by more than 1 μm [158]. This limitation has recently been circumvented by adding a chlorine source such as CCl_4 to the source gases used for growth [159]. Because of the wide range of applications for OMVPE in the development of III-V device structures, extensive and wide ranging development efforts will continue to improve the performance, reproducibility, and safety of commercial OMVPE reactors. An example of the scope of current work in the OMVPE field can be inferred from the proceedings of workshops and international conferences dedicated to this versatile crystal growth technique [160].

I CONCLUSION

This Datareview has summarised the advantages and disadvantages of the various epitaxial growth techniques and noted points of practical importance in preparing substrates. The following techniques are reviewed individually: LPE, MBE, CBE, GSMBE, chloride VPE and OMVPE.

REFERENCES

[1 H. Nelson [*RCA Rev. (USA)* vol.24 (1963) p.603-15]

[2] G.A. Antypas, R.L. Moon, W. James, J. Edgecumbe, R.L. Bell [*Gallium Arsenide and Related Compounds*, 1972, Inst. Phys. Conf. Ser. (UK) no.17 (1973) p.48-54]

[3] T.P. Pearsall, Ed. [*GaInAsP Alloy Semiconductors* (Wiley, 1982)]

[4] M.G. Astles [*Liquid-Phase Epitaxial Growth of III-V Compound Semiconductor Materials and Their Device Applications* (Adam Hilger, Bristol, UK, 1990)]

[5] A. Sasaki, Y. Takeda, N. Shikagawa, T. Takagi [*Jpn. J. Appl. Phys. (Japan)* vol.16 (1977) p.239-43]

[6] T.P. Pearsall [*IEEE J. Quantum Electron. (USA)* vol.16 (1980) p.709-20]

[7] R.L. Moon [in *Crystal Growth*, 2nd Edition, Ed. B. Pamplin (Pergamon, Oxford, 1980) p.421-57]

[8] S. Takahashi [*J. Appl. Phys. (USA)* vol.52 (1981) p.6104-8]

[9] S.H. Groves, M.C. Plonko [*Appl. Phys. Lett. (USA)* vol.38 (1981) p.1003-4]

[10] K. Nakajima, S. Yamazaki, K. Akita [*J. Cryst. Growth (Netherlands)* vol.61 (1983) p.535-45]

[11] S.E.H. Turley, P.D. Greene [*J. Cryst. Growth (Netherlands)* vol.58 (1982) p.409-12]

[12] M.B. Panish, I. Hyashi, S. Sumski [*IEEE J. Quantum Electron. (USA)* vol.5 (1969) p.210-11]

[13] L.R. Dawson [in *Solid State Chemistry* vol.7, Ed. H. Reiss (Pergamon, Oxford, 1972)]

[14] G.B. Stringfellow [*Rep. Prog. Phys. (UK)* vol.45 no.5 (1982) p.469-525]

[15] M.G. Astles, F.G.H. Smith, E.W. Williams [*J. Electrochem. Soc. (USA)* vol.120 (1973) p.1750-7]
[16] J.J. Hsieh [*Gallium Arsenide and Related Compounds*, 1976, Inst. Phys. Conf. Ser. (UK) no.33b (1977) p.74-80]
[17] T.P. Pearsall, R.W. Hopson Jr. [*J. Appl. Phys. (USA)* vol.48 (1977) p.4407-10]
[18] K. Nakajima, A. Yamaguchi, K. Akita, T. Kotani [*J. Appl. Phys. (USA)* vol.49 (1978) p.5944-50]
[19] K. Nakajima, T. Tanahashi, K. Akita, T. Yamaoka [*J. Appl. Phys. (USA)* vol.50 (1979) p.4975-81]
[20] T.P. Pearsall et al [*Gallium Arsenide and Related Compounds*, 1978, Inst. Phys. Conf. Ser. (UK) no.45 (1979) p.94-101]
[21] J.J. Hsieh [*IEEE J. Quantum Electron. (USA)* vol.17 (1981) p.11-122]
[22] L.W. Cook, M.M. Tashima, G.E. Stillman [*J. Electron. Mater. (USA)* vol.10 (1981) p.119-39]
[23] P.A. Houston [*J. Mater. Sci. (UK)* vol.16 (1981) p.2935-61]
[24] E. Kuphal [*J. Cryst. Growth (Netherlands)* vol.54 (1981) p.117-26]
[25] J.H. Marsh [Ph.D Thesis, University of Sheffield, 1982]
[26] L.W. Cook, M.M. Tashima, N. Tabatabaie, T.S. Law, G.E. Stillman [*J. Cryst. Growth (Netherlands)* vol.56 (1982) p.475-84]
[27] G.B. Stringfellow [*J. Appl. Phys. (USA)* vol.43 (1976) p.3455-60]
[28] Y. Takeda, A. Sasaki [*J. Cryst. Growth (Netherlands)* vol.45 (1978) p.257-61]
[29] J.J. Hsieh, M.C. Finn, J.A. Rossi [*Gallium Arsenide and Related Compounds*, 1976, Inst. Phys. Conf. Ser. (UK) no.33b (1977) p.74-80]
[30] K. Nakajima, K. Akita [*J. Electrochem. Soc. (USA)* vol.129 (1982) p.2603-9]
[31] E. Kuphal, D. Fritzsche [*J. Electron. Mater. (USA)* vol.12 (1983) p.743-64]
[32] V.L. Wrick, K.T. Ip, L.F. Eastman [*J. Electron. Mater. (USA)* vol.7 (1978) p.253-61]
[33] T. Amano, K. Takahei, H. Nagai [*Jpn. J. Appl. Phys. (Japan)* vol.20 (1981) p.2105-9]
[34] E. Kuphal, A. Pocker [*J. Cryst. Growth (Netherlands)* vol.58 (1982) p.133-42]
[35] M.R. Aylett, J. Haigh [*J. Cryst. Growth (Netherlands)* vol.58 (1982) p.127-32]
[36] K. Nakajima, S. Yamazaki, T. Takanohashi, K. Akita [*J. Cryst. Growth (Netherlands)* vol.59 (1982) p.572-82]
[37] J.L. Benchimol, M. Quillec, S. Slempkes [Presented at the 2nd NATO Workshop on *Materials Aspects of InP*, 28-30 March 1983]
[38] J.D. Oliver Jr., L.F. Eastman [*J. Electron. Mater. (USA)* vol.9 (1980) p.693-712]
[39] G.A. Antypas [*Appl. Phys. Lett. (USA)* vol.37 (1980) p.64-5]
[40] J. Kinoshita, H. Okuda, Y. Uematsu [*Electron. Lett. (UK)* vol.19 (1983) p.215-6]
[41] J.L. Benchimol, M. Quillec, C. Le Cornec, G. Le Roux [*Appl. Phys. Lett. (USA)* vol.36 (1980) p.454-6]
[42] Y. Takeda, M. Kuzuhara, A. Sasaki [*Jpn. J. Appl. Phys. (Japan)* vol.19 (1980) p.899-903]
[43] S. Fujita, K. Matsuda, A. Sasaki [*Jpn. J. Appl. Phys. (Japan)* vol.20 (1981) p.L889-92]
[44] O. Wada, A. Majerfeld, P.N. Robson [*J. Electrochem. Soc. (USA)* vol.127 (1980) p.2278-84]
[45] I. Umebu, P.N. Robson [*J. Cryst. Growth (Netherlands)* vol.53 (1981) p.292-6]
[46] E. Kuphal [*J. Cryst. Growth (Netherlands)* vol.54 (1981) p.117-26]
[47] N. Chand, P.A. Houston [*J. Electron. Mater. (USA)* vol.11 (1982) p.37-52]
[48] I. Umebu, P.N. Robson [*J. Cryst. Growth (Netherlands)* vol.53 (1981) p.292-5]

[49] H. Beneking, N. Grote, J. Selders [*J. Cryst. Growth (Netherlands)* vol.54 (1981) p.59-63]
[50] M. Kurihara, T. Morizumi, K. Takahashi [*Solid-State Electron. (UK)* vol.16 (1973) p.763-71]
[51] H. Nagai, Y. Noguchi [*J. Appl. Phys. (USA)* vol.49 (1978) p.450-2]
[52] A.W. Smith, L. Eaves, P.D. Greene, J.R. Hayes, A.R. Adams, L.G. Shantharama [*J. Phys. D (UK)* vol.16 (1983) p.679-86]
[53] S.B. Phatak, S.M. Bedair, s. Fujita [*Solid-State Electron. (UK)* vol.23 (1980) p.839-44]
[54] S. Fujita, M. Kuzuhara, M. Yagyu, A. Sasaki [*Solid-State Electron. (UK)* vol.25 (1982) p.359-65]
[55] P.C. Chen, K.L. Yu, S. Margarit, A. Yariv [*Appl. Phys. Lett. (USA)* vol.38 (1981) p.301-3]
[56] F. Fiedler, A. Schlachetzki [*J. Cryst. Growth (Netherlands)* vol.53 (1981) p.623-6]
[57] V. Swaminathan, A.T. Macrander [*Materials Aspects of GaAs- and InP-Based Structures* (Prentice Hall, 1991) p.78-114]
[58] R.L. Mesham [Ph.D Thesis, University of Sheffield, UK, 1979]
[59] H. Rupprecht [*Inst. Phys. Conf. Ser. (UK)* no.3 (1967) p.57-62]
[60] T.P. Lee, C.A. Burrus Jr., A.G. Dentai [*IEEE J. Quantum Electron. (USA)* vol.17 (1981) p.232-9. See also many other interesting review articles in this issue]
[61] N. Chand, P.A. Houston, P.N. Robson [*Electron. Lett. (UK)* vol.17 (1981) p.726-7]
[62] N. Chand, A.V. Syrbu, P.A. Houston [*J. Cryst. Growth (Netherlands)* vol.61 (1983) p.53-60]
[63] E.H.C. Parker, Ed. [*The Technology and Physics of Molecular Beam Epitaxy* (Plenum, New York, 1985)]
[64] A.Y. Cho, J.R. Arthur [in *Progress in Solid State Chemistry* vol.10, Eds G.A. Somorjai, J. McCaldin (Pergamon, New York, 1975)]
[65] A.Y. Cho [*J. Vac. Sci. Technol. (USA)* vol.16 (1979) p.275]
[66] K. Ploog [in *Crystals: Growth, Properties and Applications* vol.3, Ed. H.C. Freyardt (Springer, Berlin, 1980)]
[67] A. Madhukar, S.V. Ghaisas [*CRC Crit. Rev. Solid State Mater. Sci. (USA)* vol.14 (1988) p.1]
[68] C.E.C. Wood [*Phys. Thin Films (USA)* vol.11 (1979) p.35]
[69] Proceedings of the Annual US MBE Workshops published in *J. Vac. Sci. Technol. B*, as well as proceedings of the last five MBE International Conferences (MBE-3, San Fransisco, 1984, published in *J. Vac. Sci. Technol. B* vol.3 (1985); MBE-4, York, 1986, published in *J. Cryst. Growth* vol.81 (1987); MBE-5, Sapporo, 1988, published in *J. Cryst. Growth* vol.95 (1989); MBE-6, La Jolla, CA, August 1990, published in *J. Cryst. Growth* vol.111 (1991); MBE-7, Germany, to be published in *J. Cryst. Growth.*
[70] M.A. Herman, M. Sitter [*MBE: Fundamental and Current Status* (Springer-Verlag, 1989)]
[71] E.C. Larkins, J.S. Harris Jr. [in *MBE*, Eds R. Farrow, J. Arthur (Noyes Publications, NJ, in press)]
[72] N. Chand [*J. Cryst. Growth (Netherlands)* vol.97 (1989) p.415-29]
[73] N. Chand, A.M. Sergent, J.P. van der Ziel, D.V. Lang [*J. Vac. Sci. Technol. B (USA)* vol.7 (1989) p.399-404]
[74] N. Chand, S.N.G. Chu [*J. Cryst. Growth (Netherlands)* vol.104 (1990) p.485-97]
[75] V. Swaminathan, A.T. Macrander [*Materials Aspects of GaAs and InP Band Structures* (Prentice Hall, 1991) p.138-55]

[76] K.Y. Cheng, A.Y. Cho, W.R. Wagner, W.A. Bonner [*J. Appl. Phys. (USA)* vol.52 (1981) p.1015-21]

[77] K.Y. Cheng, A.Y. Cho, W.R. Wagner [*J. Appl. Phys. (USA)* vol.52 (1981) p.6328-30]

[78] K.Y. Cheng, A.Y. Cho [*J. Appl. Phys. (USA)* vol.53 (1982) p.4411-15]

[79] K. Alavi, H. Temkin, W.R. Wagner, A.Y. Cho [*Appl. Phys. Lett. (USA)* vol.42 (1983) p.254-6]

[80] I.J. Fritz, P.L. Gourley, L.R. Dawson [*Appl. Phys. Lett. (USA)* vol.51 (1987) p.1004-6]

[81] T.G. Andersson, Z.G. Chen, V.D. Kulakovskii, A. Uddin, J.T. Vallin [*Appl. Phys. Lett. (USA)* vol.51 (1987) p.752-4]

[82] P.L. Gourley, I.J. Fritz, L.R. Dawson [*Appl. Phys. Lett. (USA)* vol.52 (1988) p.377-9]

[83] J.-P. Reithmaier, M. Cerva [*Appl. Phys. Lett. (USA)* vol.54 (1989) p.48-50]

[84] B. Elman, E.S. Koteles, P. Melman, C. Jagannath, J. Lee, D. Duggor [*Appl. Phys. Lett. (USA)* vol. 55 (1989) p.1659-61]

[85] J. Woodhead [*Appl. Phys. Lett. (USA)* vol.55 (1989) p.2147-8]

[86] J.Y. Yoo, T.G. Andersson, G.L. Dunlop [*J. Appl. Phys. (USA)* vol.69 (1991) p.2224-30]

[87] D.J. Dunstan, S. Young, R.M. Dixon [*J. Appl. Phys. (USA)* vol.70 (1991) p.3038-45]

[88] W.-C. Hsu, S.-Z. Chang, W. Lin [*Jpn. J. Appl. Phys. (Japan)* vol.31 (1992) p.26-9]

[89] Y.C. Chen, P.K. Bhattacharya, J. Singh [*Appl. Phys. (Germany)* vol.57 (1990) p.692-4]

[90] B. Elman, E.S. Koteles, P. Melman, K. Ostreicher, C. Sung [*J. Appl. Phys. (USA)* vol.70 (1991) p.2634-40]

[91] S.-L. Weng, C. Webb, J.N. Eakstein [*J. Vac. Sci. Technol. B (USA)* vol.7 (1989) p.361-4]

[92] D.C. Radulescu, W.J. Schaff, L.F. Eastman, J.M. Ballingall, G.O. Bamseyer, S.D. Hersee [*J. Vac. Sci. Technol. B (USA)* vol.7 (1989) p.111-5]

[93] J.-P. Reithmaier, H. Riechert, H. Schlötterer, G. Weigmann [*J. Cryst. Growth (Netherlands)* vol.111 (1991) p.407-11]

[94] M.T. Emeny, L.K. Howard, K.P. Homewood, J.D. Lambkin, C.R. Whitehouse [*J. Cryst. Growth (Netherlands)* vol.111 (1991) p.413-7]

[95] N. Chand, E.A. Fitzgerald, M.Geva [*Appl. Phys. Lett. (USA)* vol.61 (1992) p.2893]

[96] N. Chand, J.P. van der Ziel, S.N.G. Chu, A.M. Sergent [*J. Vac. Sci. Technol. B (USA)* vol.10 (1992) p.1006-10]

[97] H.K. Choi, C.A. Wang [*Appl. Phys. Lett. (USA)* vol.57 (1990) p.321-3]

[98] N. Chand, E.E. Becker, J.P. van der Ziel, S.N.G. Chu, N.K. Dutta [*Appl. Phys. Lett. (USA)* vol.58 (1991) p.1704-6]

[99] M.T. Asom, L. Luther, G. Livescu, V. Swaminathan, S.N.G. Chu [*Proc. SPIE (USA)* vol.1676 (1992) p.107-16]

[100] E. Fitzgerald [Datareview in this book: *1.2 Lattice mismatch and dislocations in InGaAs/GaAs strained heterostructures*]

[101] E. Veuhoff, W. Pletschen, P. Balk, H. Lüth [*J. Cryst. Growth (Netherlands)* vol.55 (1981) p.30]

[102] For reviews see: W.T. Tsang [*J. Cryst. Growth (Netherlands)* vol.95 (1989) p.121]; W.T. Tsang [in *VLSI Electronics Microstructure Science* vol.21, Ed. N.G. Einspruch (Academic Press, New York, 1989) ch.6 p.255-357]; W.T. Tsang [*J. Cryst. Growth (Netherlands)* vol.105 (1990) p.1]. Also see annual conference proceedings on MBE and CBE published in *J. Cryst. Growth* and in *J. Vac. Sci. Technol. B*.

[103] V. Swaminathan, A.T. Macrander [*Materials Aspects of GaAs- and InP-Based Structures* (Prentice Hall, 1991) p.156-65]
[104] M.B. Panish, H. Temkin [*Annu. Rev. Mater. Sci. (USA)* vol.19 (1989) p.209-29]; M.B. Panish [*J. Cryst. Growth (Netherlands)* vol.81 (1987) p.249-60]; M.B. Panish, H. Temkin [*Gas Source MBE: Growth and Properties of P Containing III-V Heterostructures*, to be published by Springer Verlag, Heidelberg]
[105] C.R. Abernathy [in *Indium Phosphide and Related Materials*, Ed. A. Katz (Artech House, 1992) p.177-209]
[106] M.B. Panish [*J. Electrochem. Soc. (USA)* vol.127 (1980) p.2729-33]
[107] H. Asahi, Y. Kawamura, M. Ikeda, H. Okamoto [*J. Appl. Phys. (USA)* vol.52 (1981) p.2852-9]
[108] C.T. Foxon, B.A. Joyce, M.T. Norris [*J. Cryst. Growth (Netherlands)* vol.49 (1980) p.132-40]
[109] W.T. Tsang, J.C. Campbell [*Appl. Phys. Lett. (USA)* vol.48 (1986) p.1416-18]
[110] A.R. Calwa [*Appl. Phys. Lett. (USA)* vol.38 (1981) p.701-3]
[111] R. Chow, Y.G. Chai [*Appl. Phys. Lett. (USA)* vol.42 (1983) p.383-5]
[112] M.B. Panish, M. Temkin, S. Sumski [*J. Vac. Sci. Technol. B (USA)* vol.3 (1985) p.657-65]
[113] A.Y. Cho [*J. Appl. Phys. (USA)* vol.46 (1975) p.1733-5]
[114] R.A. Hamm, M.B. Panish, R. Nottenburg, Y.K. Chen, D. Humphery [*Appl. Phys. Lett. (USA)* vol.54 (1989) p.2586]
[115] D. Ritter, R.A. Hamm, M.B. Panish, M. Geva [unpublished]
[116] M. Weyers, J. Musolf, D. Marx, A. Kohl, P. Balk [*J. Cryst. Growth (Netherlands)* vol.105 (1990) p.383]
[117] J. Nagle, R.J. Malik, D. Gershoni [*J. Cryst. Growth (Netherlands)* vol.111 (1991) p.264-8]
[118] M. Konagai et al [*J. Cryst. Growth (Netherlands)* vol.105 (1990) p.359]
[119] C.R. Abernathy [*J. Cryst. Growth (Netherlands)* vol.107 (1990) p.982]
[120] T.J. de Lyon, N.I. Buchan, P.D. Kirchner, J.M. Woodall, G.J. Scilla, F. Cardone [*Appl. Phys. Lett. (USA)* vol.58 (1991) p.517]
[121] T.P. Chin, P.D. Kirchner, J.M. Woodall, C.W. Tu [*Appl. Phys. Lett. (USA)* vol.59 (1991) p.2865]
[122] S.A. Stockman, A.W. Hanson, G.E. Stillman [*Appl. Phys. Lett. (USA)* vol.60 (1992) p.2903]
[123] G.B. Stringfellow [*J. Cryst. Growth (Netherlands)* vol.105 (1990) p.260]
[124] D. Ritter, M.B. Panish, R.A. Hamm, D. Gershoni [*2nd Int. Conf. on Chemical Beam Epitaxy and Related Growth Techniques*, Houston, December 1989, published in *J. Cryst. Growth (Netherlands)* vol.105 (October 1990) no.1-4]
[125] A. Doi, Y. Aoyagi, S. Namba [*Appl. Phys. Lett. (USA)* vol.48 (1986) p.1787]
[126] T.H. Chiu, J.E. Cunningham, A. Robertson Jr., D.L. Malm [*J. Cryst. Growth (Netherlands)* vol.105 (1990) p.155]
[127] J.A. Long, R.A. Logan, R.F. Karlicek Jr. [in *Optical Fiber Telecommunications II*, Eds S.E. Miller, J.P. Kaminow (Academic Press, New York, 1988) ch.16]
[128] V. Swaminathan, A.T. Macrander [*Materials Aspects of GaAs- and InP-Based Structures* (Prentice Hall, Englewood Cliffs, 1991) ch.2]
[129] G.H. Olsen [in *GaInAsP Alloy Semiconductors*, Ed. T.P. Pearsall (John Wiley & Sons, New York, 1982) ch.1]
[130] M. Yoshida, H. Terao, H. Watanabe [*J. Electrochem. Soc. (USA)* vol.132 (1985) p.930]

[131] R.F. Karlicek Jr., D. Mitcham, J.C. Ginocchio, B. Hammarlund [*J. Electrochem. Soc. (USA)* vol.134 (1987) p.470]
[132] K. Tanaka, K. Nakai, O. Aoki, M. Sugawara, K. Wakao, S. Yamakoshi [*J. Appl. Phys. (USA)* vol.61 (1987) p.4698]
[133] R.F. Karlicek Jr. [*J. Cryst. Growth (Netherlands)* vol.91 (1988) p.133]
[134] Y. Miura et al [*Proc. 2nd Int. Conf. on Indium Phosphide and Related Materials* (IEEE, New York, 1990) p.104-7]
[135] J.N. Hollenhorst, D.T. Ekholm, J.M. Geary, V.D. Mattera, R. Pawelek [*Proc. SPIE, High Frequency Analog Communications* vol.995 (Soc. Photo-Optical Instrumentation Engineers, Bellingham, USA, 1989) p.53-60]
[136] G. Olsen, F. Hawrylo, D.J. Channin, D. Botez, M. Ettenberg [*IEEE J. Quantum Electron. (USA)* vol.17 (1981) p.2130]
[137] T. Yanase et al [*Electron. Lett. (UK)* vol.19 (1983) p.700]
[138] H. Temkin et al [*Appl. Phys. Lett. (USA)* vol.53 (1988) p.1156]
[139] V.S. Ban, D. Rodefeld, J.R. Flemish, K.A. Jones [*Appl. Phys. Lett. (USA)* vol.62 (1993) p.160]
[140] W.D. Johnston, M.A. DiGiuseppe, D.A. Wilt [*AT&T Tech. J. (USA)* vol.6 (1989) p.53]
[141] H.M. Cox [*J. Cryst. Growth (Netherlands)* vol.69 (1984) p.641]
[142] A. Koukitu, H. Seki [*J. Cryst. Growth (Netherlands)* vol.49 (1980) p.325]
[143] M. Yoshida, H. Watanabe [*J. Electrochem. Soc. (USA)* vol.132 (1985) p.1733]
[144] D.W. Shaw [*J. Electrochem. Soc. (USA)* vol.113 (1966) p.904]
[145] R.F. Karlicek Jr., B.P. Segner, J.D. Wynn, A.J. Becker, U.K. Chakrabarti, R.A. Logan [*J. Electrochem. Soc. (USA)* vol.137 (1990) p.2639]
[146] H.M. Manasevit [*Appl. Phys. Lett. (USA)* vol.116 (1969) p.1725]
[147] G.B. Stringfellow [*Organometallic Vapor-Phase Epitaxy: Theory and Practice* (Academic Press, San Diego, 1989)]
[148] M. Razeghi [*The MOCVD Challenge* (Adam Hilger, Philadelphia, 1989)]
[149] E.K. Byrne [*Indium Phosphide and Related Materials: Processing, Technology and Devices*, Ed. A. Katz (Artech House, Norwood, USA, 1992) ch.5]
[150] Application Note 2, American Cyanamide Company, 1989.
[151] J.L. Zilko, P.S. Davisson, L. Luther, K.D.C. Trapp [*J. Cryst. Growth (Netherlands)* vol.124 (1992) p.112]
[152] T.F. Keuch, M.A. Tischler, R. Potemski [*Appl. Phys. Lett. (USA)* vol.54 (1989) p.910]
[153] Morton International, Speciality Chemicals Group, Ethyldimethylindium Product Data Sheet, June 1990.
[154] L.W. Fannin, D.W. Webb, R.H. Pearce [*J. Cryst. Growth (Netherlands)* vol.124 (1992) p.307]
[155] K.F. Jensen, D.I. Fotiadis, T.J. Mountziaris [*J. Cryst. Growth (Netherlands)* vol.107 (1991) p.1]
[156] K. Streubel, J. Wallin, M. Amiotti, G. Landgren [*J. Cryst. Growth (Netherlands)* vol.124 (1992) p.541]
[157] R. Bhat [*J. Cryst. Growth (Netherlands)* vol.120 (1992) p.362]
[158] T. Sanada, K. Nakai, K. Wakao, M. Kuno, S. Yamakoshi [*Appl. Phys. Lett. (USA)* vol.51 (1987) p.1054]
[159] N. Nordell, J. Borglind [*Appl. Phys. Lett. (USA)* vol.61 (1992) p.22]
[160] For example, see *Proc. 6th Int. Conf. on Metalorganic Vapor Phase Epitaxy*, Eds G.B. Stringfellow, J.J. Coleman (North Holland, 1992)

8.2 n- and p-type doping in InGaAs

Y.C. Chen and P.K. Bhattacharya

March 1993

A INTRODUCTION

The ability to obtain controllable and reproducible n- and p-type doping is crucial in the fabrication of high performance semiconductor devices. Since InGaAs-based materials grown on InP are important for electronic and optoelectronic applications, intensive efforts have been made for the establishment of well-behaved dopants for InGaAs. Ideally, a dopant suitable for the growth of high performance devices should have: (1) a very high maximum carrier concentration; (2) a very low diffusion coefficient; (3) negligible surface segregation; (4) negligible effects on morphology at high doping levels; and (5) very slight compensation. Apparently, very few dopant species for InGaAs meet all these criteria.

Studies on InGaAs doping are based, more or less, on the experiences with GaAs. The characteristics of dopants in InGaAs are, however, different from those in GaAs. We review here the work done on n- and p-type doping in InGaAs. Unless the composition is described explicitly, the ternary composition is that lattice-matched to InP, i.e., $In_{0.53}Ga_{0.47}As$. TABLE 1 summarises the characteristics of the main dopants for InGaAs grown by various techniques. Deep-level impurities in InGaAs are discussed by Bhattacharya elsewhere in this book [1].

B n-TYPE DOPING

B1 Silicon (Si)

Si is one of the most important and most convenient n-type dopants in InGaAs and other III-V semiconductors. It produces a very shallow donor level (~ 5 meV below the conduction band edge) and a small diffusion coefficient. Maximum electron concentrations of 5.0×10^{19} and 6.1×10^{19} cm^{-3} have been obtained in InGaAs grown by molecular beam epitaxy (MBE) at substrate temperatures of 420°C and 370°C, respectively [2]. As a group IV element Si is, in principle, an amphoteric dopant in III-V compounds. It behaves as a donor when it occupies a group III site and as an acceptor when it occupies a group V site. However, in InGaAs, it preferentially occupies group III sites, showing negligible self-compensation.

For InGaAs grown by MBE, gas-source MBE (GSMBE), metalorganic MBE (MOMBE) and chemical beam epitaxy (CBE), Si is usually introduced from an elemental Si source evaporated from a Knudson effusion cell. Disilane has also been used to introduce Si doping in InGaAs grown by metalorganic vapour phase epitaxy (MOVPE) [3]. However, the incorporation efficiency is very low (< 10^{-3}).

TABLE 1 n- and p-type dopants for InGaAs.

Dopant	Maximum Doping Concentration	Source	Growth Technique	Comments
n-type				
Si	6.1×10^{19} cm^{-3} at $T_{sub} = 370°C$	Si, Si_2H_6	MBE, MO- or hydride-based epitaxy	Almost ideal
Sn	$\sim 10^{20}$ cm^{-3}	TESn	MO- or hydride-based epitaxy, LPE	Beyond 10^{20} cm^{-3}, additional Sn is not active
S	$\leq 10^{18}$ cm^{-3} for $T_{sub} = 625°C$, $\sim 10^{20}$ cm^{-3} for $T_{sub} = 525°C$	H_2S	MO- or hydride-based epitaxy	
Se		H_2Se	MO- or hydride-based epitaxy	Memory effects
Te	1×10^{18} cm^{-3}	Te, H_2Te	VPE, LPE	
p-type				
Be	5.6×10^{20} cm^{-3} at $T_{sub} = 365°C$	Be	MBE, MO- or hydride-based epitaxy	Diffuses very rapidly above $\sim 10^{19}$ cm^{-3}
Zn	3×10^{19} cm^{-3}	DEZn, In-Zn alloy	MO- or hydride-based epitaxy, LPE	Zn redistribution occurs in the presence of a highly Si-doped neighbouring layer
Cd	2×10^{19} cm^{-3}			Doping obtained by diffusion
C	3×10^{19} cm^{-3}	CCl_4	MO- or hydride-based epitaxy	Low diffusion coefficient
Mn	3×10^{18} cm^{-3}	Mn	LPE	

B2 Tin (Sn)

Sn is a widely used shallow donor in all the epitaxial techniques. Panish et al [4] have investigated the incorporation of Sn in InGaAs grown by hydride-source MBE at 450°C. Their results show that Sn is a well-behaved donor in InGaAs to a doping concentration of $\sim 10^{20}$ cm^{-3}, although increased compensation is observed with increasing doping level. They also show that at a total concentration beyond about 10^{20} cm^{-3}, additional Sn is not electrically active. The epitaxial quality remains high without noticeable morphology changes, defects, or precipitates, to a concentration of 10^{21} cm^{-3}. Pinzone et al [5] have also achieved an n-type carrier concentration of 6.1×10^{19} cm^{-3} in InGaAs grown by MOCVD using tetraethyltin (TESn).

It is clearly established that Sn segregates at or near the growing surface during MBE growth of GaAs. In the case of InGaAs, Kawaguchi and Nakashima [6] have shown that in the $n = 10^{16}$ - 10^{17} cm^{-3} range, Sn introduced by the use of TESn during MOMBE at 490°C does not exhibit surface segregation.

B3 Sulphur (S)

S as a donor is available as a stable mixture of H_2S in H_2 and may be conveniently introduced in vapour-phase epitaxy (VPE), where H_2 is generally used as the carrier gas. S is essentially completely ionised at room temperature. Logan et al [7] have demonstrated that S can be easily incorporated into InGaAs at levels $\leq 10^{18}$ cm^{-3} during MOVPE growth at 625°C, where crystal morphology is optimal, and at levels as high as $\sim 10^{20}$ cm^{-3} to form low-resistance n-contact layers with growth at 525°C.

B4 Selenium (Se)

In the growth of InGaAs by vapour phase techniques, n-type doping can be accomplished by the use of Se from a H_2Se source and some success has been achieved [8,9].

B5 Tellurium (Te)

Te is a shallow donor with an activation energy of about 10 meV.. Te metal has been used to produce n-type InGaAs in hydride VPE growth and carrier concentrations in the range 1×10^{16} to 1×10^{18} cm^{-3} have been obtained [10]. Te is also used as an n-dopant in LPE.

B6 Chromium (Cr)

Cr-doping has been successfully used to grow high-resistivity GaAs. However, in InGaAs grown by liquid phase epitaxy (LPE), Cr-doping seems to produce donor-like behaviour and the electron concentration increases monotonically with increased addition of Cr to the melt [11].

C p-TYPE DOPING

C1 Beryllium (Be)

Be is a very important shallow acceptor in III-V semiconductors. For MBE growth, Be is preferred to other p-type dopants (Zn, Mg, Cd) because it exhibits a sticking coefficient close to 1. Doping concentrations in excess of 10^{19} cm^{-3} can be achieved without degradation of the morphology under proper growth conditions. This is very useful for the growth of the p^+ base region of n-p-n heterojunction bipolar transistors (HBTs), the p^+ contact layer of heterostructure lasers and the gate layer of junction field effect transistors (JFETs). The highest doping in InGaAs grown by beam epitaxy techniques at 500°C has ranged from 2 to 5×10^{19} cm^{-3} [12-14]. Hole concentration as high as 5.6×10^{20} cm^{-3} has been reported by Hamm et al [15] for InGaAs grown by GSMBE at a growth temperature of 365°C.

A well-known problem associated with Be as a p-dopant is its high diffusivity in InGaAs and GaAs. Scott et al [16] have measured the effective diffusion coefficient of Be in InGaAs. At doping levels $> 5 \times 10^{16}$ cm^{-3}, Be has a diffusion coefficient of $(1.9 \pm 0.6) \times 10^{-15}$ cm^2s^{-1} at 680°C and an activation energy of 0.45 ± 0.2 eV. At doping levels $< 5 \times 10^{16}$ cm^{-3}, the diffusion coefficient at 680°C is $(9 \pm 3) \times 10^{-16}$ cm^2s^{-1}. Maier and Selders [17] have reported a diffusion coefficient of 1×10^{-11} cm^2s^{-1} for Be in InGaAs at 800°C. The large diffusion coefficient of Be is detrimental for HBTs. In this device, Be can diffuse from the p-type base region into the emitter, lowering the emitter injection efficiency. The migration of Be at elevated temperatures is due to fast interstitial-substitutional diffusion mechanisms [18,19].

Be is very efficient at gettering oxygen-containing gas species, forming electrically inactive centres such as BeO in growing layers [20]. In MOMBE or CBE, there are also problems arising from the reaction of hot elemental Be with decomposing metalorganics to form Be_2C in the effusion cell. The formation of stable Be_2C lowers the Be vapour pressure, giving rise to difficulties in obtaining high doping concentrations.

C2 Zinc (Zn)

Zn is a shallow acceptor (binding energy ~ 25 meV) in InGaAs [21]. Diethylzinc (DEZn) can be used as a dopant source during MOMBE, GSMBE or CBE growth. Hamm et al [22] have reported a very low Zn incorporation efficiency ($< 10^{-3}$) for MOMBE growth of InGaAs at 480°C - 510°C using DEZn. However, they were able to obtain doping levels from $p = 1 \times 10^{17}$ to 3×10^{19} cm^{-3}. Benzaquen et al [21] have also reported Zn-doping levels in the range 3.22×10^{15} to 1.95×10^{20} cm^{-3} in $In_{0.15}Ga_{0.85}As$ grown by MOVPE. In LPE growth, Zn doping can be achieved by using either In-Zn alloy or Zn-doped GaAs as the dopant source in the growth melt [23-25].

Zn has an unfavourably large diffusion coefficient (3×10^{-9} cm^2s^{-1} at 500°C in InGaAsP [26]). Kurishima et al [27] have reported that a highly doped n^+ subcollector layer (Si concentration $> 10^{19}$ cm^{-3}) can induce considerable broadening of the Zn doping profile in the base region of InP/InGaAs HBT structures. Similar phenomena have been observed previously in GaAs/AlGaAs HBT structures, where the redistribution of Zn is caused by the presence of a highly Si-doped emitter layer [28].

C3 Cadmium (Cd)

Cd has been introduced into InGaAs by diffusion to form heavily-doped p-type layers [29,30]. Highly doped (2×10^{19} cm^{-3}) shallow ($x_j \simeq 0.1 - 0.4$ m) p^+-n junctions have been reported [30]. The diffusion coefficient of Cd at 600°C is measured to be $(1 - 2) \times 10^{-13}$ cm^2s^{-1} [29,30].

C4 Carbon (C)

Heavily C-doped p-type InGaAs layers were grown successfully only very recently [31,32]. Early attempts to produce C-doped $In_xGa_{1-x}As$ resulted in either highly compensated n-type materials [33] or p-type conduction for $x < 0.8$ and n-type conduction for $x > 0.8$ [34]. SIMS studies by Tokumitsu et al [34] indicate that the p to n conversion occurring at high In compositions may be due to limited C incorporation during growth rather than self-compensation. Recently, Chin et al [31] have achieved a hole concentration as high as 3×10^{19} cm^{-3} for $In_{0.5}Ga_{0.5}As$ using CCl_4 during growth by GSMBE. They found that low growth temperatures produced the most heavily doped layers, and short post-growth anneal leads to an increase in the hole concentration. Stockman et al [32] have reported C doping of $In_xGa_{1-x}As$ ($0 < x \leq 0.53$) grown by low-pressure MOCVD using CCl_4. A hole concentration of 1×10^{20} cm^{-3} was obtained for $x < 0.12$ and 1×10^{19} cm^{-3} for $In_{0.53}Ga_{0.47}As$. The presence of CCl_4 in the growth ambient was found to affect the In content in the epitaxial layer.

Pearton et al [35] have measured the diffusion coefficient of C in InGaAs and found it to be less than 3.3 x 10^{-14} cm^2s^{-1} at 800°C, which is significantly lower than that of other p-type dopants. Consequently, the use of C as a p-type dopant in InGaAs-based HBTs seems to be a potential solution to base dopant redistribution problems encountered with Be and Zn.

C5 Manganese (Mn)

Mn is a p-type dopant in $In_xGa_{1-x}As$ with an activation energy of 77 meV for $0 < x < 0.25$ [36] and 52.5 ± 2.5 meV for x = 0.53 [37]. Chand et al [37] have reported hole concentrations in the range 1.5 x 10^{17} to 3 x 10^{18} cm^{-3} in InGaAs grown by LPE. They have also measured the diffusion coefficient of Mn in InP at 640°C and found it to be 2.6 x 10^{-12} cm^2s^{-1}.

D CONCLUSION

We presented the properties of several n- and p-type dopants in InGaAs. At present, Si and Be are still the most convenient and controllable donor and acceptor species for MBE growth of InGaAs. However, for MOMBE, GSMBE, CBE and other vapour-phase-based growth techniques, C may evolve as a promising acceptor species.

REFERENCES

[1] P. Bhattacharya [Datareview in this book: *6.3 Deep levels in InGaAs/GaAs and InGaAsP/InP*]
[2] T. Fujii, T. Inata, K. Ishii, S. Hiyamizu [*Electron. Lett. (UK)* vol.22 (1986) p.191-2]
[3] E. Woelk, H. Beneking [*J. Appl. Phys. (USA)* vol.63 (1988) p.2874-6]
[4] M.B. Panish, R.A. Hamm, L.C. Hopkins, S.N.G. Chu [*Appl. Phys. Lett. (USA)* vol.56 (1990) p.1137-9]
[5] C.J. Pinzone, N.D. Gerrard, R.D. Dupuis, N.T. Ha, H.S. Luftman [*J. Appl. Phys. (USA)* vol.67 (1990) p.6823-9]
[6] Y. Kawaguchi, K. Nakashima [*J. Cryst. Growth (Netherlands)* vol.95 (1989) p.181-4]
[7] R.A. Logan, T. Tanbun-Ek, A.M. Sergent [*J. Appl. Phys. (USA)* vol.65 (1989) p.3723-5]
[8] L. Hollan, M. Boulou, J.P. Chane [*J. Electron. Mater. (USA)* vol.10 (1981) p.193-212]
[9] C.R. Lewis, M.J. Ludowise, W.T. Dietze [*J. Electron. Mater. (USA)* vol.13 (1984) p.447-61]
[10] N. Susa, Y. Yamauchi, H. Ando, H. Kanbe [*J. Cryst. Growth (Netherlands)* vol.58 (1982) p .527-33]
[11] M.V. Rao, P.K. Bhattacharya [*J. Appl. Phys. (USA)* vol.57 (1985) p.333-7]
[12] K.Y. Cheng, A.Y. Cho, W.A. Bonner [*J. Appl. Phys. (USA)* vol.52 (1981) p.4672-5]
[13] A. Le Corre et al [*Appl. Phys. Lett. (USA)* vol.51 (1987) p.1597-9]
[14] W.T. Tsang, A.F.J. Levi, E.G. Burkhardt [*Appl. Phys. Lett. (USA)* vol.53 (1988) p.983-5]
[15] R.A. Hamm, M.B. Panish, R.N. Nottenburg, Y.K. Chen, D.A. Humphrey [*Appl. Phys. Lett. (USA)* vol.54 (1989) p.2586-8]

[16] E.G. Scott, D. Wake, G.D.T. Spiller, G.J. Davies [*J. Appl. Phys. (USA)* vol.66 (1989) p.5344-8]
[17] M. Maier, J. Selders [*J. Appl. Phys. (USA)* vol.60 (1986) p.2783-7]
[18] D.L. Miller, P.M. Asbeck [*J. Appl. Phys. (USA)* vol.57 (1985) p.1816-22]
[19] J.N. Miller, D.M. Collins, N.J. Moll [*Appl. Phys. Lett. (USA)* vol.46 (1985) p.960-2]
[20] M. Gauneau et al [*J. Appl. Phys. (USA)* vol.66 (1989) p.2241-7]
[21] R. Benzaquen, A.P. Roth [*J. Appl. Phys. (USA)* vol.72 (1992) p.4288-94]
[22] R.A. Hamm, D. Ritter, H. Temkin, M.B. Panish, M. Geva [*Appl. Phys. Lett. (USA)* vol.58 (1991) p.2378-80]
[23] T.P. Pearsall, G. Beuchet, J.P. Hirtz, N. Visentin, M. Bonnet [in *Proc. Int. Symp. on Gallium Arsenide and Related Compounds* 1980 (Inst. Phys. Conf. Ser. no.56, Institute of Physics, London, 1981) p.639-49]
[24] M.M. Tashima, L.W. Cook, G.E. Stillman [*Appl. Phys. Lett. (USA)* vol.39 (1981) p.960-1]
[25] M.V. Rao [*J. Appl. Phys. (USA)* vol.58 (1985) p.4313-6]
[26] G.J. van Gurp, T. van Dongen, G.M. Fontijn, J.M. Jacobs, D.L.A. Tjaden [*J. Appl. Phys. (USA)* vol.65 (1989) p.553-60]
[27] K. Kurishima, T. Kobayashi, U. Gosele [*Appl. Phys. Lett. (USA)* vol.60 (1992) p.2496-98]
[28] P. Enquist, J.A. Hutchby, T.J. de Lyon [*J. Appl. Phys. (USA)* vol.63 (1988) p.4485-93]
[29] S. Aytac, A. Schlachetzki [*J. Cryst. Growth (Netherlands)* vol.64 (1983) p.169-73]
[30] D.M. Shah, W.K. Chan, R. Bhat, H.M. Cox, N.E. Schlotter, C.C. Chang [*Appl. Phys. Lett. (USA)* vol.56 (1990) p.2132-4]
[31] T.P. Chin, P.D. Kirchner, J.M. Woodall, C.W. Tu [*Appl. Phys. Lett. (USA)* vol.59 (1991) p.2865-7]
[32] S.A. Stockman, A.W. Hanson, G.E. Stillman [*Appl. Phys. Lett. (USA)* vol.60 (1992) p.2903-5]
[33] M. Kamp et al [*J. Cryst. Growth (Netherlands)* vol.95 (1989) p.154-7]
[34] E. Tokumitsu et al [*J. Cryst. Growth (Netherlands)* vol.120 (1992) p.301-5]
[35] S.J. Pearton, W.S. Hobson, A.P. Kinsella, J. Kovalchick, U.K. Chakrabarti, C.R. Abernathy [*Appl. Phys. Lett. (USA)* vol.56 (1990) p.1263-5]
[36] S.B. Phatak, S.M. Bedair, S. Fujita [*Solid-State Electron. (USA)* vol.23 (1980) p.839-44]
[37] N. Chand, P.A. Houston, P.N. Robson [*Electron. Lett. (UK)* vol.17 (1981) p.726-7]

8.3 Selective etching of InGaAs and its heterostructures

I. Adesida

May 1993

A INTRODUCTION

The fabrication of heterostructure devices in III-V compound semiconductors requires the availability of selective etchants. These are needed to selectively remove one epitaxial layer over another. Extensive investigations have been conducted on the development of selective wet chemical etching processes for the fabrication of heterostructure lasers in the InP/InGaAsP/InGaAs material system. These have resulted in the availability of various solutions primarily based on acids which are capable of etching InGaAs on InP and vice versa. Heterostructure materials based on AlGaAs/InGaAs, InAlAs/InGaAs, and AlGaSb/InAs have more recently been developed for high speed heterostructure field-effect transistor (HFET) applications. We discuss here selective etching of InGaAs with special reference to the fabrication of HFETs. The term HFET is used generically and includes such devices as the modulation-doped field-effect transistors (MODFETs).

A schematic of an HFET layer structure is shown in FIGURE 1. The epitaxial materials commonly used in the different layers of each family of HFET are shown in TABLE 1. Although the fabrication procedure of HFETs may vary from laboratory to laboratory, it is possible to identify three common etching steps in the realisation of $In_xGa_{1-x}As$-channel HFETs. These are the mesa etch, gate recess etch, and channel sidewall recess. The mesa etch, which is for device isolation, requires good etch depth control. This is because metal gate discontinuity resulting from breakage at the mesa edge can occur, especially for devices with ultra-small gatelengths. The gate recess etch is needed to etch through the cap layer to expose the underlying Schottky barrier layer for gate metal contact. The cap layer which is usually a thin (< 20 nm) layer of n^+-InGaAs in some of these HFETs is used to facilitate good ohmic contacts to the heterostructure. A selective etching of this layer with respect to the Schottky layer is highly desirable in order to obtain uniform characteristics in the fabricated devices. A tight control of threshold voltage, V_t, is essential for high performance circuits. The mesa etch results in sidewalls where the device channel is exposed and comes in contact with the gate metallisation running up the mesa. Although the gate/sidewall contact area is very small in comparison with the total gate area, the low barrier height of metals with InGaAs constitutes a current leakage path from the gate to the channel. It has been shown that for InAlAs/InGaAs HFETs, sidewall leakage results in excessive gate leakage current and degraded breakdown characteristics. To

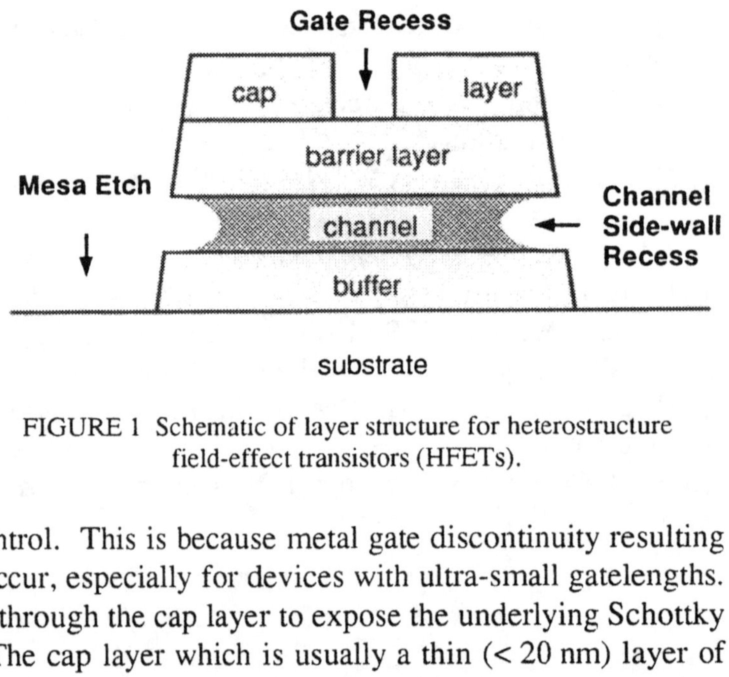

FIGURE 1 Schematic of layer structure for heterostructure field-effect transistors (HFETs).

avoid this problem, Bahl and del Alamo [1] proposed the selective etching of the InGaAs channel sidewall to prevent any gate metal contact to the channel. It is seen that selective etchants are essential for gate recess and channel sidewall recess.

TABLE 1 Epitaxial materials for the layers (see FIGURE 1) for heterostructure field-effect transistors in different material systems.

Heterostructure	Cap layer	Barrier layer	Channel	Buffer layer
InAlAs/InGaAs	$In_{0.53+x}Ga_{0.47-x}As$	$In_{0.52-y}Al_{0.48+y}As$	$In_{0.53+x}Ga_{0.47-x}As$	InAlAs InP
InAlAs/InP	$In_{0.53+x}Ga_{0.47-x}As$	$In_{0.52-y}Al_{0.48+y}As$	InP	InAlAs InP
AlGaAs/InGaAs	GaAs	$Al_yGa_{1-y}As$	$In_xGa_{1-x}As$	GaAs
AlGaSb/InAs	GaSb	$Al_yGa_{1-y}Sb$	InAs	$Al_yGa_{1-y}Sb$

The discussion here deals with the selective etchants needed for these processes in the fabrication of HFETs where the cap layer and/or channel materials are epitaxial $In_xGa_{1-x}As$ layers with x being non-zero. As seen from TABLE 1, the etching steps largely involve the removal of $In_xGa_{1-x}As$ over other materials such as InP and InAlAs. In our discussion, InGaAs and InAlAs will refer to the lattice-matched $In_{0.53}Ga_{0.47}As$ and $In_{0.52}Al_{0.48}As$, respectively, unless otherwise stated. Both wet (Section B) and dry (Section C) etching processes will be discussed. The implementation of these etching steps usually requires photoresists or electron beam resists as the masks; therefore, the etchants adopted must not attack resists.

B WET ETCHING

In general, wet chemical etching of III-V compound semiconductors relies heavily on oxidation-reduction reactions. In such reactions, most etchants include an oxidising agent, a complexing agent, and a dilutant such as water. The most common oxidising agents are Br_2 and H_2O_2, while the most common complexing agents are H_3PO_4, H_2SO_4, HF, and NH_4OH. The etching mechanism involves the oxidation of the semiconductor through reactions with an oxidising agent. The oxidised layer is not soluble in water and can be removed only with the aid of a complexing agent. Details on the physical and chemical mechanisms involved in wet etching are discussed by Mukherjee and Woodard [2].

B1 Citric Acid:H_2O_2

FIGURE 2 shows the results of Tong et al [3] on the etch rates of InGaAs, InAlAs and InP in citric acid/hydrogen peroxide solutions. Citric acid monohydrate was mixed 1:1 by weight in deionised water. Citric acid/H_2O_2 solutions were prepared by adding 30% H_2O_2 to citric acid in the desired proportions just before etching. The selectivity of InGaAs over InAlAs is the etch rate of InGaAs divided by the etch rate of InAlAs. As deduced from FIGURE 2, the selectivity varies from a maximum of 25 at r = 1 to a value of 1 at r = 15, where r is the ratio of citric acid to H_2O_2. The selectivity of InGaAs on InP is over 100 for all solution ratios investigated. InGaAs etch profiles obtained in solutions with r = 2 and r = 10 are slightly

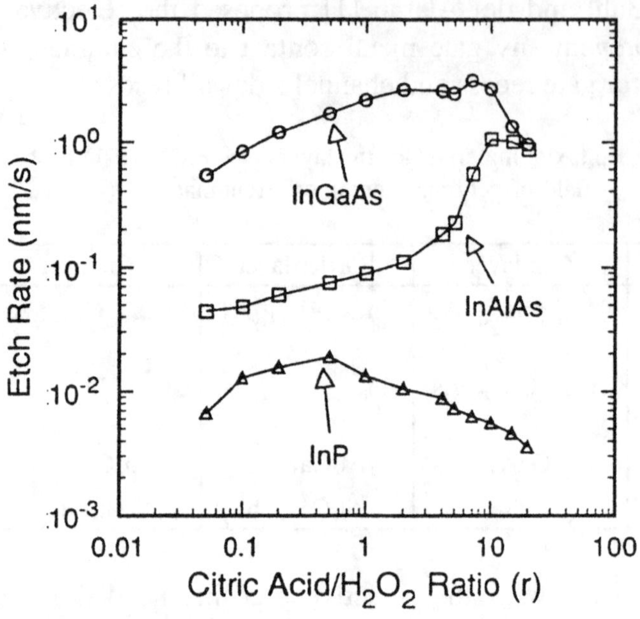

FIGURE 2 Etch rates of $In_{0.53}Ga_{0.47}As$, $In_{0.52}Al_{0.48}As$, and InP in citric acid/H_2O_2 solutions.

tapered along both [110] and [1$\bar{1}$0] directions. In a solution with r = 2, etch rates increased with temperature. The derived activation energies for InGaAs, InAlAs, and InP are 16, 18.9, and 17.4 kcal/mol, respectively. Different ratios of this solution have been used to perform mesa etch, channel sidewall recess etch, and gate recess etch [4].

B2 Organic Acids

Although the NH_4OH-H_2O_2 system is used extensively to selectively etch GaAs on AlGaAs, the etch does not form soluble complexes with indium and therefore cannot be used to etch indium-based compounds. The etch rate of InGaAs in a 1:30 solution of NH_4OH:H_2O_2 at 4°C is as low as 1.25 nm/min and that of InAlAs is 0.25 nm/min [5].The etch rate of $In_{0.18}Ga_{0.82}As$ in NH_4OH:H_2O_2 with a pH of 7.0 at 21°C is 6.7 nm/min [6]. Broekaert and Fonstad [7,8] alleviated this etch rate limitation by adding organic acids which readily form soluble complexes with indium. They have investigated an extensive number of acids, some of which are listed in TABLE 2. The preparation of the acids into aqueous solutions is also shown in TABLE 2 with the corresponding pH value to which each has been adjusted using ammonia. Etch rates of InGaAs and InAlAs and the corresponding selectivities as determined in mixtures of organic acids with hydrogen peroxide by Broekaert and Fonstad [8] are summarised in TABLE 3. Selectivities of both materials with respect to AlAs have also been measured. InGaAs to AlAs selectivities are very high, exceeding 1000 for succinic acid (SA). Dimethylsuccinic acid (DSA) and SA have demonstrated high selectivities for the etching of InGaAs on InAlAs. Bahl and del Alamo [1] measured a selectivity of 23 for InGaAs on InAlAs using 6:1 SA:H_2O_2 when buffered to a pH of 5.5 using NH_4OH. For a 10:1 ratio of the same solution at the same pH, a selectivity of 50 was obtained as shown in TABLE 3. This 6:1 mixture has been successfully applied to selective etching of the InGaAs channel sidewall. For DSA, the selectivity is over 125. The mechanism for the selectivity obtained for InGaAs over InAlAs or AlAs is due to the slow removal of oxidised Al.

TABLE 2 The concentration and pH of some of the acid ammonia solutions that have been utilised in etching InGaAs and InAlAs. Acids marked with asterisks need ammonia for complete dissolution of the acid [7,8].

Label	Acid added per litre of H_2O	pH, by adding ammonia
SA	200 g succinic*	4.2
AA	200 g methylsuccinic*	5.2
MSA	250 g methylsuccinic*	3.8
DSA	100 g 2, 2 dimethyl-succinic*	4.0
PTA	200 g 1, 2, 3 propane tricarboxylic	1.75
BTA	200 g 1, 2, 3, 4 butane tetracarboxylic	1.85

TABLE 3 Etch rates and selectivity of $In_{0.53}Ga_{0.47}As$ and $In_{0.52}Al_{0.48}As$ in some acid solutions [7,8].

Acid:H_2O_2 ratio (pH)	InGaAs etch rate (nm/min)	InAlAs etch rate (nm/min)	InGaAs to InAlAs selectivity
AA 25:1	30	0.4	75
MSA 25:1	40	0.5	80
DSA 25:1	25	<0.2	>125
PTA 25:1 (1.75)	50	1	50
BTA 25:1 (4)	50	<0.5	>100
SA 10:1 (5.5)	150	3	50

B3 H_2SO_4:H_2O_2:H_2O

Nishitani and Kotani [9] have shown that this is a very slow etch for (100) InP which also provides a smooth etched surface. The etch rate is dependent on the H_2O_2 content; however, the overall etch rate for (100) InP is very low. For selective etch applications, this system is used for the removal of InGaAs on InP. Various volume ratios (1:1:5 [10], 1:10:220 [11]) of the solution have been used to perform gate recess etch with InP as an etch stop. No etch rates were given. A 1:1:10 solution at 38°C etches InGaAs at 1860 nm/min with no measurable etching of InP [12]. Otsubo et al [13] have shown the etchant solutions with high H_2SO_4 content attack photoresists.

B4 H_3PO_4:H_2O_2:H_2O

This system is used for the selective etching of InGaAs on InP. It also etches InAlAs, and therefore can only be used as a selective etch where InP is utilised as the etch stop. Various volume ratios (1:1:20 [14], 1:1:8 [15], 38:1:1 [16], 1:1:38 [12], and 1:1:40 [17]) have been used for gate recess applications. The 1:1:8 solution etches InGaAs at a rate of 470 nm/min while the etch rates of InGaAs and InAlAs in the 1:1:38 solution are 132 nm/min and 210 nm/min, respectively. A 3:1:150 solution has been shown to etch InGaAs on $In_{0.75}Ga_{0.25}P$ with a selectivity of 20 at 15°C.

B5 HCl:H_2O

This etchant is normally used for InP etching. Recent work by Sauer and Chough [18] has shown that it can be used to selectively etch InAlAs over InGaAs. For a 3:1 solution, the etch rate of InAlAs is 648 nm/min while the etch rate of InGaAs is much below 0.33 nm/min. This

3:1 solution also etches InAlGaAs quaternaries selectively. For InAlGaAs with an AlAs mole fraction of x = 0.20, lattice-matched to InP, no etching occurs. However, for InAlGaAs with x = 0.34, the etch rate is ~110 nm/min.

C DRY ETCHING

Selective dry etching is used primarily for the gate recess process in the fabrication of HFETs. This process is well established for GaAs/AlGaAs HFETs. A variety of processes based on reactive ion etching (RIE) in CCl_2F_2 and similar gases have been developed for the removal of GaAs on AlGaAs [19]. It is only recently that work on dry selective etching involving indium-based compounds has yielded dividends. To date, the processes available are for the selective etching of InGaAs on InAlAs.

C1 CH_4:H_2

The reactive ion etching characteristics of InGaAs and InAlAs in CH_4:H_2 plasmas have been investigated by Lauterbach et al [20] and Andideh [21]. These experiments were carried out in parallel plate reactors operating at a radio frequency of 13.56 MHz. For the etching conditions of 1:2 CH_4:H_2, power density of 0.32 W/cm^2, and a self-bias of -470 V, Lauterbach et al [20] obtained etch rates of 15 nm/min and 2.5 nm/min for InGaAs and InAlAs, respectively. Andideh [21] obtained etch rates of 32 nm/min and 4.2 nm/min for InGaAs and InAlAs, respectively, at the etching conditions of 1:3 CH_4:H_2, power density of 0.44 W/cm^2, pressure of 60 mtorr, and a self-bias voltage of -420 V. Although devices have been fabricated using these processes for the gate recess etch [20,21], these are not viable processes because of the low selectivities (< 8), high self-generated bias, and polymer formation which is always a by-product of methane-based plasmas. Processes involving high self-bias voltages must be avoided in gate recessing to prevent the degradation of HFETs.

C2 CH_3Br

Kuroda et al [22] have investigated a photochemical dry etching process involving CH_3Br and a low-pressure mercury lamp (λ = 185 nm and 254 nm) for the etching of InGaAs and InAlAs. The CH_3Br introduced into a high vacuum etching chamber was decomposed by UV light from a mercury lamp which also activates the reaction at the substrate. At a substrate temperature of 70°C, the etch rates of InGaAs and InAlAs are 17 nm/min and 0.7 nm/min, respectively, resulting in a selectivity of about 25. Since the etch mechanism is purely chemical, the possibility of ion-induced damage is avoided. The etch stop mechanism is the formation of aluminium oxide which inhibits etching. Highly uniform InAlAs/InGaAs MODFETs have been fabricated using this process.

C3 HBr

Agarwala et al [23] have demonstrated selective reactive ion etching of InGaAs on InAlAs in HBr plasmas. Etching was conducted in a parallel plate reactor operating at 13.56 MHz. For the etching conditions of 120 mtorr chamber pressure, RF net power density of 0.23 W/cm^2, and a plasma self-bias voltage of 100 V, the etch rates of InGaAs and InAlAs are 11 nm/min and 0.067 nm/min, respectively. The associated selectivity is over 160 which is the highest reported to date for this material system. Etch rates at other bias voltages are shown in TABLE 4. As should be expected, selectivity is degraded at higher bias voltages due to the physical etching component of RIE [24]. This etch process has been successfully applied to the fabrication of InAlAs/InGaAs MODFETs.

TABLE 4 Etch rates and selectivity of $In_{0.53}Ga_{0.47}As$ and $In_{0.52}Al_{0.48}As$ in various dry etching processes.

Etch gases	Plasma self-bias voltage (-V)	Etch rates (nm/min)		InGaAs to InAlAs selectivity	Ref
		InGaAs	InAlAs		
CH_4/H_2 (RIE)	470	15	2.4	6	[20]
	420	32	4.2	7	[21]
CH_3Br (Photochemical)	-	17	0.7	25	[22]
HBr (RIE)	100	11	0.067	160	[24]
	200	26	0.50	50	
	300	73	1.37	50	

D CONCLUSION

The selective etching of InGaAs and its heterostructures using wet and dry etching techniques has been discussed. Although the discussion was particularly focused on the application of selective etching to the fabrication of HFETs, the processes enumerated above are useful for the fabrication of other heterostructure devices. More work needs to be done in this area, especially in dry etching, in order to permit flexibility in the processing of InGaAs and its heterostructures.

REFERENCES

[1] S. Bahl, J. del Alamo [*IEEE Electron Device Lett. (USA)* vol.13 (1992) p.195-7]

[2] S. Mukherjee, D. Woodard [in *Gallium Arsenide Materials, Devices, and Circuits* Eds M.J. Howes, D.V. Morgan (Wiley-Interscience, 1985) p.119]

[3] M. Tong, K. Nummila, A. Ketterson, I. Adesida, L. Aina, M. Mattingly [*J. Electrochem. Soc. (USA)* vol.139 (1992) p.L91-3]

[4] M. Tong, K. Nummila, A. Ketterson, I. Adesida, C. Caneau, R. Bhat [*IEEE Electron Device Lett. (USA)* vol.13 (1992) p.525-7]

[5] M. Kamada [private communication]

[6] Ch. Greus, A. Forchel, J. Straka, K. Pieger, M. Emmerling [*J. Vac. Sci. Technol. B (USA)* vol.9 (1991) p.2882-5]

[7] T.P. Broekaert, C.G. Fonstad [*IEEE Trans. Electron Devices (USA)* vol.39 (1992) p.533-6]

[8] T.P. Broekaert, C.G. Fonstad [*J. Electrochem. Soc. (USA)* vol.139 (1992) p.2306-9]

[9] Y. Nishitani, T. Kotani [*J. Electrochem. Soc. (USA)* vol.126 (1979) p.2269-71]

[10] K. Matsuda, M. Kubo, K. Ohnaka, J. Shibata [*IEEE Trans. Electron Devices (USA)* vol.35 (1988) p.1284-8]

[11] D.R. Greenberg, J. del Alamo, R. Bhat [*IEEE Electron Device Lett. (USA)* vol.13 (1992) p.137-9]

[12] A. Ketterson, I. Adesida [unpublished]
[13] M. Otsubo, T. Oda, H. Kumabe, H. Miki [*J. Electrochem. Soc. (USA)* vol.123 (1976) p.676-80]
[14] E.F. Schubert, W.T. Tsang, M.D. Feuer, P.M. Mankiewich [*IEEE Electron Device Lett. (USA)* vol.9 (1988) p.145-7]
[15] J.C. Gammel, H. Ohno, J.M. Ballantyne [*IEEE J. Quantum Electron. (USA)* vol.17 (1981) p.269-72]
[16] A.S.H. Liao, B. Tell, R.F. Leheny, T.Y. Chang [*Appl. Phys. Lett. (USA)* vol.41 (1982) p.280-2]
[17] A. Mesquida Kusters, A. Kohl, R. Muller, V. Sommer, K. Heime [*IEEE Electron Device Lett. (USA)* vol.14 (1993) p.36-9]
[18] N.J. Sauer, K.B. Chough [*J. Electrochem. Soc. (USA)* vol.139 (1992) p.L10-11]
[19] K. Hikosaka, T. Mimura, K. Joshin [*Jpn. J. Appl. Phys. (Japan)* vol.20 (1981) p.L847-9]
[20] Ch. Lauterbach, H. Albrecht, M. Beschoner, R. Gessner, M. Schier [*Proc. 3rd Int. Conf. Indium Phosphide and Related Materials*, Wales, UK (IEEE, USA, 1991) p.610]
[21] E. Andideh [Ph.D. thesis, University of Illinois, USA, 1990]
[22] S. Kuroda, K. Imanishi, N. Harada, K. Hikosaka, M. Abe [*IEEE Electron Device Lett. (USA)* vol.13 (1992) p.105-7]
[23] S. Agarwala, I. Adesida, C. Caneau, R. Bhat [*Appl. Phys. Lett. (USA)* vol.62 to be published in May 31st, 1993 issue]
[24] S. Agarwala, I. Adesida, C. Caneau, R. Bhat [to be presented at the *37th Int. Symp. on Electron, Ion, Photon Beams*, San Diego, USA, June 1993]

8.4 X-ray characterisation of InGaAs

C.R. Wie

January 1993

A INTRODUCTION

A compositional change in the InGaAs alloy causes a large variation in both lattice constant and refractive index (or structure factor), and therefore both lattice spacing and refractive index can be a sensitive tool in X-ray characterisation of InGaAs. Here, we discuss several structural properties of InGaAs obtained from X-ray rocking-curve analysis (rocking-curve FWHM and X-ray penetration depth (Section B), critical layer thickness and strain relaxation (Section C), elastic strain and composition (Section D), and parameters for lattice-matched or strained InGaAs quantum well (section E)).

B ROCKING-CURVE FWHM AND X-RAY PENETRATION DEPTH

For a single InGaAs heteroepitaxial layer, the theoretical intrinsic width of a CuKα 004 rocking-curve profile limited by dynamical diffraction and instrumental convolution (4-reflection Si(220) monochromator) may be given as

$$\text{FWHM}(\text{"arc}) = 19.928 - 31.379y + 41.415y^2 - 45.025y^3 + 27.912y^4, \quad R^2 = 1.000 \tag{1}$$

where y = log(h) with h being the $In_{0.5}Ga_{0.5}As$ layer thickness in μm. Eqn (1) is a polynomial fit to theoretical FWHM of rocking curves simulated using dynamical diffraction theory and convoluted with the monochromator function. For the entire composition range $0 \le x \le 1$, Eqn (1) is valid within 1"arc for $0.35\ \mu m \le h \le 8.0\ \mu m$ and within 4 or 5 "arc for $0.1\ \mu m \le h < 0.35\ \mu m$. Type and number of the monochromator crystals have only a small influence on its value. FWHM decreases monotonically with layer thickness, being about 165" at 0.1 μm, 84" at 0.2 μm, 43" at 0.4 μm, 24" at 0.8 μm, and 13" at 2.0 μm. For thicknesses greater than about three times the extinction depth (i.e. > 2 μm for $In_{0.5}Ga_{0.5}As$), the CuKα FWHM remains approximately constant at about 10". For high-quality layers lattice-matched to InP, the experimental data up to 1.5 μm in the layer thickness seem to agree with Eqn (1) within a factor of two [1]. The intrinsic width of the diffraction profile depends on X-ray wavelength. To compare the rocking-curve FWHMs obtained with a different X-ray wavelength, a constant conversion factor may be used: for example, FWHM(MoKα) ≈ 0.40 FWHM(CuKα) and FWHM(FeKα) ≈ 1.43 FWHM(CuKα) for both bulk crystals (see TABLE 1, column 2) and a thin epitaxial layer.

TABLE 1 Angular width of total reflection (≈ rocking curve FWHM) and X-ray penetration depth for 004 reflection in bulk crystals. λ = 1.54051 Å (Cu), 1.93597 Å (Fe), and 0.70926 Å (Mo).

	Width of total refl. ('')	Absorption depth (μm)	Extinction depth (μm)
GaAs	8.2(Cu); 12(Fe); 3.5(Mo)	15.9(Cu); 10.6(Fe); 7.84(Mo)	0.74(Cu); 0.72(Fe); 0.69(Mo)
$In_{0.5}Ga_{0.5}As$	8.9(Cu); 13(Fe); 3.7(Mo)	7.77(Cu); 5.27(Fe); 8.78(Mo)	0.67(Cu); 0.66(Fe); 0.65(Mo)
InAs	9.6(Cu); 14(Fe); 3.9(Mo)	5.38(Cu); 3.67(Fe); 9.82(Mo)	0.61(Cu); 0.61(Fe); 0.61(Mo)

For other wavelengths, the intrinsic width can be estimated using the fact that it is approximately proportional to

$$\lambda / \sqrt{4\,d^2 - \lambda^2}$$

where λ is wavelength and d is lattice plane spacing. Values which are substantially larger than Eqn (1) may be attributed to poor layer quality, strain relaxation and wafer bending. TABLE 2 shows experimental rocking-curve FWHM data for a partially strain-relaxed 3 μm-thick $In_xGa_{1-x}As$ layer on GaAs(001), grown by MOCVD (rows 1 and 2) [2] and a 2.0 μm-thick layer by MBE (rows 3 and 4) [3]. Interpretation of FWHM data of partially-relaxed strained InGaAs layers must consider the following two points: (1) preferential propagation of threading dislocations into either side of the mismatched interface during strain relaxation; and (2) the X-ray penetration depth.

TABLE 2 FWHM in "arc of experimental CuKα 004 rocking-curves. First and second rows: a 3.0 μm thick MOCVD-grown $In_xGa_{1-x}As$ layer on GaAs(001) [2]; and third and fourth rows: a 2.0 μm thick MBE-grown InGaAs layer [3].

x	0	0.07	0.13	0.19	0.27	0.37	0.48	0.59	0.69	0.76	0.84	0.9	0.94	1
fwhm	24.7	248	362	465	817	881	827	944	749	630	547	582	562	278

x	0.1	0.2	0.3	0.4	0.5	0.63	0.75	0.9	1
fwhm	371	680	969	1,196	1,134	1,155	1,381	856	371

(1) As for the preferential propagation of threading dislocations in $In_xGa_{1-x}As$/GaAs, a transmission electron microscopy (TEM) study of layers with x up to x = 0.5 [4] shows the threading dislocations being present predominantly in the GaAs substrate rather than in the InGaAs epilayer for small values of x (x < 0.18), in both epilayer and substrate for medium values (0.18 < x < 0.28), and predominantly in the InGaAs epilayer for large values (0.28 < x < 0.5). This could partially account for the increasing FWHM with increasing x up to 0.5 in TABLE 2. For 0 < x < 0.5, the increasing FWHM is also due to the increasing lattice-mismatch that will lead to more misfit and threading dislocations. The decreasing FWHM with x beyond x = 0.5 is not clearly understood. TEM data for x > 0.5 are not available to my knowledge. A similar trend in the FWHM vs. x data exists in the partially-relaxed Ge_xSi_{1-x}/Si samples [5].

(2) For perfect crystals, the exact penetration depth at any given angle can be calculated from the Darwin-Prins theory [6]. The extinction depth is the relevant penetration depth in interpreting the FWHM data because penetration length at an incident angle lying within the Bragg profile equals the extinction length. The X-ray signal comes nearly 65% from the first extinction depth and over 85% from the first two extinction depths.

TABLE 1 shows that the 004 extinction depth in InGaAs ranges from 0.61 μm to 0.74 μm depending on radiation and composition. Therefore, over 95% of the X-ray signal comes from the region within about 2 μm from the surface. A larger FWHM value is expected in the 2 μm-thick layer than in the 3 μm-thick layer because the InGaAs layer responsible for the X-ray signal is more defective in the 2 μm samples than in the 3 μm samples. Listed in TABLE 3 are FWHM values for an MBE-grown InGaAs layer on GaAs(001) obtained with MoKα (rows 1 and 2, h = 0.2 μm, [7]) and with CuKα (rows 3, 4 and 5, [8]). If we convert the MoKα value to a value equivalent to CuKα in the same way as for the intrinsic width of a perfect crystal (i.e. divide the MoKα FWHM by 0.4), those significantly relaxed 0.2 μm layers (the last three compositions in TABLE 3, rows 1 and 2) give an FWHM which is 4 to 7 times larger than a corresponding value in TABLE 2. Thus, the rocking-curve FWHM of partially-relaxed strained InGaAs layers strongly depends on the layer thickness due to a non-uniform depth-distribution of defect density and the small X-ray extinction depth.

TABLE 3 FWHM values of MBE-grown InGaAs on GaAs(001) from [7] (first two rows) and from [8] (last three rows). The superscript r represents films with strain relief well above 10^{-4}, see [12].

x	0.05	0.08	0.12	0.15	0.2	thickness 0.2 μm
MoKα	62"	62"	634"^{r}	922"^{r}	1248"^{r}	

x	0.07	0.07	0.14	0.14	0.25	0.25
h	0.2 μm	1.5 μm	0.1 μm	1.0 μm	0.03 μm	1.0 μm
CuKα	123"	338"^{r}	315"	646"^{r}	662"	1243"^{r}

The FWHM data may be further broken down according to

$$\mathrm{FWHM} = \sqrt{\beta_i^2 + \beta_r^2 + \beta_d^2} \qquad (2)$$

where β_i is the intrinsic FWHM given by Eqn (1) which includes the instrumental broadening, β_r is due to sample curvature, and β_d is due to defects in the crystal [9]. The 004 CuKα values can range up to well over 1000" arc for β_d, from 10" to over 100" for β_r depending on the residual elastic strain (or wafer bending), and from 200" down to 10" for β_i depending on layer thickness. In summary, the rocking-curve FWHM of lattice-matched or strained pseudomorphic InGaAs layers is shown to depend strongly on layer thickness (especially below 0.5 μm, see Eqn (1)) and X-ray wavelength (TABLE 1, column 2). These points are seldom mentioned in the discussion of rocking curve FWHM in the literature, but are important in comparing different samples. In partially strain-relaxed InGaAs layers, an important consideration should be given to the distance between the mismatched interface and the sample surface. This is because of the short X-ray penetration depth (0.61 - 0.74 μm for

004 reflection in InGaAs) and a better structural quality farther away from the mismatched interface.

C CRITICAL LAYER THICKNESS AND STRAIN RELAXATION

The critical thickness h_c of a single strained InGaAs layer, for relaxation of strain via formation of misfit dislocations, may be given approximately as

$$h_c = A \mid x - x_0 \mid^{-b} \tag{3}$$

TABLE 4 lists values of parameters, A and b, obtained from fitting Eqn (3) to the values calculated using the single-kink mechanism, anisotropic elastic constants and the equilibrium theory of Matthews and Blakeslee [10]. This simple formula gives a good fit to the calculated data as the R^2 value indicates for both GaAs(001) and InP(001) substrates.

TABLE 4 The parameter values of Eqn (3) for a strained InGaAs layer.

Substrate	x_0	A (Å)	b	R^2	x range
GaAs	0	8.16	1.24	1	$0 < x < 0.4$
InP	0.53	10.27	1.23	1	$0 < x-x_0 < 0.4$
InP	0.53	7.79	1.31	1	$0 < x_0-x < 0.4$

Critical thickness according to the model by Matthews and Blakeslee (MB) is the thickness at which misfit dislocations are just formed by elongation at the interface through glide of existing threading dislocations. The critical thickness determined from the change in Bragg peak separation, however, is the layer thickness at which the strain is already relieved by at least 10^{-4}. Strain relief at $h = h_c$ given by the MB model is much less than 10^{-4}. The plastic flow model by Dodson and Tsao [11] gives a proper account for the critical thickness measured from the X-ray rocking-curves. This model can also predict quantitatively the amount of strain relief at any given layer thickness and lattice mismatch for InGaAs [12,13]. X-ray rocking curve data and optical micrographs [14] or X-ray data combined with TEM pictures [15] show a significant difference in strain relaxation in the two <110> in-plane directions for the InGaAs system. Difference in the in-plane mismatch (or the strain relief) is manifested as a different cross-hatched line density in the two directions [14,15].

D LAYER COMPOSITION AND ELASTIC STRAIN

For a pseudomorphic InGaAs layer, whether strained or lattice-matched, the composition deduced from perpendicular lattice-spacing by applying Vegard's law and linear elasticity theory is usually accurate. A careful study of lattice-matched AlGaAs/GaAs samples has shown that the composition deduced from X-ray lattice-spacing data is accurate for the AlGaAs ternary alloy (even for samples slightly strain-relieved by about 3×10^{-5}) [16]. The lattice constant varies more rapidly with composition in the InGaAs alloy (about 56 times more rapidly than in AlGaAs), and therefore the compositions obtained from the rocking-curve data are more reliable for this alloy. However, this simple method of Bragg peak separation can result in a significant underestimation of layer composition if the InGaAs layer is thinner than [17]

$$h\ (\text{Å}) = 3.9\ \frac{(1-\nu)a_s}{(1+\nu)(a_f - a_s)} \tag{4}$$

where ν is the Poisson ratio of InGaAs and a is the cubic unit cell lattice constant for substrate (s) and fully-relaxed epilayer (f). However, h in Eqn (4) is significant (i.e. h > a few 100 Å) only for layers nearly lattice-matched to InP. For an InGaAs layer thicker than h of Eqn (4), the InAs mole fraction (x) and elastic strains (ε_{xx} and ε_{zz}) may be found from the Bragg peak separation ($\Delta\theta_B = \theta_{B,InGaAs} - \theta_{B,sub}$) using:

$$x = x_0 - A\ \frac{\sqrt{4d^2 - \lambda^2}}{\lambda}\ \Delta\theta_B(\text{"arc}) \tag{5a}$$

$$\varepsilon_{zz} = -B\ \frac{\sqrt{4d^2 - \lambda^2}}{\lambda}\ \Delta\theta_B(\text{"arc}),\ \varepsilon_{xx} = -C\varepsilon_{zz} \tag{5b}$$

GaAs substrate: $x_0 = 0$; $A = 3.559 \times 10^{-5}$; $B = 2.298 \times 10^{-6}$; $4d^2 = 7.990$ Å^2; $C = 1.110$.
InP substrate: $x_0 = 0.5315$; $A = 3.54 \times 10^{-5}$; $B = 2.404 \times 10^{-6}$; $4d^2 = 8.610$ Å^2; $C = 1.017$.

Here, the surface is taken as x-y plane and surface normal as z-axis with the reflecting lattice planes (004) parallel to surface.

For pseudomorphic strained InGaAs layers, the Poisson ratio ν, deduced from X-ray measured lattice spacing and known composition, always comes close to the value calculated from elastic constants. For partially relaxed strained InGaAs layers, however, the Poisson ratio ν deduced from the X-ray measured $a_\perp$ and $a_{||}$ and the known layer composition (or known a_f) using [14]:

$$\frac{a_f - a_s}{a_s} = \frac{1-\nu}{1+\nu}\frac{a_\perp - a_s}{a_s} + \frac{2\nu}{1+\nu}\frac{a_{||} - a_s}{a_s} \tag{6}$$

substantially deviates from that calculated from elastic constants. The apparent Poisson ratio from Eqn (6) is given in TABLE 5 using the X-ray data from [7]. A large deviation in ν from 0.31 ~ 0.35, the value expected from elastic constants, is seen in TABLE 5. This shows that use of Eqn (6) for estimation of layer composition (from Vegard's law) and elastic strain, (e.g. $\varepsilon_{xx} = (a_{||} - a_f)/a_f$), with the Poisson ratio given by elastic constants, is unjustified in partially-relaxed layers. This problem needs further investigation.

TABLE 5 Apparent Poisson ratio ν estimated from Eqn (6) using X-ray data of [7] for partially strain relaxed 0.2 μm InGaAs layers.

x	0.05	0.08	0.12	0.15	0.2
$\varepsilon_\perp(10^{-3})$	6.58	9.94	10.6	13.1	16.8
$\varepsilon_\parallel(10^{-3})$	0.48	0.48	2.23	3.03	6.73
ν	0.32	0.27	0.15	0.13	0.14

E SINGLE AND MULTIPLE InGaAs QUANTUM WELLS

The interface structure of multiple quantum wells (MQW) and superlattices (SL) is important in their electrical and optical properties such as low leakage current, nonradiative recombination, and electron scattering. For an InGaAs/InP SL nominally lattice-matched to the InP(001) substrate, a large strain is expected in the interfacial biatomic layers even in the ideally abrupt interface. By comparing experimental and calculated high resolution X-ray diffraction profiles, Vandenburg et al [18] have reported a negative strain $\varepsilon_\perp = -1.7\%$ with d = 1.443 Å at the InP → InGaAs interface and a positive strain $\varepsilon_\perp = +1.9\%$ with d = 1.495 Å at the InGaAs → InP interface. They report that these values are an arithmetic average of the lattice spacing of InGaAs and InGaP for the former, and an average of InAs and InP for the latter. An alternative interpretation may be that, considering the fact that both arsenic and phosphorus shutters are open during the MBE growth of the interfacial group V atomic layer, these negative and positive strains are equivalent to an $As_{0.37}P_{0.63}$ atomic layer for the InP → InGaAs interface and an $As_{0.63}P_{0.37}$ atomic layer for the InGaAs → InP interface. The (004) crystal planes in [001] are therefore

$$\text{--- -P-In-}As_{0.37}P_{0.63}\text{-InGa-As- --- -InGa-}As_{0.63}P_{0.37}\text{-In-P- ---}$$

The perpendicular mismatch values are $\varepsilon_\perp = +2.5\%$ for In-$As_{0.37}P_{0.63}$, $\varepsilon_\perp = -4.2\%$ for $As_{0.37}P_{0.63}$-InGa, $\varepsilon_\perp = +4.3\%$ for InGa-$As_{0.63}P_{0.37}$, and $\varepsilon_\perp = -2.5\%$ for $As_{0.63}P_{0.37}$-In. Lyons et al [19] report a 9 - 10 Å wide graded interfacial layer, a conclusion drawn from matching the intensities of various superlattice peaks between experimental and calculated diffraction data. Because of the nearly lattice-matched nature of layers in the InGaAs/InP system and because of a large difference in atomic radius between arsenic and phosphorus, the high resolution X-ray diffraction is sensitive even to an interfacial atomic monolayer.

A strained InGaAs/GaAs MQW structure on GaAs(001) has important application in high efficiency laser diodes. All three basic layer parameters (ternary composition and two layer thicknesses) can be found from the X-ray rocking curve data. Wang et al [20] report that the width and peak intensity of the main SL orders and the asymmetry of interference fringes around the main SL peaks are good indicators of strain relaxation in this strained MQW structure. They also report that the relaxation process is governed by a single kink mechanism and the MB theory. For these samples, the photoluminescence signal was not sensitive to the strain relaxation at all [20].

F CONCLUSION

The X-ray rocking curve technique provides many structural properties of InGaAs. We have discussed some of these structural properties. It is pointed out that the very widely employed figure of merit, the rocking-curve FWHM, must be interpreted carefully by considering the X-ray wavelength, extinction depth, InGaAs layer thickness, and defect depth distribution. This point is particularly important in comparing the FWHM data obtained with different radiation on different samples. A theoretical formula, Eqn (1), is given for the intrinsic FWHM (004 CuKα). Experimental FWHM data, TABLES 2 and 3, of partially strain-relaxed InGaAs layer on GaAs(001), obtained from several papers, were presented along with discussion. A simple formula, Eqn (3), was given for the critical thickness of InGaAs based on the Matthews-Blakeslee theory [10]. Experimental X-ray data on critical thickness of InGaAs can be well explained by the Dodson-Tsao model [11,12]. The InGaAs layer composition and elastic strain can be reliably obtained using Eqn (5) if the layer is pseudomorphic and thicker than h of Eqn (4). For partially strain-relaxed InGaAs layers, Eqn (6) may not be accurate for finding the cubic unit cell lattice constant. This needs further examination. Presently, there is ongoing research on the very thin InGaAs layers, lattice-matched (such as InGaAs/InP) or strained (such as InGaAs/AlGaAs, see [21]). Investigation with atomic monolayer scale resolution is particularly interesting and requires further research. This is because the present state of research on X-ray determination of interfacial structures with a monolayer resolution is largely qualitative and somewhat arbitrary. This may be seen from the analysis of InGaAs/InP SL samples done by the same authors [22]. The ambiguity underscores the need for further work in the dynamical simulation and modelling study of InGaAs/InP and other SL and MQW structures.

REFERENCES

[1] E. Kuphal, A. Pocker, A. Eisenbach [*J. Appl. Phys. (USA)* (1993) to be published]

[2] D.I. Westwood, D.A. Woolf, R.H. Williams [*J. Cryst. Growth (Netherlands)* vol.98 (1989) p.782-92]

[3] S. Fujita et al [*J. Cryst. Growth (Netherlands)* vol.95 (1989) p.224]

[4] V. Krishnamoorthy, P. Ribas, R.M. Park [*Appl. Phys. Lett. (USA)* vol.58 (1991) p.2000]

[5] J.M. Baribeau, T.E. Jackman, D.C. Houghton, P. Maigne, M.W. Denhoff [*J. Appl. Phys. (USA)* vol.63 (1988) p.5738]

[6] R.W. James [*The Optical Principles of The Diffraction of X-rays* (Ox Bow Press, Woodbridge, CT, 1948), p.55-66]

[7] K. Kamigaki, H. Sakashita, H. Kato, M. Nakayama, N. Sano, H. Terauchi [*Appl. Phys. Lett. (USA)* vol.49 (1986) p.1071]

[8] P.J. Orders, B.F. Usher [*Appl. Phys. Lett. (USA)* vol.50 (1987) p.980]

[9] J. Lee, W.E. Mayo, T. Tsakalakos [*J. Electron. Mater. (USA)* vol.21 (1992) p.867]

[10] J.W. Matthews, A.E. Blakeslee [*J. Cryst. Growth (Netherlands)* vol.27 (1974) p.118]

[11] B.W. Dodson, J.Y. Tsao [*Appl. Phys. Lett. (USA)* vol.51 (1987) p.1325]

[12] C.R. Wie [*J. Appl. Phys. (USA)* vol.65 (1989) p.2267]

[13] Y.W. Choi, C.R. Wie, K.R. Evans, C.E. Stutz [*J. Appl. Phys. (USA)* vol.68 (1990) p.1303]

[14] C.R. Wie, H.M. Kim, K.M. Lau [in *Micro-Optoelectronic Materials*, SPIE Proc. vol.877 (1988) p.41]

[15] K.L. Kavanagh et al [*J. Appl. Phys. (USA)* vol.64 (1988) p.4843]

[16] B.K. Tanner, A.G. Turnbull, C.R. Stanley, A.H. Kean, M. McElhinney [*Appl. Phys. Lett. (USA)* vol.59 (1991) p.2272]

[17] C.R. Wie [*J. Appl. Phys. (USA)* vol.66 (1989) p.985]

[18] J.M. Vandenberg, M.B. Panish, H. Temkin, R.A. Hamm [*Appl. Phys. Lett. (USA)* vol.53 (1988) p.1920]

[19] M.H. Lyons, E.G. Scott, M.A.G. Halliwell [*Microscopy of Semiconducting Materials*, Inst. Phys. Conf. Ser. no.100 (1989) p.473]

[20] C.A. Wang, S.H. Groves, J.H. Reynolds, D.R. Calawa [*Electron. Mater. Conf. 1992*, June 24-26, MIT, Cambridge, MA, USA, paper no.P7]

[21] Y.W. Choi, H.M. Kim, C.R. Wie [*Mat. Res. Soc. Symp. Proc. (USA)* vol.240 (1992) p.627]

[22] J.M. Vandenberg, S.N.G. Chu, R.A. Hamm, M.B. Panish, D. Ritter, A.T. Macrander [*Mat. Res. Soc. Symp. Proc. (USA)* vol.240 (1992) p.141]

CHAPTER 9

DEVICE APPLICATION OF InGaAs AND RELATED HETEROSTRUCTURES

9.1 InGaAs photodetectors

T.P. Pearsall

February 1993

A INTRODUCTION

$Ga_{0.47}In_{0.53}As$ is a direct bandgap semiconductor with a fundamental energy gap of 0.75 eV at 295 K [1]. This material is usually grown as an epitaxial thin film on InP. Although there are efforts to develop bulk GaInAs crystals, these are still in the research stage [2]. GaInAs with a mole fraction of InAs equal to 0.53 can be grown lattice-matched directly on InP [3]. By increasing the mole fraction of InAs, the absorption edge can be extended to longer wavelengths accompanied by the introduction of lattice mismatch strain and misfit dislocations [4]. Following the successful epitaxial synthesis of this semiconductor, it soon became clear that it would be widely used in photodetection applications in the near infra-red [6-9]. Today, GaInAs photodiodes are standard items in the optoelectronics offerings of many electronic device suppliers.

The principal photodiode characteristics that determine device performance are spectral response, quantum efficiency, speed, capacitance, dark current, reliability and manufacturability [10]. GaInAs is a material that happily represents a combination of materials properties that leads to near optimum performance in all of these categories. The only competition for GaInAs in the near infra-red ($1\ \mu m < \lambda < 2\ \mu m$) is the Ge photodiode. Ge detectors do not offer the same level of performance with regard to spectral response, speed, and noise. However, they are still considerably less expensive than GaInAs detectors, and Ge detectors are still widely used in non-critical applications where photodetector cost represents a major consideration in system design [11]. The spectral range of useful photoresponse for Ge photodiodes is shown in FIGURE 1(a).

Starting from the minority-carrier diffusion equation, it is straightforward to calculate the response of a GaInAs photodiode absorbing G_L photons/s. The excess carrier density $\Delta n(x)$ obeys the minority-carrier diffusion equation:

$$\frac{\partial \Delta n(x)}{\partial t} = D_n \bullet \frac{\partial^2 \Delta n(x)}{\partial x^2} - \frac{\Delta n(x)}{\tau_n} + G_L = 0 \qquad (1)$$

The boundary condition at the p-n junction is specified by Boltzmann statistics:

$$\frac{\Delta n(x=0)}{n_o} = \exp(eV/kT) - 1 \qquad (2)$$

The solution to the differential equation is:

$$\Delta n(x) = [n_o \{\exp(eV/kT) - 1\} - G_L \bullet \tau_n]\bullet\exp(-x/L_n) + G_L \bullet \tau_n \qquad (3)$$

and the current-voltage relationship is derived using the expression for the diffusion current: $J = I/A = eD_n \bullet dn/dx$.

$$I = \frac{eAD_n}{L_n} \bullet \left[\frac{n_i^2}{N_A} \bullet \{\exp(eV/kT) - 1\} - G_L \tau_n \right] \tag{4}$$

$$\tau_n = (L_n)^2/D_n$$

Eqn (4) shows that the current response of semiconductor photodiodes is <u>exactly</u> linear in the incident light intensity. Deviations from linearity occur only in the high injection limit. The minority carrier lifetime τ in Eqn (4) has been measured for both n- and p-type material [12]. Other features of Eqn (4) such as the time-response and the dark current will be discussed in the following sections.

B SPECTRAL RESPONSE

The spectral response of GaInAs, like that of all semiconductor photodiodes, is determined by its bandgap. The response to photons with energy below 0.75 eV is essentially zero. The spectral response for above bandgap photons is determined by the absorption coefficient. In a direct bandgap material like GaInAs, the absorption coefficient rises to 10^4 cm^{-1} for energies only a few kT above the bandgap energy. Therefore, just above the bandgap, all photons are absorbed within a few microns of the surface. For photon energies considerably above the bandgap, near 1.0 eV, the absorption coefficient is even higher, ~ 10^5 cm^{-1}, so that a higher percentage of photons is absorbed near the surface. The photo-generated electron-hole pairs tend to recombine at the surface reducing the photocurrent of minority carriers reaching the p-n junction. This surface recombination reduces the response as the wavelength becomes shorter. This feature can be seen in FIGURE 1(b). Also shown in FIGURE 1(b) is the response of the photodiode when the Ga is covered with a wide bandgap window. The reduction of quantum efficiency due to surface recombination is nearly eliminated resulting in a more uniform response as a function of optical wavelength.

The quantum efficiency of a photodetector is the ratio of electrons collected in the external circuit to the photons incident on the detector surface:

$$\eta = \frac{\text{no. electrons out}}{\text{no. photons in}} = \frac{I\phi}{P_{opt}} \bullet \frac{\hbar\omega}{e} = R \bullet \frac{\hbar\omega}{e} \tag{5}$$

The quantity R is called the responsivity and has units of amps/watt. The responsivity is related to the quantum efficiency. For example, a photodetector with a quantum efficiency of 100% operating at $\lambda = 1.24$ μm has a responsivity of 1 amp/watt. The maximum possible quantum efficiency for a GaInAs photodiode is limited to about 75% because of optical reflection at the air/semiconductor interface. By adding an antireflection coating, the quantum efficiency can be increased to 100% at the optical wavelength of interest [13].

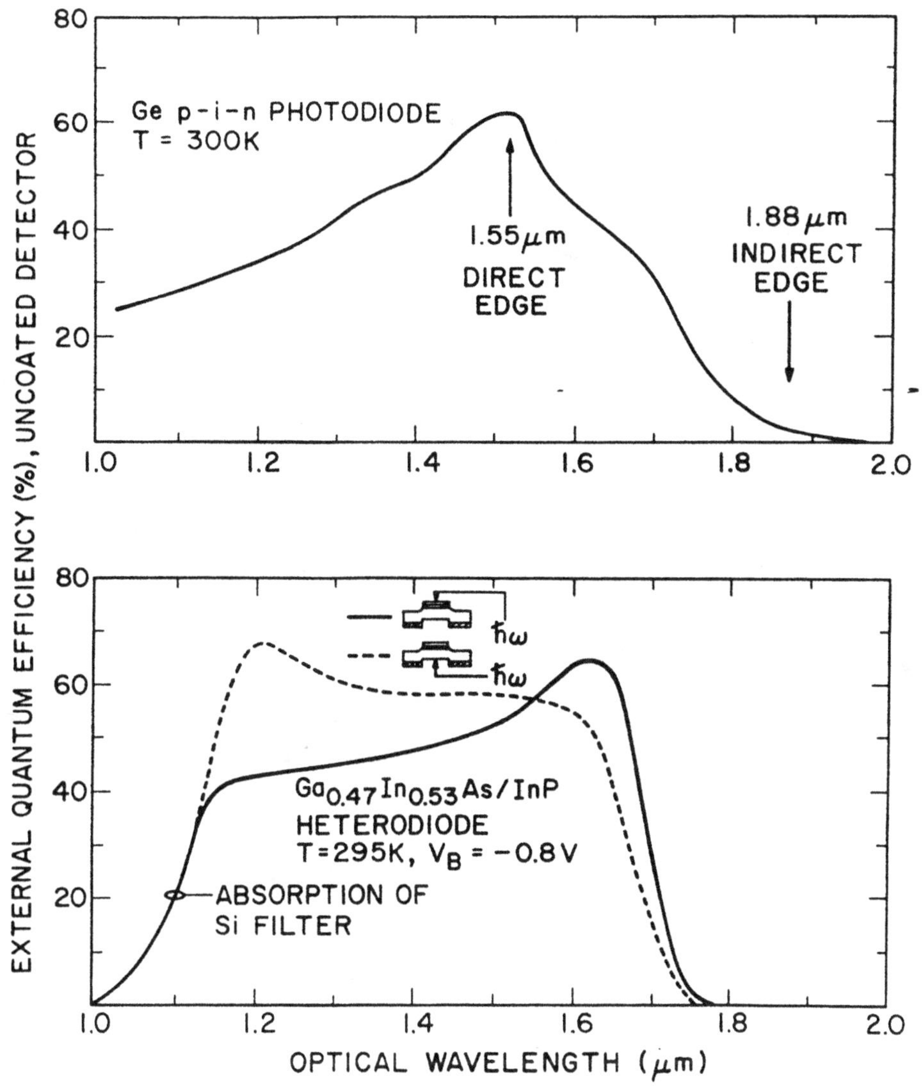

FIGURE 1 (a) Spectral response of a Ge photodiode at T = 295 K. The response shows two features, a slow-rising photoresponse at 1.8 μm and a second sharp rise near 1.5 μm. These features are the result of indirect gap absorption and direct gap absorption respectively. (b) Spectral photoresponse of a GaInAs photodiode. The sharp rise in the response near 1.6 μm reflects the 295 K bandgap (0.75 eV) of this material which is grown lattice matched on InP.

By increasing the InAs mole fraction above 0.53 required for 295 K lattice matching to InP, the bandgap is reduced. At the same time the compressive stress applied by an InP substrate tends to increase the bandgap. Overall, the bandgap is reduced. Martinelli et al [14] have commercialised these pseudomorphic photodiodes for applications in the 1.7 μm < λ < 2.7 μm range. In FIGURE 2 we show the optical spectral response of these photodiodes. Alternatives in this wavelength range have been developed using InAs and PbS.

The growth of strained GaInAs both on GaAs and InP has been reviewed in some detail by Schaff et al [5].

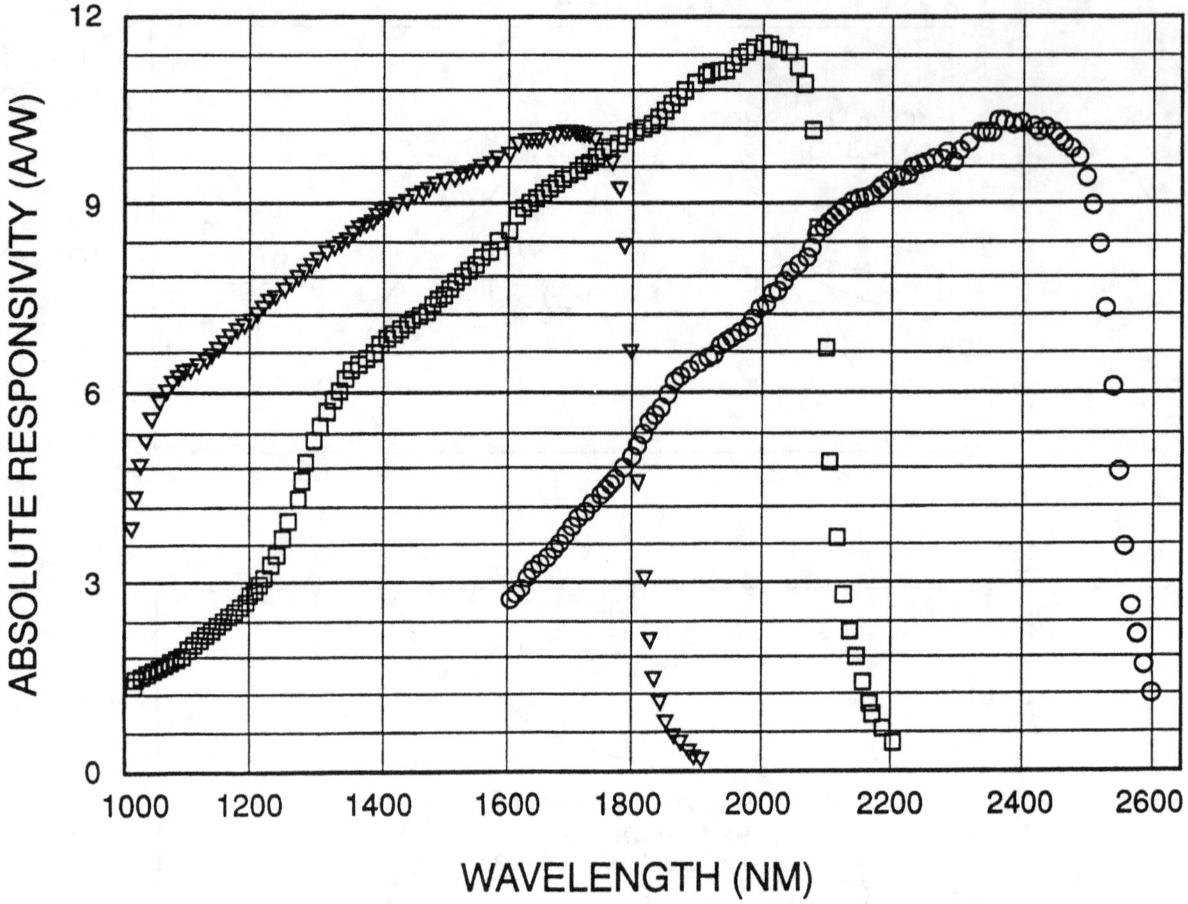

FIGURE 2 Extended wavelength response of strained GaInAs photodiodes grown on InP but containing a higher InAs mole fraction than is required for a good lattice match with the substrate. These photodiodes are currently marketed by EPITAXX, Inc. (reproduced by permission).

C DARK CURRENT

The principal source of photodetector dark current arises from generation and recombination at defect centres in the depletion region [6]:

$$I_{G\text{-}R} = \left[\frac{eAWn_i^2}{\tau_{eff}}\right] \bullet [\exp(eV/kT) - 1] \tag{6}$$

where W = width of the depletion region.

Note that this contribution to the dark current depends on the depletion width which increases as the square-root of the reverse bias voltage. The effective recombination time in the depletion region, τ_{eff}, varies over many orders of magnitude depending on where the recombination levels lie in the forbidden gap. For high quality n-type material they lie only a few meV from the band edge and τ_{eff} is at its maximum of approximately 10^{-6} s [6]. There is a secondary contribution from the same kind of generation-recombination current from centres lying outside the depletion region, but within a diffusion-length of the junction. J_{Diff} is given by the pre-factor in the I-V characteristic:

$$I_{Diff} = eAn_i^2\left[\frac{D_n}{L_nN_A} + \frac{D_p}{L_pN_D}\right] = kTAn_i^2\left[\frac{\mu_n}{L_nN_A} + \frac{\mu_p}{L_pN_D}\right] \quad (7)$$

The diffusion current is minimised in the p^+n configuration. The minority carrier electron mobility exceeds that for holes by a factor of about 20. This undesirable feature is offset by more heavily doping the p-side. The n^+p configuration would be expected to have a diffusion-current density about 4 orders of magnitude higher.

All sources of current in a photodetector, including the signal photocurrent, contribute to the noise power in photon detection, by generating a shot-noise power of $2e(I)BR_L$, where B is the bandwidth, and R_L is the load resistance. As long as the photo-current is larger than the dark current, the photo-current will be the dominant contributor to the noise power. The usual objective is to keep the level of shot-noise power below the equivalent noise power of the amplifier. This noise power is shown in FIGURE 3 as the dark current equivalent for the often used standard of one error in 10^9 bits. This figure shows that at 1 GHz, the dark current should be at least one order of magnitude less than 10^{-6} A in order not to degrade the signal-to-noise ratio of the detection circuit. This is easily achieved in GaInAs photodiodes whose diameter is 1 mm or less [13]. It is also easily achieved by a Ge photodiode [11]. At lower bandwidths, particularly below 1 MHz, the dark current level becomes much more critical. In this frequency range, the GaInAs photodiode is expected to be less noisy than a Ge photodiode, with the shot noise power being reduced by several orders of magnitude. In many kinds of lower frequency, low-light-level sensing applications, the GaInAs photodiode is a clear choice over a Ge detector.

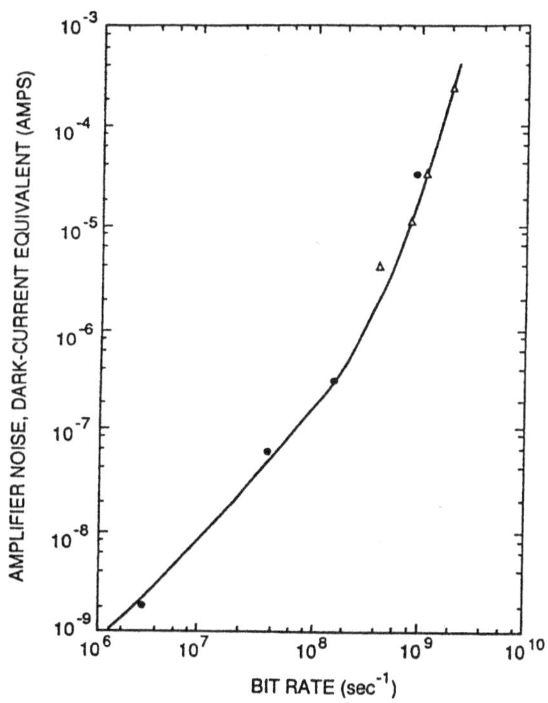

FIGURE 3 The noise power equivalent dark current for photodetectors shown as a function of communications bit rate for a digital communications system. This figure compares the noise associated with state of the art preamplifiers to the noise associated with a certain level of dark current. Thus, to be assured that the shot noise associated with the dark current is much less than the amplifier noise at 100 Mbits/s, then the dark current should be less than about 10^{-8} A. (• = FET front-end, Δ = bipolar front-end. Bit error rate is 10^{-9}.)

D SPEED

The response time of a photodiode is determined by three factors: the diffusion time associated with the transport of carriers photogenerated outside the depletion region, the drift time across the depletion region and the RC time constant associated with charging the external circuit. The diffusion time, τ_{Diff}, is given by:

$$\tau_{Diff} = \frac{L^2}{D} = \frac{e\,L^2}{kT\mu} \tag{8}$$

In Eqn (8), L is the shorter of the diffusion length or the thickness of the absorption layer. Since it requires only 2 μm to absorb nearly all of the incident optical flux, this is the appropriate length to use. The diffusion time is also determined by the minority carrier mobility. The response time is shorter for photons absorbed in p-type material since electrons have the higher mobility. For GaInAs lattice-matched to InP, this leads to a diffusion-limited response time of 200 ps with near ideal quantum efficiency. This figure is confirmed in experiment [15]. Absorption in n-type material would lead to a diffusion-limited time about 20 times longer.

Faster response requires that the photons be absorbed directly in the depletion region of the photodiode. This is accomplished most easily by designing a high bandgap window (InP, GaInAsP, or AlInAs) on either side of the depletion region. The transit time across the depletion region depends on the depletion region width which is a function of bias voltage and carrier concentration. When the wider bandgap InP substrate is used for the window, the diode is operated in the fully-depleted punch-through mode in order to avoid a diffusion tail from photons absorbed in the undepleted GaInAs. The desired width of the depletion region is several times the absorption length, typically 2 μm. The doping of this region must be low, less than 10^{15} cm^{-3}, so that the absorption region can be fully depleted with an applied bias of 5 volts or less. The transit time across the region is:

$$t_{Drift} = \frac{L}{v_{sat}} = 10 \text{ ps} \tag{9}$$

where the saturated drift velocity of 2 x 10^7 cm/s has been assumed. Commercially available GaInAs p-i-n detectors based on these design principles are available from Hewlett-Packard [16]. The rise time of these detectors is 13 ps in good agreement with Eqn (9). A faster photodiode response can be obtained only by sacrificing quantum efficiency [17]. Under these conditions, the speed is determined by the width of the depletion region (W), while the quantum efficiency is proportional to exp($-\alpha$W). Reducing the depletion region width by half will lower the quantum efficiency by only about 25%. A typical approach in GaAs design is to use the interdigitated metal-semiconductor-metal (MSM) detector configuration with submicron spacing between the contacts. Fast GaInAs MSM detectors with picosecond response times have been demonstrated using this approach [18,19].

Achieving the ultimate speed requires that the RC time constant is less than any other transport-related delays. The intrinsic resistance and capacitance of the photodetector are:

$$R = \frac{\rho \bullet L}{A},\ C = \frac{\varepsilon\varepsilon_0 \bullet A}{W} \text{ and } RC = \frac{\rho \bullet L \bullet \varepsilon\varepsilon_0}{W}$$

So, in principle, the RC time constant of the photodiode is independent of the surface area of the diode. The capacitance can be lowered by reducing the doping of the n^- region, and by applying a reverse bias. However, as the doping of the n^- region is lowered, its contribution to the series resistance goes up. This undesirable feature can be eliminated if the n^- region is fully depleted. For design of high-speed performance, the n-doping is less than 10^{15} cm^{-3}, and the desired 2 μm depletion width can be attained with less than 5 volts applied reverse bias.

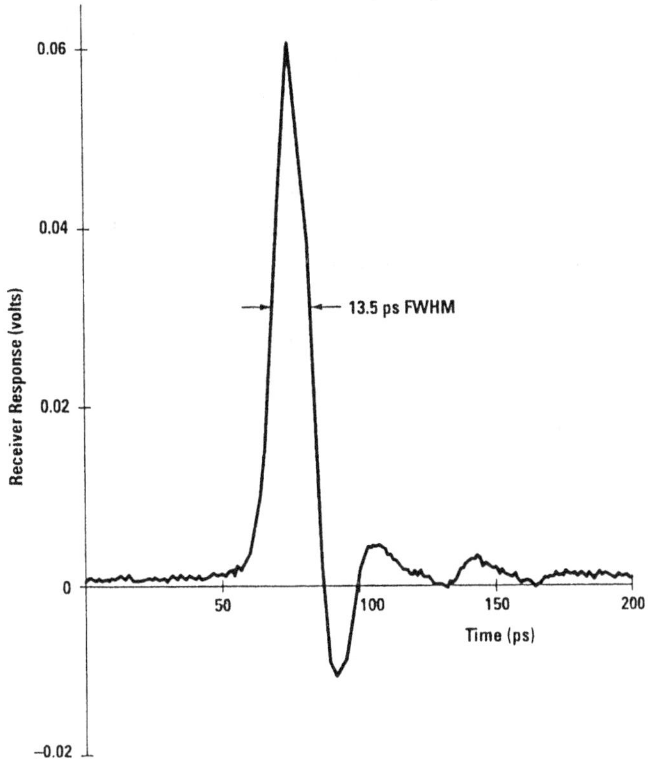

FIGURE 4 The response time of a GaInAs p-i-n photodiode is ultimately limited by the electron drift velocity across the depletion region. In a properly designed device the ultimate limit is about 10 ps without sacrificing quantum efficiency. In this figure, the response of a Hewlett-Packard HP83440 p-i-n detector is shown with a measured response time of 13.5 ps, which is close to this limit (from King et al [16], © 1993 Hewlett-Packard Company, reproduced with permission).

In determining the overall response time due to RC effects, the resistance of the external circuit, typically dominated by the input resistance of the following pre-amplifier stage, needs to be taken into account. In most cases this resistance is larger than the series resistance of the photodiode. This means that photodiode response becomes slower as its surface area increases - the result of the product of the internal capacitance and the external resistance. The larger this resistance, the lower the associated noise of the following preamplifier. There is obviously an important compromise between the reduction of noise, permitting higher sensitivity, and the bandwidth of the detector circuit. In addition, the capacitance of the photodetector contributes an AC noise contribution similar to that of thermal noise developed in the load resistor. This contribution will be discussed in more detail in Section F. The upper limit of capacitance is set by the amplifier noise power. This limit fixes the photodiode area, because n-doping levels cannot be practically reduced below 10^{14} cm^{-3}, and bias voltage greater than -5 volts is not standard practice. Assuming a 50 Ω charge resistance, the capacitance should be less that 0.02 pF in order not to degrade the minimum 10 ps response mentioned earlier. To achieve these characteristics, the photodetector should have an area less than 4 x 10^{-6} cm^2, or a diameter less than about 20 μm.

E AVALANCHE GAIN

Carrier avalanche multiplication can be used as a means to achieve internal gain in a photodiode. It results in an increase in the signal-to-noise ratio when the dark current is much less than the amplifier equivalent noise current. For GaInAs photodiodes, the dark current typically exceeds the amplifier noise current for bandwidths less than 1 MHz. So avalanche gain is a useful feature only above this operating bandwidth.

Reverse breakdown in pn junctions occurs when there is a large increase in the free carrier population in the depletion region. This increase has two sources: one is direct Zener tunnelling of electrons from the valence band to the conduction band and the other is avalanche multiplication of carriers that reach the depletion region by diffusion [13]. In materials with E_g less than 0.5 eV tunnelling dominates, and in materials with E_g greater than 1.0 eV avalanche multiplication dominates. In GaInAs reverse breakdown is a combination of the two [20]. Pure, usable avalanche is observed only in high purity ($N_D \leq 10^{15}$ cm^{-3}) material [21] or at temperature substantially less than 295 K.

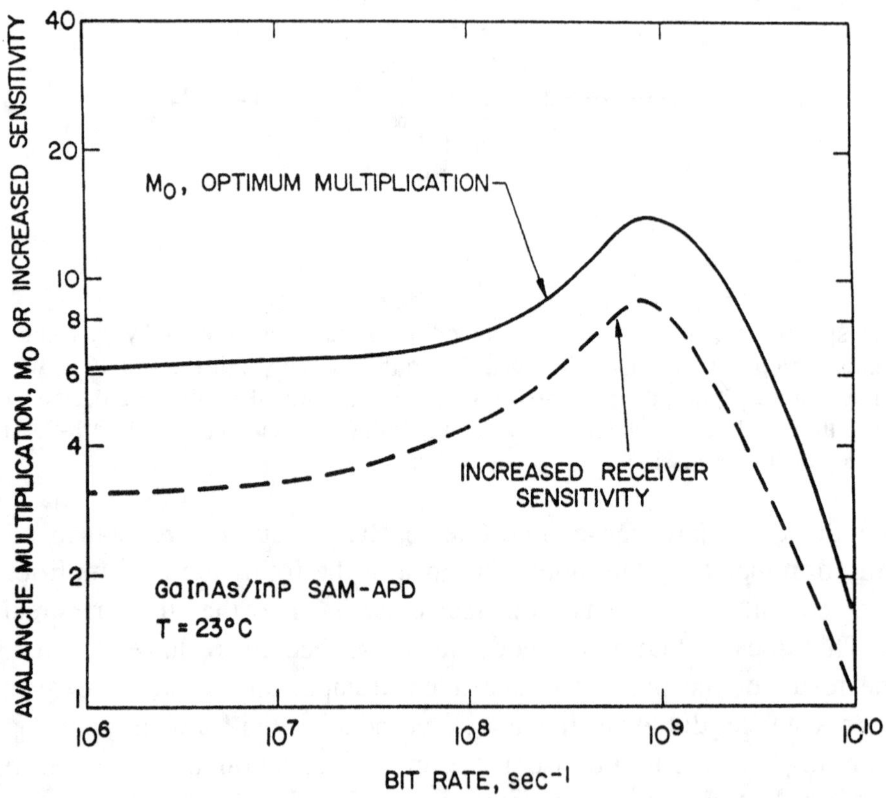

FIGURE 5 The optimum avalanche gain as a function of bit rate for GaInAs photodiodes. At the highest bit rates, the gain must be reduced to preserve the bandwidth. Eventually the gain reaches unity and for frequencies higher than this limit, a p-i-n photodiode must be used.

Successful avalanche photodiode design to avoid tunnelling breakdown was demonstrated by Nishida who used the lower bandgap materials for optical absorption, but created avalanche breakdown in higher bandgap InP [22]. This kind of structure is called a separate absorption and multiplication (SAM) photodiode [23]. The amount of useful avalanche gain that can be obtained is just that which raises the photodiode noise level up to that of the amplifier. In a 'perfect' photodiode with no dark current, the photocurrent shot noise will limit the gain. The avalanche gain process introduces its own noise as well. In most applications, the optimum avalanche gain is 10 or less.

The noise introduced by avalanche gain comes from the creation of electrons, which move one way, and holes which move in the opposite direction in the depletion region. Both kinds of carriers create additional electrons and holes throughout the depletion region extending the time duration of the signal that caused the avalanche process to begin. Therefore, the gain is obtained at the expense of speed of response or bandwidth. This feature puts an upper limit on the frequency range where avalanche gain can be used [13]. In FIGURE 5 the useful range of bandwidth for avalanche gain is shown.

F NOISE

The basic performance characteristic for a photodetector is the signal to noise ratio. The desired level of signal detection quality (for example a bit error rate of 10^{-9}) determines the signal to noise ratio. Then it is a straightforward matter to calculate the required optical power given the noise levels in the detection system. Noise comes from three sources: the shot noise from the optical signal, noise from the photodetector, and amplifier noise [13]. The photodetector-generated noise is composed of a shot noise component related to the DC dark current, and an AC component related to the capacitance of the diode. Both are important in determining the minimum detectable optical signal at a given bandwidth. In FIGURES 3 and 6 we show the minimum detectable optical power for a photodetection system expressed in terms of diode dark current and diode capacitance for a digital communication system [13]. Thus in order for the detection system not to be limited in detection sensitivity by the properties of the photodiode at 100 Mbits/s, the dark current must be less than 10^{-8} A and the capacitance must be less than 0.2 pF in order to have a negligible effect on detector performance.

Most of the desirable performance features of GaInAs photodiodes are achieved by growing a very high purity n-type GaInAs to form the depletion region. The n-region is most often grown by chemical vapour deposition in order to achieve the highest purity material. Liquid phase epitaxy can be used, but the wafer size is typically 2 orders of magnitude smaller. Molecular beam epitaxy produces wafers of equal area and quality, but with a lower throughput. Chemical beam epitaxy cannot currently produce material with the required low doping density. The necessary heavy p-doping for the contact region cannot be incorporated during growth without degrading the quality of the low-doped n-region. Hence, p-doping is incorporated afterwards by diffusion [10], or by ion-implantation [24].

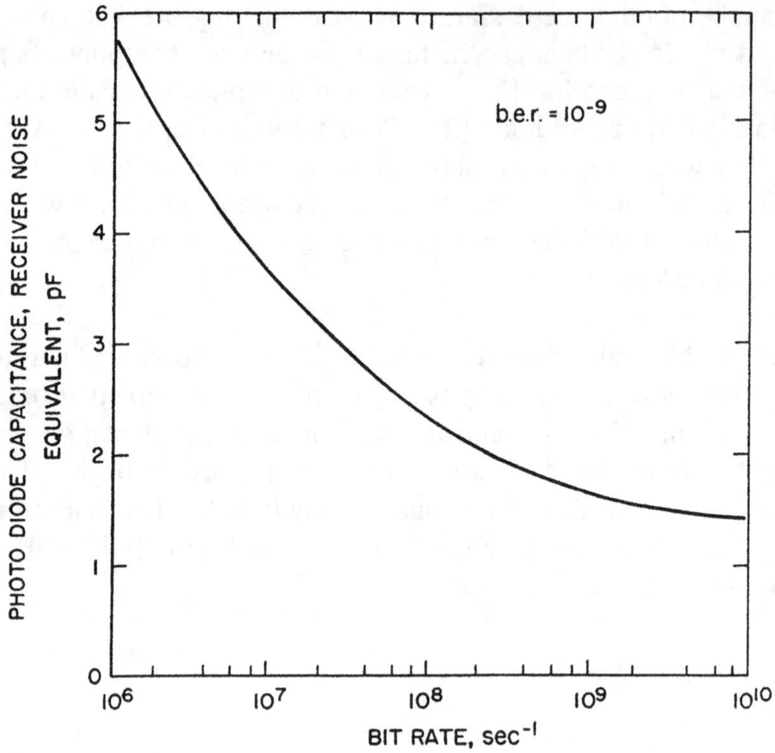

FIGURE 6 Noise-equivalent capacitance for GaInAs photodiodes as a function of bit rate. While the demands on dark current are relaxed as the detection bandwidth increases, the requirements on capacitance become more stringent. At a bit rate of 100 MHz, the capacitance must be well below 1 pF in order to have a negligible effect on photodetection performance.

G ARRAYS AND INTEGRATED STRUCTURES

The simplest and most widely-used integrated photonic device is the photodetector array. Array detectors have many applications from motion detection, to imaging, to parallel data collection. The basic array consists of a series of photodiodes which are interfaced to a switching network and a shift register for read-out. In imaging applications, it is helpful to have a second array of 'dummy' photodiodes so that differential measurements of received intensity can be made. Linear array detectors incorporating these design features are manufactured by EPITAXX as linear 1 x 30, 1 x 64, 1 x 128, and 1 x 256 arrays [25]. They are incorporated in EG&G optical multichannel analyser systems for long-wavelength spectroscopy.

GaInAs p-i-n structures are normally composed of several compositional layers. As such they are easily integrated with structurally similar devices such as bipolar transistors or lasers. The heterojunction phototransistor which is such a combination is a simple integrated device that has long been known for high gain, but long response times. Recent calculations modelled by Leu et al [26] show a theoretical f_T as high as 10 GHz which would, if verified experimentally, offer some improvement in frequency response typical of these devices. It falls far short, however, of the performance that can now be achieved with discrete p-i-n devices (150 GHz) and bipolar amplifiers (~ 100 GHz) using the same materials.

MSM photoconductive detectors are based on a metal semiconductor Schottky barrier that is inherently low ($\phi_B \sim 0.3$ eV), so high resistivity GaInAs is essential for an acceptable on-off conductivity ratio. MSM photodetectors are essentially planar structures, and are most easily integrated with other planar devices such as FETs, strip-line circuits and optical waveguides [27]. The advantages of the MSM structure, extremely fast response and simplicity of fabrication, can both be maintained in this kind of integration scheme. In most integrated photonics applications the emphasis is on these two features: speed and robust fabrication processes, rather than on high sensitivity/low-noise performance. This latter area is a concern principally for spectroscopy, imaging, or optical fibre communications for which the p-i-n diode was originally developed.

H CONCLUSION

This Datareview has derived the current response of a GaInAs diode and reviewed the present understanding of spectral response, dark current, speed of response, avalanche gain and signal-to-noise ratio. Arrays and integrated structures have been briefly discussed.

REFERENCES

[1] T.P. Pearsall [*IEEE J. Quantum Electron. (USA)* vol.16 (1980) p.709-20]

[2] T. Kusunoki, C. Takenaka, K. Nakajima [*J. Cryst. Growth (Netherlands)* vol.112 (1991) p.33-8]

[3] T.P. Pearsall, R.W. Hopson Jr. [*J. Appl. Phys. (USA)* vol.48 (1977) p.4407-9]

[4] A.W. Mabbitt, J.W. Burgess, D. Jenkins, R. Nicklin [*Gallium Arsenide and Related Compounds*, Vienna, 1980 (Inst. Phys., Bristol, 1981) p.371-8]

[5] W.J. Schaff P.J. Tasker, M.C. Foisy, L.F. Eastman [*Semicond. Semimet. (USA)* vol.33 Ed. T.P. Pearsall (Academic Press, Boston, 1992) ch.2 p.73-138]

[6] T.P. Pearsall, M. Papuchon [*Appl. Phys. Lett. (USA)* vol.33 (1978) p.640-2]

[7] T.P. Pearsall [in *Optical Fibre Communications*, Eds M.J. Howes, D.V. Morgan (John Wiley, Chichester, UK, 1980) ch.3 p.107-64]

[8] Y. Matsushima, K. Sakai [in *GaInAsP Alloy Semiconductors*, Ed. T.P. Pearsall (John Wiley, Chichester, UK, 1982) ch.16 p.413-36]

[9] J.C. Campbell, W. Tsang, G. Qua, J. Bowers [*Appl. Phys. Lett. (USA)* vol.51 (1987) p.1454-6]

[10] P.A. Beck [*Hewlett-Packard J. (USA)* vol.44 (1993) p.49-51]

[11] T. Kaneda [*Semicond. Semimet. (USA)* vol.22, Ed. W.T. Tsang (Academic Press, New York, 1985) ch.3]

[12] C.H. Henry, R.A. Logan, F.R. Merritt, C.G. Bethea [*Electron. Lett. (UK)* vol.21 (1984) p.358-9]

[13] T.P. Pearsall, M.A. Pollack [*Semicond. Semimet. (USA)* vol.22, Ed. W.T. Tsang (Academic Press, New York, 1985) ch.2]

[14] R.A. Martinelli, T.J. Zamerowski, P.A. Longeway [*Appl. Phys. Lett. (USA)* vol.53 (1988) p.989-91]

[15] T.P. Pearsall, R.A. Logan, C.G. Bethea [*Electron. Lett. (UK)* vol.19 (1983) p.611-12]

[16] R. King, D.M. Braun, S.W. Hinch, K. Schubert [*Hewlett-Packard J. (USA)* vol.44 (1993) p.83-6]

[17] Y.G. We et al [*Appl. Phys. Lett. (USA)* vol.58 (1991) p.2156-8]

[18] O. Wada, H. Nobuhara, H. Hamaguchi, T. Mikawa, A. Tackeuchi, T. Fujii [*Appl. Phys. Lett. (USA)* vol.54 (1989) p.16-17]

[19] R. Loepfe, A. Schaelin, H. Melchior, M. Blaser, H. Jaeckel, G.L. Bona [*Appl. Phys. Lett. (USA)* vol.52 (1988) p.2130-32]

[20] T.P. Pearsall [*Electron. Lett. (UK)* vol.16 (1980) p.771-3]

[21] Y. Takeda, A. Sasaki [*Jpn. J. Appl. Phys. (Japan)* vol.20 suppl.20-1 (1981) p.189-92]

[22] K. Nishida, K. Taguchi, Y. Matsumoto [*Appl. Phys. Lett. (USA)* vol.35 (1979) p.251-3]

[23] S. Yamazaki, K. Nakajima, Y. Kishi [*Fujitsu Sci. Tech. J. (Japan)* vol.20 (1984) p.329-83]

[24] T. Shirai, T. Mikawa, T. Kaneda [*Fujitsu Sci. Tech. J. (Japan)* vol.20 (1984) p.303-28]

[25] G.H. Olsen, A.M. Joshi, V.S. Ban [*Proc. SPIE - Int. Soc. Opt. Eng. (USA)* vol.1540 (1991) p.596-605]

[26] L.Y. Leu, J.T. Gardner, S.R. Forrest [*J. Appl. Phys. (USA)* vol.69 (1991) p.4818-22]

[27] P. Cinguino, F. Genova, C. Rigo, C. Cacciatore, A. Stano [*Appl. Phys. Lett. (USA)* vol.50 (1987) p.1515-17]

9.2 InGaAs DH and quantum well lasers

N.K. Dutta

March 1993

A INTRODUCTION

The quaternary alloy $In_xGa_{1-x}As_yP_{1-y}$ has been grown epitaxially over single crystalline InP using liquid phase epitaxy (LPE), vapour phase epitaxy (VPE), chemical beam epitaxy (CBE) and metalorganic chemical vapour deposition (MOCVD) growth techniques [1]. The material grown is of high quality to allow highly reliable low threshold operation for lasers fabricated using this material.

The ternary alloy $In_{0.53}Ga_{0.47}As$ has the same lattice constant as InP and has also been grown using the previous techniques. Ternary alloy compositions of $In_xGa_{1-x}As$ which differ slightly (< 1%) in lattice constant from that of InP have been grown by CBE and MOCVD growth techniques. The epitaxial layer in this case is under strain. Lasers have been fabricated using both strained and lattice-matched $In_xGa_{1-x}As$ grown over InP. In the presence of lattice mismatch induced strain, high quality (dislocation free) material can be grown only if the grown layer thickness is sufficiently small. For thicknesses above a certain value (critical thickness), the material quality degrades through the formation of dislocations. This necessitates strained layer lasers to have a quantum well active region. The alloy $In_xGa_{1-x}As$ has also been grown on GaAs substrates. The lattice mismatch between $In_xGa_{1-x}As$ and GaAs increases with increasing x reaching a value of ~1.5% for x = 0.2. For x larger than ~0.3, the material quality degrades due to the presence of strain induced dislocations, so that high quality lasers can no longer be fabricated using this system. The laser fabricated using the $In_{0.2}Ga_{0.8}As$ alloy composition, which emits near 0.98 μm, is of particular interest as a pump laser for erbium doped glass fibre optical amplifiers. This Datareview describes the fabrication and performance characteristics of InGaAs/InP and InGaAs/GaAs lasers.

B InGaAs/InP DH LASERS

The InGaAsP alloy is a direct bandgap semiconductor when grown over InP. Under forward bias, electrons and holes are injected in the conduction and valence bands respectively; when they recombine, light at the bandgap energy E_g is emitted. The bandgap of $In_xGa_{1-x}As_yP_{1-y}$ lattice-matched to InP has been measured using photoluminescence. The bandgap is represented by the expression

$$E_g \text{ (in eV)} = 1.35 - 0.72y + 0.12y^2 \text{ at 300 K with } x \sim 0.45y \quad (1)$$

The bandgap of InGaAs lattice-matched to InP is 0.75 eV at 300 K.

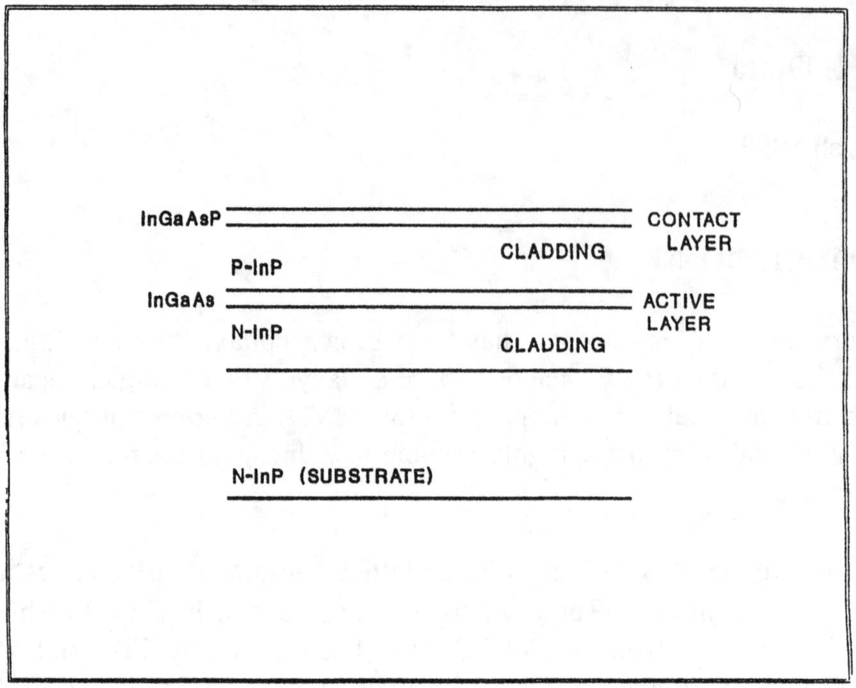

FIGURE 1 Schematic of the layer structure of an InGaAs/InP double heterostructure laser.

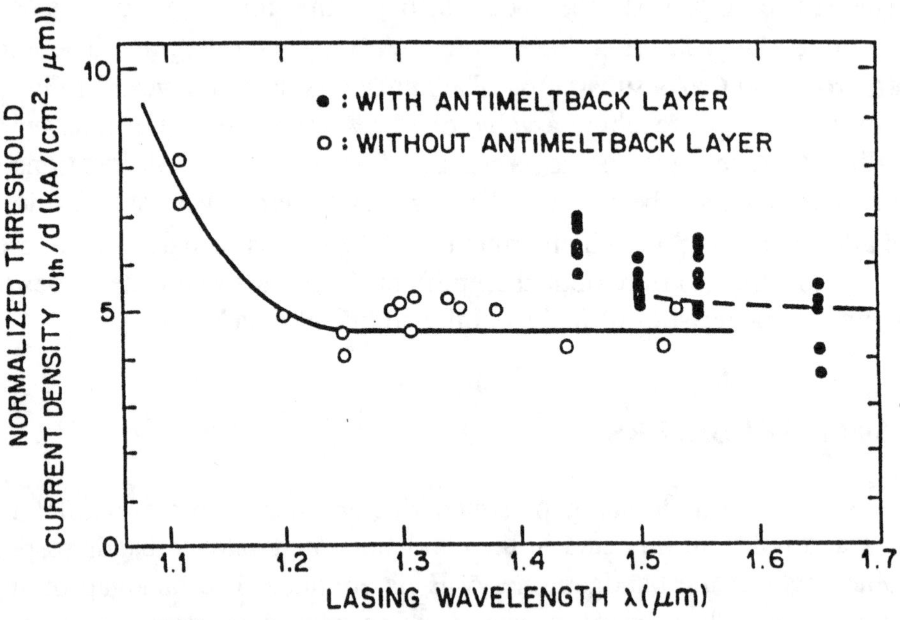

FIGURE 2 Experimental data for normalised threshold current density for broad area InGaAsP lasers emitting at different wavelengths. After [2].

Semiconductor lasers have been fabricated using InGaAs as the active (light emitting) region. The typical layer structure is shown in FIGURE 1. It consists of a double heterostructure (DH) with the active layer bounded by n-InP and p-InP cladding layers followed by a p-InGaAs contact layer. If the structure shown in FIGURE 1 is grown by LPE, an additional layer above the InGaAs active layer is needed in order to protect it from melting back during subsequent growth of the p-InP layer. This layer is InGaAsP (generally ~ 0.1 μm thick) and has a composition with bandgap in the 1.1 to 1.3 μm range. The threshold current density of a laser depends on the thickness of the active layer. The normalised threshold current density for broad area InGaAsP lasers emitting at different wavelengths is shown in FIGURE 2. A typical threshold current of an InGaAs/InP double heterostructure laser which utilises index guiding for optical and electrical confinement is 10 mA at room temperature.

C InGaAs/InP QW LASERS

A class of lasers that have been extensively studied over the last few years are the quantum well (QW) lasers which have a very thin (typically 20 Å to 100 Å) active region. The fabrication of quantum well lasers requires very high perfection of materials growth and interfaces. Such high perfection has been achieved using MOCVD (metalorganic chemical vapour deposition) growth techniques for InP-based materials. As a result it has been possible to fabricate high performance unstrained and strained InGaAsP/InP quantum well lasers and examine the performance improvement for system applications. For GaAs-based materials, QW lasers can be fabricated using both molecular beam epitaxy (MBE) and MOCVD growth techniques. The principal advantages of QW lasers over regular double heterostructure (DH) lasers are (i) lower threshold current density, (ii) lower optical loss which can result in low threshold current and higher power, and (iii) lower dynamic linewidth under modulation which results in longer repeater spacing for optical fibre transmission systems [1].

The parameters for optimisation of a quantum well laser structure are the well and barrier widths, barrier composition and grading for a graded index separate confinement heterostructure (GRIN-SCH) design. The schematic of QW laser structures with different barrier layer composition is shown in FIGURE 3. Best results are obtained for lasers with high energy barriers.

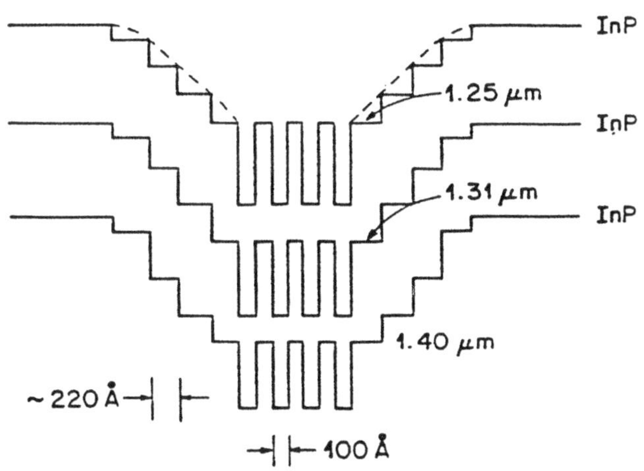

FIGURE 3 Schematic of three GRIN MQW designs. The bandgap of the barrier layer is varied from 1.25 μm to 1.40 μm. After [3].

Index guiding along the junction plane is necessary in order to control the light emission pattern of a semiconductor laser. The schematic of an index guided buried heterostructure (BH) laser is shown in FIGURE 4. The active region width is ~1 μm. The laser utilises

Fe-doped InP layers for confinement of the current to the active region and for providing lateral index guiding. The light vs. current characteristics of an MQW BH laser at different temperatures is shown in FIGURE 5.

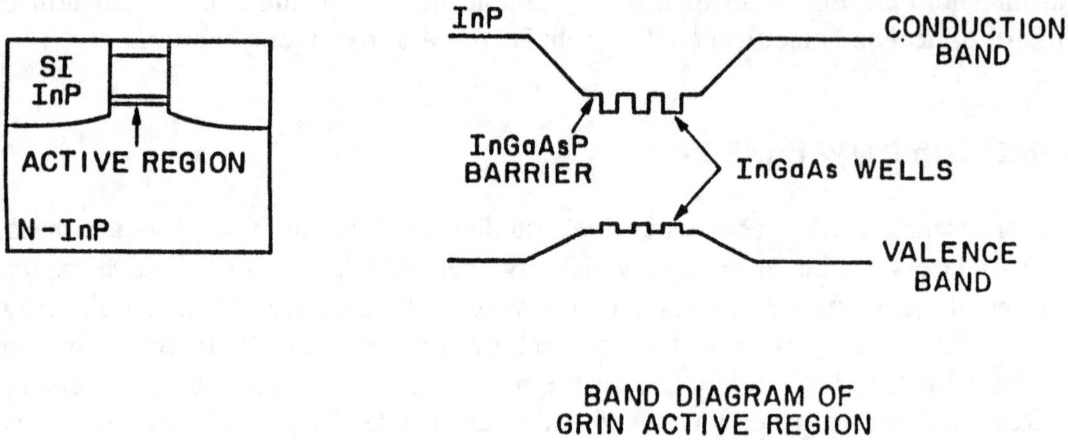

FIGURE 4 Schematic of a buried heterostructure laser utilising semi-insulating Fe-doped InP current confining layer.

Buried heterostructures were fabricated using the MQW active layer designs shown in FIGURE 3. The MQW designs are identical except for the composition of the barrier layer. They all have four 100 Å thick InGaAs wells and 100 Å thick barrier layers confined between two step graded index regions (GRIN)..

The threshold current density J_{th} depends on the confinement of the optical mode and hence varies with the design parameters of the MQW active region such as well thickness, number of wells, etc. The optical gain g in a QW is approximately related to the current density J by

$$g = a(J - J_0) \tag{2}$$

where a is the gain constant and J_0 is the transparency current density. The threshold gain g_{th} is given by

$$\Gamma g_{th} = \alpha + \frac{1}{L} \ln\left(\frac{1}{R}\right) \tag{3}$$

where L is the cavity length, α is the total loss and R is the facet reflectivity. From Eqns (2) and (3), the threshold current density J_{th} is given by

$$J_{th} = J_0 + \left(\alpha + \frac{1}{L} \ln\left(\frac{1}{R}\right)\right)/a\Gamma \tag{4}$$

Eqn (3) shows J_{th} decreases with increasing cavity length and for small losses it equals the transparency current density J_0 as $L \rightarrow \infty$. Threshold current density of 380 A/cm^2 has been reported for 2 mm long lasers with cleaved facets [4].

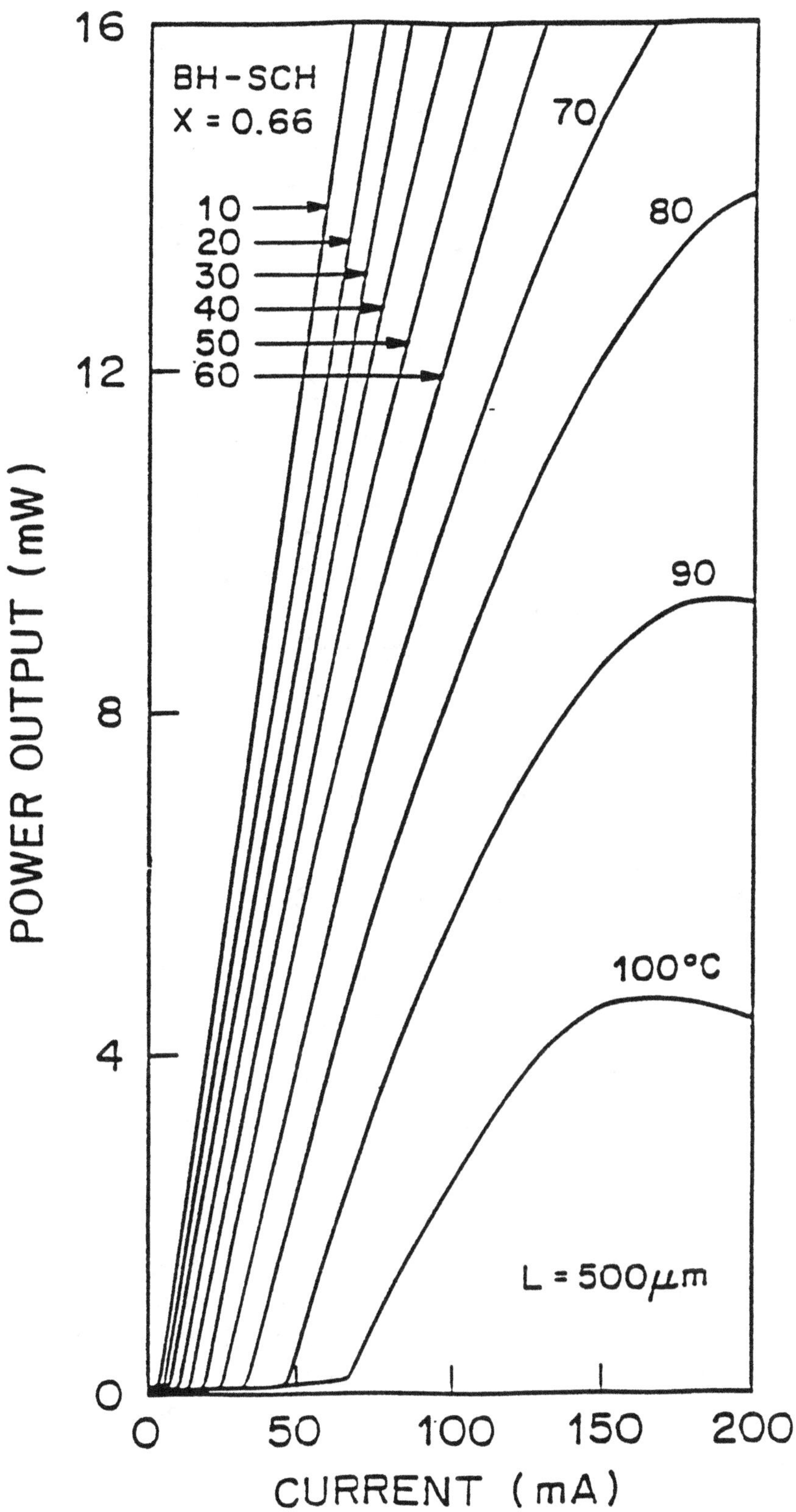

FIGURE 5 Light vs. current characteristics of an MQW BH laser at different temperatures.

Eqn (4) shows that the threshold current density of a laser is primarily determined by two components: (i) transparency current density, J_0 and (ii) current needed to overcome losses. The latter is small if α is small, such as that for a QW laser, and can be further reduced using high reflectivity coatings. The low absorption loss of QW structures allows the fabrication of very low threshold lasers [5]. The light vs. current characteristics of an MQW BH laser with

90% and 70% facet coatings are shown in FIGURE 6. The laser has a threshold current of 1.1 mA at 25°C and 0.9 mA at 10°C.

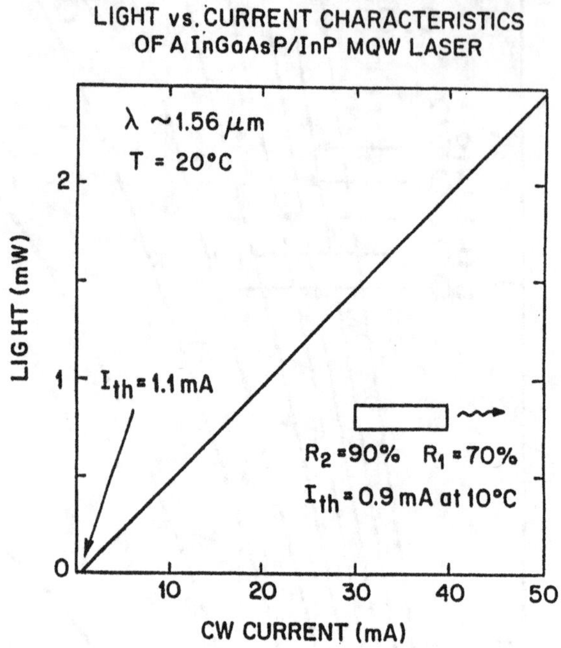

FIGURE 6 Light vs. current characteristics of an MQW BH laser with 90% and 70% facet coatings. The cavity length is 200 μm.

D InGaAs/InP STRAINED LAYER QW LASERS

The alloy $In_{0.53}Ga_{0.47}As$ has the same lattice constant as InP. Lasers with an $In_{0.53}Ga_{0.47}As$ active region have been grown on InP by the MOCVD growth technique [5]. Recently, it has been shown that excellent material quality is also obtained for $In_xGa_{1-x}As$ alloys grown over InP by MOCVD for non-lattice-matched compositions. In this case, the laser structure generally consists of one or many $In_xGa_{1-x}As$ quantum well layers with InGaAsP barrier layer whose composition lattice matches that of InP. For $x < 0.53$, the active layer in these lasers is under tensile stress and for $x > 0.53$, the active layer is under compressive stress.

Superlattice structures of InGaAs/InGaAsP with tensile and compressive stress have been grown by both MOCVD and chemical beam epitaxy (CBE) growth techniques over an n-InP substrate. A threshold current density of 300 A/cm^2 has been reported for a compressively strained quantum well laser which is 25 percent lower than the value obtained for a similar lattice-matched MQW structure [3]. The threshold current of InGaAs/InP lasers as a function of In concentration (x) is shown in FIGURE 7. Lasers with compressive strain have lower threshold current than that for lasers with tensile strain. This can be explained by shifting of the light hole and heavy hole bands under stress.

It has been predicted theoretically that strained MQW lasers can have higher bandwidth due to larger gain coefficient. Experimental results on small signal bandwidth of λ ~1.5 μm strained MQW lasers have been reported. The measured maximum value of ~20 GHz at 20 mW of output power is comparable to that for regular lasers. This suggests further optimisation in device structure or MQW design may be necessary to achieve higher bandwidth.

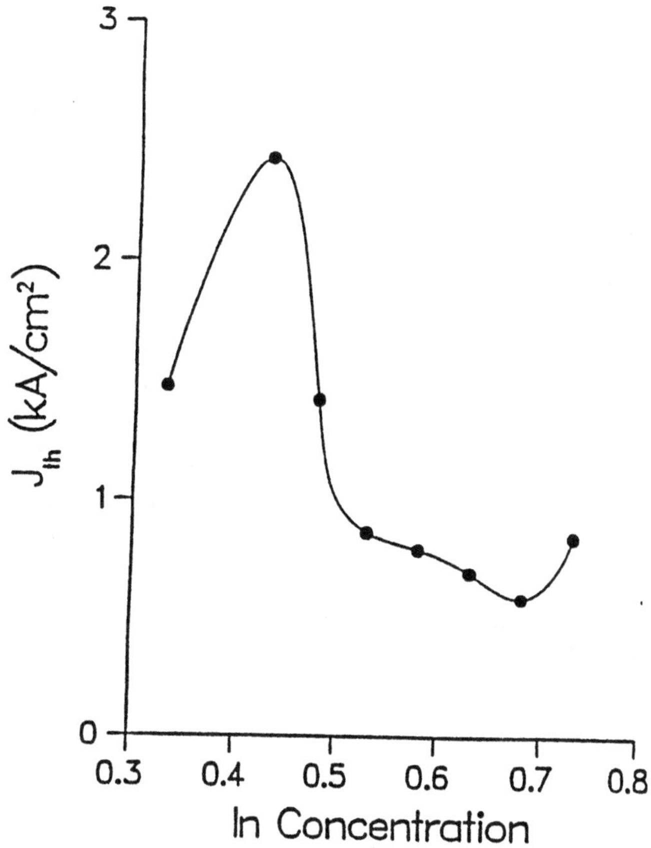

FIGURE 7 Threshold current of strained $In_xGa_{1-x}As/InP$ MQW lasers as a function of In concentration x. After [7].

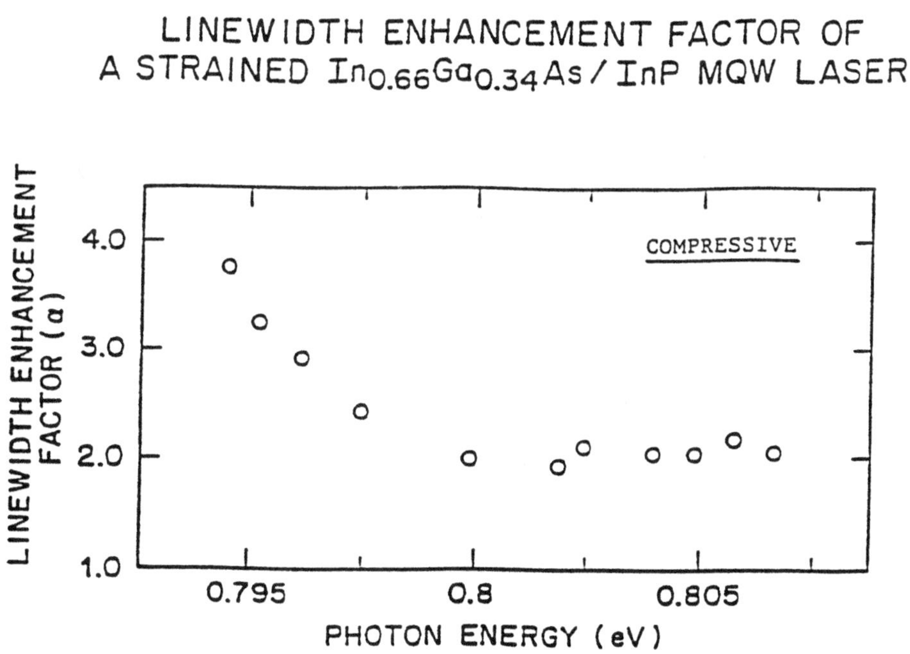

FIGURE 8 Measured linewidth enhancement factor (α) as a function of photon energy.

The frequency chirp under modulation of strained MQW lasers has been measured. The lasers exhibit somewhat smaller chirp than an unstrained MQW laser. The smaller chirp is due to the smaller linewidth enhancement factor (α) of these strained MQW lasers. The measured α as a function of photon energy of a compressively strained, MQW laser is shown in FIGURE 8. The α value at the lasing wavelength for strained lasers is smaller than the value observed for unstrained MQW lasers. The α value depends on the injected carrier density in both unstrained and strained MQW lasers and therefore a small variation of α with cavity length and MQW active region design is expected.

E InGaAs/GaAs STRAINED LAYER QW LASERS

The $In_xGa_{1-x}As$ alloy has been grown over GaAs using both MOCVD and MBE growth techniques. The typical layer structure for a laser consists of the following. An n-GaAs buffer layer is grown over the (100) oriented n-GaAs substrate. This is followed by an n-$Al_xGa_{1-x}As$ ($x \sim 0.4$) cladding layer, a graded index separate confinement heterostructure active region, a p-$Al_xGa_{1-x}As$ ($x \sim 0.4$) cladding layer and then a p-GaAs contact layer. The schematic of the QW active region for an $In_{0.2}Ga_{0.8}As$/GaAs laser is shown in FIGURE 9. This laser has three $In_{0.2}Ga_{0.8}As$ wells surrounded by GaAs barrier layers. Also shown in FIGURE 9 is the schematic of a ridge waveguide laser structure. The lasing action takes place in the region of the ridge which provides a weak lateral index guiding ($\Delta n \sim 10^{-3}$) to the optical mode.

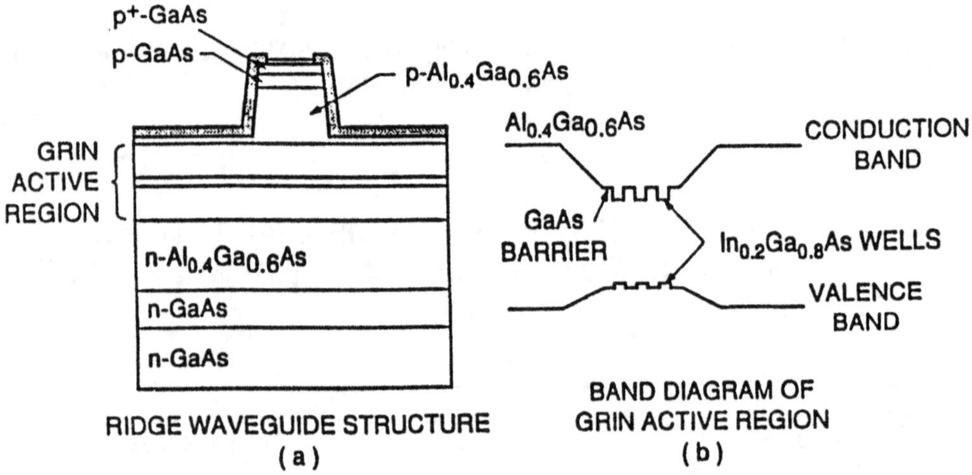

FIGURE 9 Schematic of a ridge waveguide $In_{0.2}Ga_{0.8}As$/GaAs strained quantum well laser.

The emission wavelength of the laser is primarily determined by the composition of the InGaAs active region. As the In composition increases, the emission wavelength increases (FIGURE 10) but the lattice mismatch and hence the strain also increases. If the strain is too high, the grown materials may have too many dislocations which makes its luminescence characteristics poor. High quality lasers have not been reported for $x > 0.3$ for the $In_xGa_{1-x}As$/GaAs material system.

Semiconductor lasers emitting near 0.98 μm fabricated using InGaAs/GaAs strained QW structures are of considerable interest as pump lasers for erbium doped glass fibre amplifiers. Single mode output power of greater than 200 mW has been demonstrated using a ridge waveguide laser structure. Pulsed and CW output powers greater than 4 W and 1 W have been demonstrated using wide stripe or array geometry [8] (FIGURE 11).

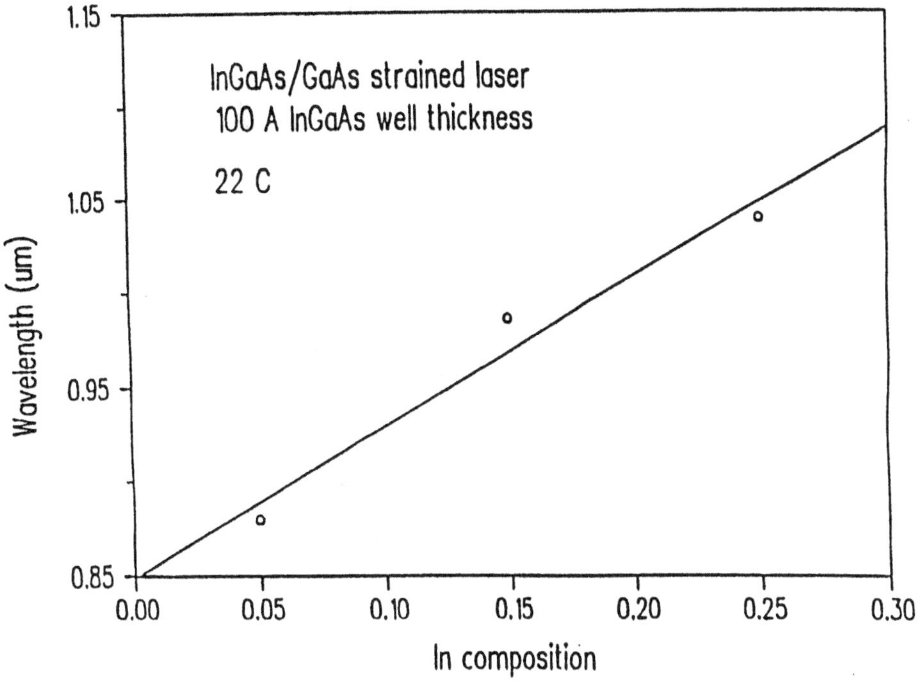

FIGURE 10 Emission wavelength as a function of In composition (x) for $In_{1-x}Ga_xAs/GaAs$ strained quantum well laser.

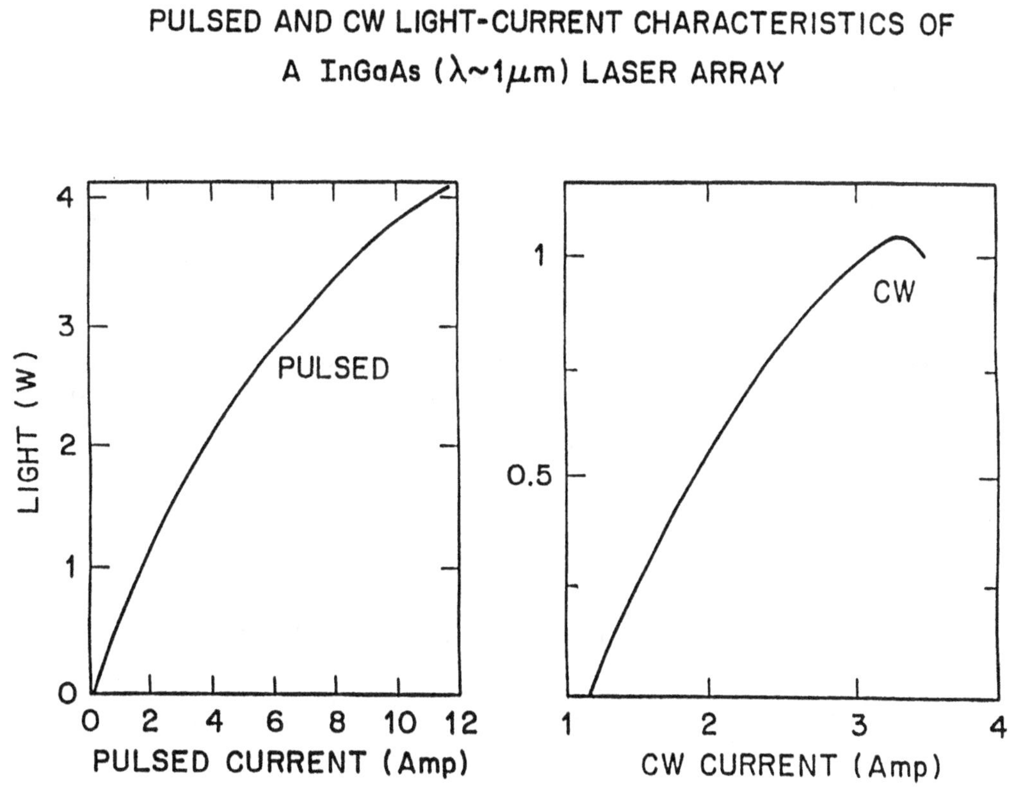

FIGURE 11 CW and pulsed light vs. current characteristics of an InGaAs/GaAs (λ ~1 μm) laser array. After [8].

Frequency chirp of strained and unstrained QW lasers has been investigated. The result at 1 Gb/s modulation is shown in FIGURE 12. The strained QW laser (InGaAs/GaAs) exhibits the lowest chirp or dynamic linewidth under modulation. The lower chirp under modulation of strained QW lasers is consistent with smaller alpha factor for regular DH, strained and unstrained QW lasers as shown in TABLE 1.

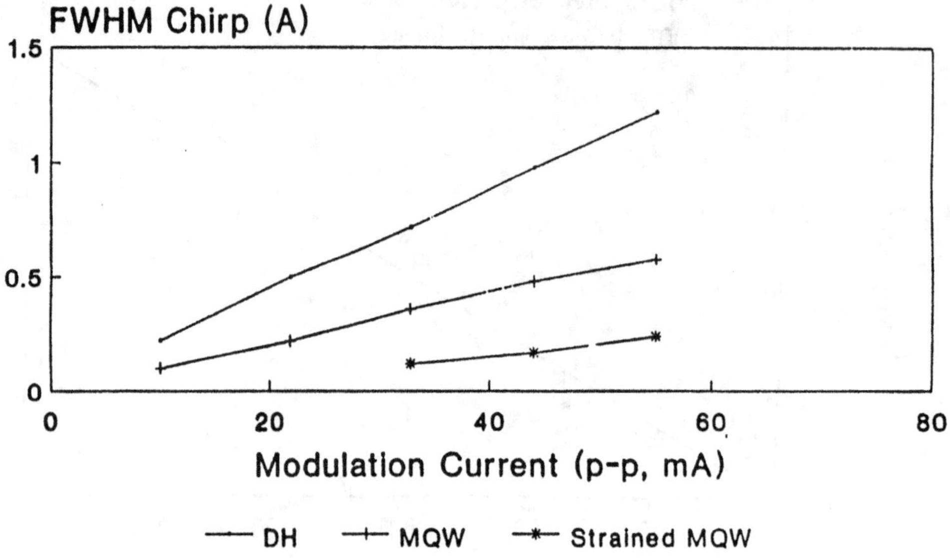

FIGURE 12 Measured full width at half maximum of the spectral width (chirp) plotted as a function of modulation current.

TABLE 1 Correlation between chirp and linewidth enhancement factor (α). Chirp is proportional to α.

	α-factor	FWHM chirp at 50 mA
DH laser	5.5	1.2 A
MQW laser	3.5	0.6 A
Strained MQW InGaAs laser	1.0	0.2 A
Strained MQW InGaAsP laser	2.0	0.4 A

The InGaAsP MQW laser technology is well advanced. Lasers with good reliability have been fabricated. These lasers are being used in lightwave transmission systems operating at data rates of 2.4 Gb/s and higher. Even lower chirp possible with strained MQW lasers makes them potentially useful for systems with larger regenerator spacing.

F CONCLUSION

Quantum well and double-heterostructure lasers with InGaAs active regions have emerged as important devices for optical communication. This Datareview describes the design and performance characteristics of InGaAs/InP double heterostructure, InGaAs/InP lattice-matched, InGaAs/InP strained and InGaAs/GaAs strained quantum well lasers.

REFERENCES

[1] See for example G.P. Agrawal, N.K. Dutta [*Long Wavelength Semiconductor Lasers* (Van Nostrand, New York, 1986)]

[2] S. Arai, T. Suematsu, Y. Itaya [*IEEE J. Quantum Electron. (USA)* vol.16 (1980) p.197]

[3] H. Temkin, T. Tanbun-Ek, R.A. Logan, J.A. Lewis, N.K. Dutta [*Appl. Phys. Lett. (USA)* vol.56 (1990) p.1222]

[4] W.T. Tsang, L. Yang, M.C. Wu, Y.K. Chen, A.M. Sergent [*Electron. Lett. (UK)* vol.26 (1990) p.2033]

[5] H. Temkin, N.K. Dutta, T. Tanbun-Ek, R.A. Logan, A.M. Sergent [*Appl. Phys. Lett. (USA)* vol.57 (1990) p.1610]

[6] For a review see P.J.A. Thijs, L.F. Triemeijer, P.I. Kuindersma, J.J.M. Binsma, T. Van Donger [*IEEE J. Quantum Electron. (USA)* vol.27 (1991) p.1426]

[7] D.T. Nichols et al [*IEEE J. Quantum Electron. (USA)* vol.8 (1992) p.1239]

[8] N.K. Dutta, J. Lopata, J.D. Wynn, D.L. Sivco, A.Y. Cho [*Electron. Lett. (UK)* vol.26 (1990) p.1816]

9.3 InGaAs channel field effect transistors

D. Pavlidis

February 1993

A INTRODUCTION

Field Effect Transistors (FETs) using InGaAs channels present various advantages in terms of carrier transport and device characteristics with respect to more conventional approaches based for example on GaAs. We review here the main device characteristics according to the material used for the donor layer, lattice-matched or strained (pseudomorphic) approach used for the channel and type of device, i.e. High Electron Mobility Transistor (HEMT), Heterostructure-Insulated-Gate FET (HIGFET).

B DEVICE STRUCTURE

InGaAs channel FETs can be built either on GaAs or InP substrates. A generic cross section for HEMTs is given in FIGURE 1.

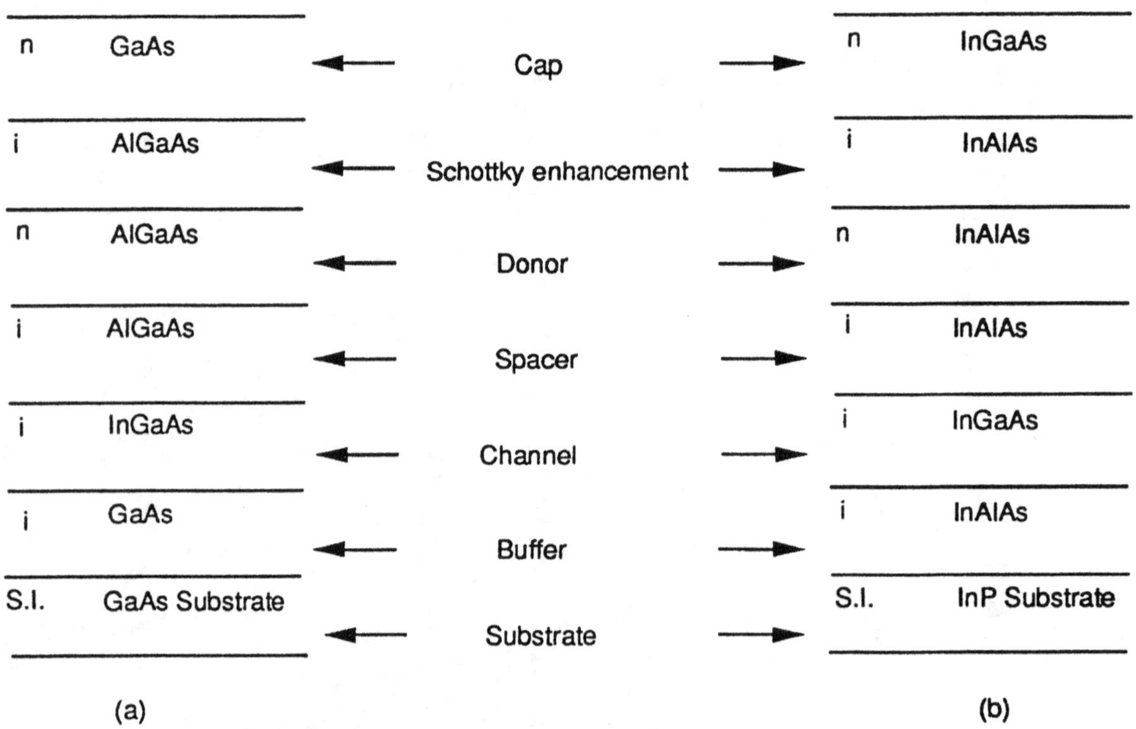

FIGURE 1 Generic cross-section of (a) GaAs- and (b) InP-based HEMTs.

The $In_xGa_{1-x}As$-channel of GaAs-based devices is lattice mismatched with respect to the other layers and results in strained (pseudomorphic) designs with maximum x-values in the range of 0.20 - 0.25. In the case of InP-based devices $In_xGa_{1-x}As$-channel FETs exist in either lattice matched (x = 0.53) or strained (x > 0.53) configuration.

By varying the position of doping in the device designs shown in FIGURE 1, one obtains other device types such as:

(a) Heterostructure Insulated Gate FETs (HIGFETs), where none of the layers of FIGURE 1 is doped; n- or p-channel operation can be achieved by selecting appropriate types of implants for contacting the channel through the source end drain terminals.

(b) Heterostructure FETs (HFETs) where the donor layer of FIGURE 1 is kept undoped and doping is introduced directly in the channel.

In the following we will refer to the wide bandgap material below the gate as a 'donor layer' although in certain instances, i.e. HIGFETs, HFETs, no doping is employed in this layer.

We will concentrate primarily on the properties of HEMTs and will only review briefly the characteristics of other device types such as HIGFETs.

C InGaAs-CHANNEL FETs WITH AlGaAs DONOR LAYERS

Strained AlGaAs/InGaAs designs permit the achievement of a high ΔE_c value while using low Al content AlGaAs layers. This results in increased sheet-carrier density (n_s) while avoiding DX-related problems. The latter occur in high-Al content AlGaAs layers and produce undesired device characteristics i.e. Persistent Photoconductivity (PPC) [1], threshold voltage shifts [2], I-V collapse [3] and bulges in '1/f'-noise characteristics [4]. Using 20% In in the InGaAs channel and an $Al_{0.2}Ga_{0.8}As$ donor layer, ΔE_c is, for example, 0.34 eV, while the ΔE_c obtained from the $Al_{0.2}Ga_{0.8}As$/GaAs combination is only 0.20 eV. The larger ΔE_c proves also to be very useful in providing better confinement of the 2DEG. In addition, InGaAs has a lower electron effective mass and thus better mobility and velocity.

Typical AlGaAs/InGaAs HEMTs consist of $In_xGa_{1-x}As$ channels with x_{max} ~ 0.20 - 0.25 and a thickness properly selected (~150 Å) not to exceed the critical thickness so that dislocation formation can be avoided. The spacer end donor AlGaAs layers use ~ 0.15 to 0.25 Al-content. The corresponding ΔE_c is in this way sufficiently large and of the order of 0.29 eV to 0.39 eV for $In_{0.2}Ga_{0.8}As$ channels. Unlike lattice-matched AlGaAs/GaAs HEMTs pseudomorphic designs employ different materials for the channel and better buffer layers (InGaAs/GaAs instead of GaAs alone). This results in a quantum well like conduction band profile instead of the conventional triangular well and permits better confinement of the 2DEG in the channel.

1 μm gate length AlGaAs/InGaAs HEMTs reported in 1985 [5] have shown better performance than the best AlGaAs/GaAs HEMTs at that time. The extrinsic and intrinsic g_m's were 270 mS/mm and 350 μS/mm at 300 K respectively, the output conductance was as low as 700 μS/mm and the current density was 290 mA/mm. The cutoff frequency (f_T) of these devices was also good and of the order of 20 GHz, a value which exceeded by ~ 25% AlGaAs/GaAs HEMTs with comparable geometry. Since then, excellent progress has been made in the areas of material growth and device design; this, coupled with the ability to fabricate very small gates down to ~ 0.1 μm, allowed AlGaAs/InGaAs HEMTs to achieve

state-of-the-art device performance well into the millimetre frequency range. Some of these results are reviewed next.

C1 DC and RF Performance

Since the first report of a submicron (0.25 μm) AlGaAs/InGaAs HEMT in 1986 [6] significant progress has been made. TABLE 1 shows the improvement through time of the major device characteristics. As shown in this table, the two most important RF figures of merit, f_T and f_{max}, can exceed 100 GHz and 200 GHz respectively with today's submicron technology. It is worth noting that the f_{max} value of 350 GHz is the best ever reported for GaAs-based HEMTs. A novel double-recess gate technique [7] was used in this case to increase the g_m/g_o and C_{gs}/G_{gd} ratios and thus the f_{max}. In addition to the impressively high gain performance, AlGaAs/InGaAs HEMTs have also shown high current density and breakdown voltage. Typical maximum current densities are of the order of ~ 500 - 600 mA/mm and values to 1.1 A/mm have been reported using double quantum-well designs [8]. The maximum drain-source voltage that could be applied on these devices before breakdown occurs was of the order of ~7 V. These attractive features have made AlGaAs/InGaAs HEMTs an excellent choice for power applications at millimetre-wave frequencies. State-of-the-art results and the various device structures explored for optimum power performance will be discussed next.

TABLE 1 Summary of the progress in device performance for submicron gate-length pseudomorphic AlGaAs/InGaAs HEMTs.

Year	L_g (mm)	g_m,intr (mS/mm)	f_T (GHz)	f_{max} (GHz)	Company/Laboratory
1986	0.25	495	70	200	University of Illinois
1987	0.1	540 - 740	-	-	GE
1988	0.15	640	100	350	GE
	0.15	650	152	230	Cornell University
	0.15 - 0.25	500	98	-	HP
1989	0.08	740 - 920	-	270	GE
	0.25	625	90	-	GE
	0.20	530	55	150	TI
1990	0.10	700	120	290	TRW
	0.20	650	82	305	COMSAT
	0.15	800	135	-	GE
1991	0.15	660	-	-	GE
	0.10	778	130	290	TRW

C2 Power Performance

An important factor for high power density characteristics is the maximum sheet carrier density, n_s, that can be obtained with a device. For single-heterojunction AlGaAs/InGaAs HEMTs, typical maximum values obtained are ~ 2.5 x 10^{12} cm^{-2}. In addition to the top AlGaAs donor layer used in the single-heterojunction HEMTs, a second AlGaAs donor layer is placed beneath the InGaAs channel in the case of AlGaAs/InGaAs DHHEMTs. This provides additional carriers into the channel, as well as an increase of the carrier confinement in the InGaAs channel. By using this approach, n_s values as high as 2.4 x 10^{12} cm^{-2} were obtained and the output power of 0.15 μm x 150 μm HEMTs measured at 94 GHz was 57 mW with a PAE of 16% and an associated gain of 2.0 dB [9].

In spite of the fact that DHHEMTs result in high n_s values, there are some drawbacks associated with this approach. First, it is difficult to grow a good inverted heterointerface between the InGaAs channel and the bottom AlGaAs layer [10]. As a result, the electron transport may be degraded and this can reflect in the overall device performance. Second, for a DHHEMT, the gate is effectively further away from the 2DEG. This results in a smaller gate-to-channel aspect ratio and enhanced short channel effects especially in the submicron regime. Finally, the channel width has to be properly designed so that a single peak is observed in the g_m versus V_g characteristics; undesirable double transconductance peaks occur otherwise. In view of these, an alternative approach, namely the doped channel HEMT (DCHEMT), is used to enhance the n_s.

A DCHEMT resembles in design any other SHHEMT except for the channel which is doped; doping can either be uniform or planar. FIGURE 2 shows the cross-section of an asymmetrical planar-doped channel AlGaAs/InGaAs HEMT [11]. Using this approach, an n_s value as high as 4.5×10^{12} cm^{-2} has been achieved. Furthermore, the g_m-V_g characteristics are broader and allow in this way better linear power performance and intermodulation distortion characteristics. Finally, the aspect ratio is also higher with this approach compared to that of the DHHEMTs and results in better performance for reduced gate length devices. Using this approach, Saunier et al at TI [12] have demonstrated power densities as high as 1 W/mm with 2.9 dB gain and 25% power added efficiency at 60 GHz. Smith et al at GE [11] have also obtained an output power of 32 mW, a PAE of 15% and a gain of 3.0 dB at 94 GHz using 0.25 mm x 75 mm pseudomorphic DCHEMTs. Streit et al at TRW [13] have further improved the power performance by using 0.1 mm x 160 mm asymmetric planar-doped channel pseudomorphic HEMTs; the measured output power was 62.7 mW at 94 GHz with a PAE of 13.2% and a gain of 4.0 dB.

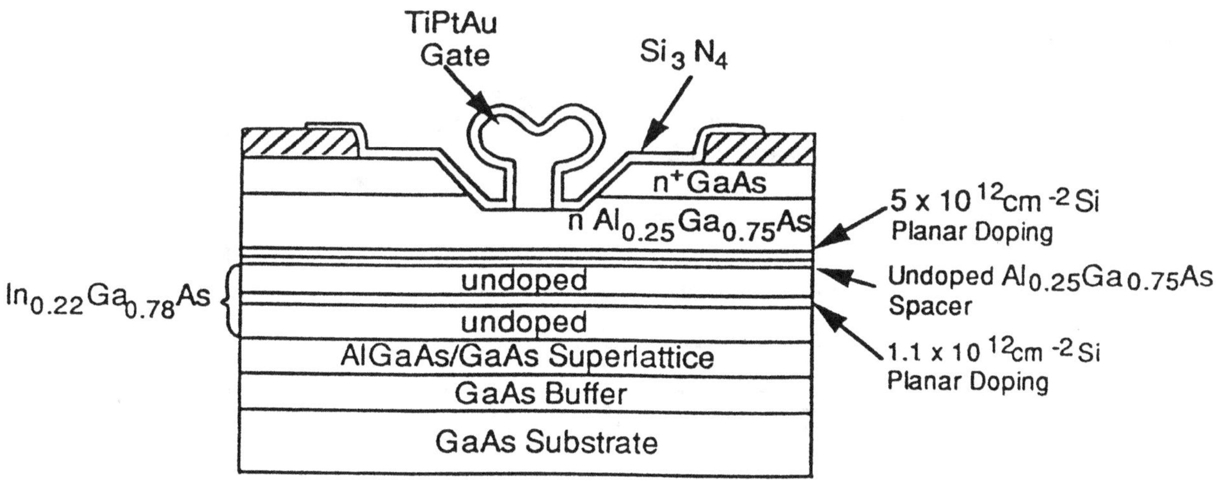

FIGURE 2 Device cross-section of an asymmetrical planar-doped channel AlGaAs/InGaAs HEMT [11] (© IEEE 1989).

By incorporating these novel approaches, namely multi-heterojunction and doped-channel, PHEMTs have shown excellent power performance superior to GaAs FETs at frequencies below 20 GHz and demonstrated great potential for power applications at even higher frequencies (20-100 GHz). These are illustrated in FIGURES 3 and 4, which summarise the best reported transistors' PAE and output power results together with the projected improvement for PHEMTs by 1994 [14].

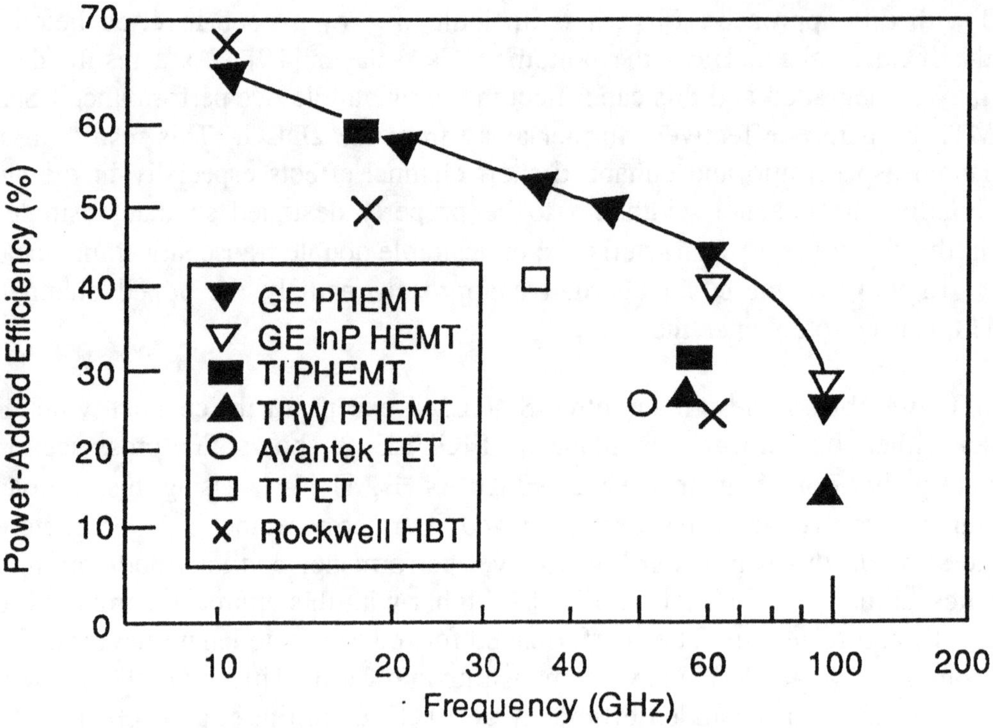

FIGURE 3 State of-the-art power added efficiency results for HEMTs and HBTs with output power typically less than 100 mW [14] (© IEEE 1992).

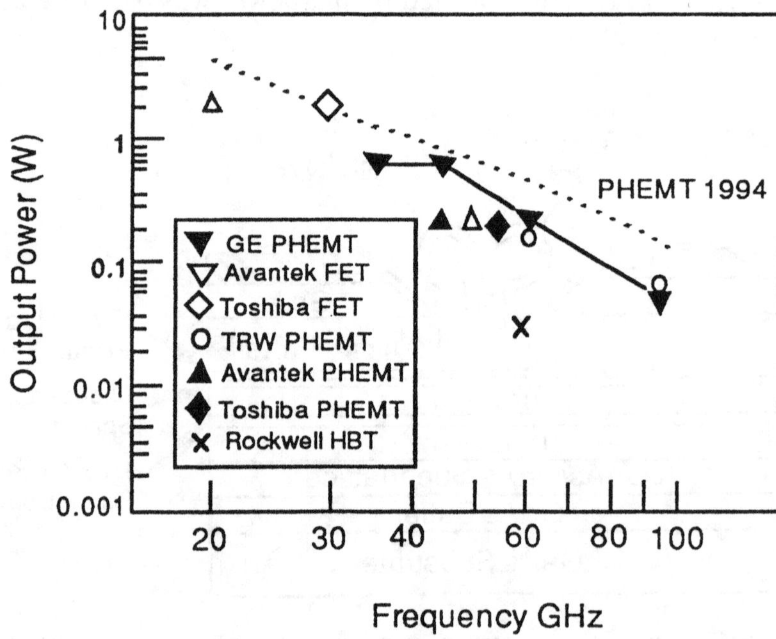

FIGURE 4 State-of-the-art output power results for HEMTs and HBTs and the projected improvement by 1994 for PHEMTs [14] (© IEEE 1992).

C3 Low Noise Performance

In addition to having superior power performance, AlGaAs/InGaAs HEMTs have also demonstrated excellent low noise characteristics at microwave and millimetre-wave frequencies. In conventional GaAs MESFETs, a trade-off usually exists between low noise and high power performance. For example, higher channel doping is required to increase the power density but this degrades the noise performance. This trade is, however, less stringent in the case of AlGaAs/InGaAs HEMTs. Henderson et al reported, for example, a 25 μm AlGaAs/InGaAs HEMT which combines acceptable characteristics for both high power (0.43 mW/mm) and low noise (NF_{min} = 2.4 dB) at 60 GHz [6]. GE has also demonstrated state-of-the-art power (139 mW) and low noise (NF_{min} = 1.8 dB, GaAs = 6.4 dB) characteristics with 0.15 μm gate-length double-heterojunction AlGaAs/InGaAs HEMTs [9] at 60 GHz. This unique feature of simultaneously good high power and low noise performance makes AlGaAs/InGaAs HEMTs very attractive for integrating both low noise and power components in a single MMIC chip.

Special efforts made on optimising the noise performance in AlGaAs/InGaAs HEMTs resulted in a minimum noise figure of 2.4 dB with an associated gain of 5.4 dB at 94 GHz [15] by GE. TRW has also reported a minimum noise figure of 2.1 dB with an associated gain of 6.3 dB at 93.5 GHz [16]. These state-of-the-art low noise characteristics have clearly demonstrated the potential of AlGaAs/InGaAs HEMTs for mm-wave low noise applications.

C4 Reliability of AlGaAs/InGaAs HEMTs

The results of reliability for AlGaAs/InGaAs HEMTs are very encouraging. FIGURE 5 shows the high-temperature DC life test results for a 0.15 μm gate-length AlGaAs/InGaAs HEMT [17]. The extrapolated DC lifetime of the device exceeds 10^6 hours at 150°C baseplate temperature which is comparable to the conventional AlGaAs/GaAs HEMT. Although the device used has a highly strained InGaAs channel with 30% InAs mole fraction, the DC lifetime results do not seem to indicate any degradation of the device characteristics due to the pseudomorphic channel.

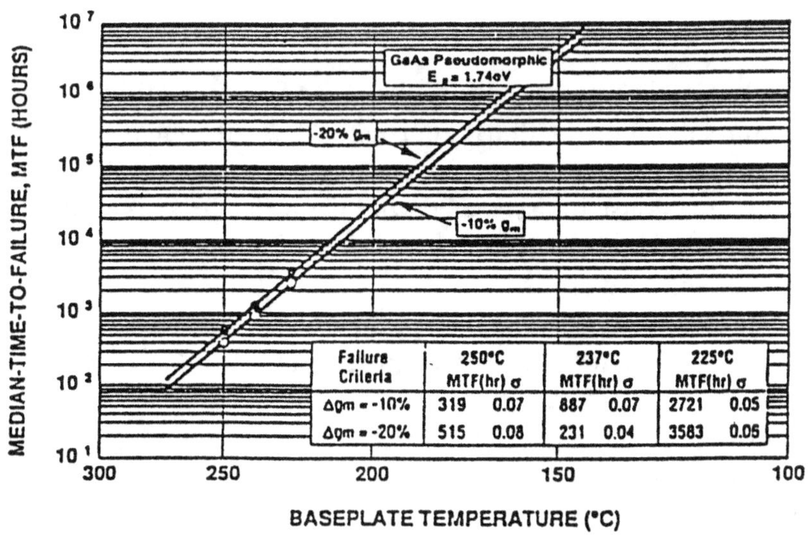

FIGURE 5 High-temperature DC life test results for a 0.15 mm x 50 mm gate AlGaAs/InGaAs HEMT; Si_3N_4 is used for passivation [17] (© IEEE 1991).

D InGaAs CHANNEL FETs WITH InAlAs DONOR LAYERS

Devices of this type are based on InP substrates. We distinguish here lattice-matched and strained designs with x = 0.53 and x > 0.53 for the $In_xGa_{1-x}As$ channel respectively. For each of the above two types of design one can have different device types such as HEMTs, HIGFETs and HFETs.

D1 Lattice-Matched FETs

D1.1 Lattice-matched HEMTs

The main features of $In_{0.52}Al_{0.48}As/In_{0.53}Ga_{0.47}As$ heterostructures compared to AlGaAs/GaAs HEMTs are:

(a) High n_s values due to the large ΔE_c (~ 0.5 eV [18]) of InAlAs/InGaAs and much higher doping achieved in InAlAs (> 1 x 10^{19} cm^{-3} for InAlAs vs. ~ 2 x 10^{18} cm^{-3} for AlGaAs); n_s values of 4 x 10^{12} cm^{-2} have for example been reported [19].

(b) High peak velocity exceeding by ~ 20% that of GaAs channels as a result of smaller electron effective mass and larger Γ-L valley separation.

(c) High mobility, typically ~ 11000 cm^2/V s instead of ~ 8000 cm^2/V s for GaAs based heterostructures.

(d) Reduced trapping and DX-related problems in InAlAs compared with AlGaAs layers. PPC, V_{th}-shift and I-V collapse problems can therefore be reduced.

The following are typical features of InGaAs/InAlAs HEMTs which are considered as drawbacks in their proper operation. Special design, material and technology precautions need therefore to be taken to control these characteristics.

(i) <u>Kink effect</u>

This manifests in the I_{ds} - V_{ds} characteristics of the HEMT as an additional current component which does not permit saturation of the I_{ds} current. It is attributed to traps in the InAlAs buffer [20] which capture highly energetic electrons and then release them by impact ionisation. Due to the time constant associated with the traps the kink effect is present only at low frequencies where one finds an increased value of output conductance. Methods to reduce the kink effect include the use of:

- Superlattice InAlAs/InGaAs buffers making use of InGaAs which has a lower trap concentration than InAlAs [20].
- Low temperature buffers with the InAlAs layer grown at ≤ 200°C and followed by annealing at 500°C [21]. Layers grown in this way have traps with very short capture time and can consequently trap very efficiently electrons injected to the buffer, minimising the possibility of their release and contribution to excess I_{ds} current.

(ii) Output conductance

InAlAs/InGaAs HEMTs show excessively high output conductance compared for example with AlGaAs/GaAs devices. This is due to energetic electrons leaving the channel and being injected in the adjacent layers. Methods to eliminate this problem make use of special designs to improve carrier confinement. The following are examples of such approaches:

- Use of double-heterojunction designs where the built-in electric field by the bottom heterojunction prevents carrier injection into the buffer. This approach has been demonstrated in 1 μm and submicron devices [22,23].
- Use of p-doped InAlAs buffer to increase the barrier height to electron injection [24].
- Use of undoped InGaAs cap layers to reduce the electric field at the gate drain region and minimise electron injection [25]. Double-recessed gate technology offers similar features [7].

(iii) Breakdown voltage

A low drain-source breakdown voltage is observed (typically 3 to 4 V) due to the high likelihood of impact ionisation in the low bandgap InGaAs channel. Undoped cap designs and double recess gate technology permit the electric field in the gate-drain region to be reduced and thus improve the gate-drain breakdown voltage [25]. Another possibility is the use of double heterojunctions where lower doping and higher Al content is used in the top donor layer compared with single heterojunction devices, while the current is still maintained reasonably high. Devices built with the latter approach demonstrated high breakdown voltage (~ 9 V) and good measurable peak characteristics (40% PAE and 288 mW i.e. 0.96 W/mm at 12 GHz) [26].

(iv) Gate leakage

Gate leakage can be excessive due to: (a) early impact ionisation in the low bandgap InGaAs channel and (b) contact of the channel at the mesa crossover of the gate feeder. Methods to reduce gate leakage include:

- Isolation by ion-implantation so that the mesa crossover is avoided. The gate breakdown improves in this case from 3 V to 10 V and the forward turn-on voltage increases from 0.3 to 0.55 V [21].
- Use of undoped InGaAs caps to reduce the electric field and decrease carrier tunnelling through the gate [25].
- Use of Pt gates to enhance the Schottky barrier height to ~ 0.82 eV [27].
- Use of high Al content Schottky layers to improve barrier height (~ 0.9 eV); $In_{0.37}Al_{0.63}As$ has for example been used successfully for this purpose [28].

(v) Sidegating

Sidegating is more important in InAlAs than in AlGaAs/GaAs HEMTs due to the high trap density of conventional InAlAs buffers and the low Schottky barrier of InGaAs [29]. To reduce sidegating the following techniques can be employed:

- InAlAs/InGaAs superlattice buffers to filter defects away from the channel [20].
- Low temperature grown InAlAs buffers to produce electrically inactive layers [21].
- Air-bridge gate crossovers of mesas [30] or ion-implantation for isolation [21].

The electrical characteristics of InAlAs/InGaAs HEMTs have been by far the best among all other types of FETs in terms of f_T, f_{max} and noise performance. Impressively high f_T's of 250 GHz [31] and f_{max}'s [32] have been reported and these values are the best ever obtained with FETs. The lowest NF_{min} reported to date has also been obtained using InAlAs/InGaAs HEMTs. FIGURE 6 shows GE's measured room temperature noise figure data using 0.15 μm InAlAs/InGaAs HEMTs. As shown by this figure, lattice matched InAlAs/InGaAs HEMTs offer excellent noise performance over the entire measured frequency range up to 94 GHz and show better noise characteristics than lattice-matched or pseudomorphic GaAs-based HEMTs; the best reported NF_{min} to date is 0.8 dB at 63 GHz [19] and 1.4 dB with 6.6 dB associated gain at 94 GHz [15]. These excellent noise characteristics undoubtedly make InAlAs/InGaAs HEMTs very attractive for mm-wave low noise applications.

Finally, InAlAs/InGaAs HEMTs have not only shown excellent results in discrete form but have also demonstrated their potential in high frequency integrated circuits applications. The first working 5 - 100 GHz distributed amplifier [33], W-band oscillator [34], W-band mixer [35] and 90 - 180 GHz doubler [36] in monolithic integrated forms all utilise InAlAs/InGaAs HEMT technology.

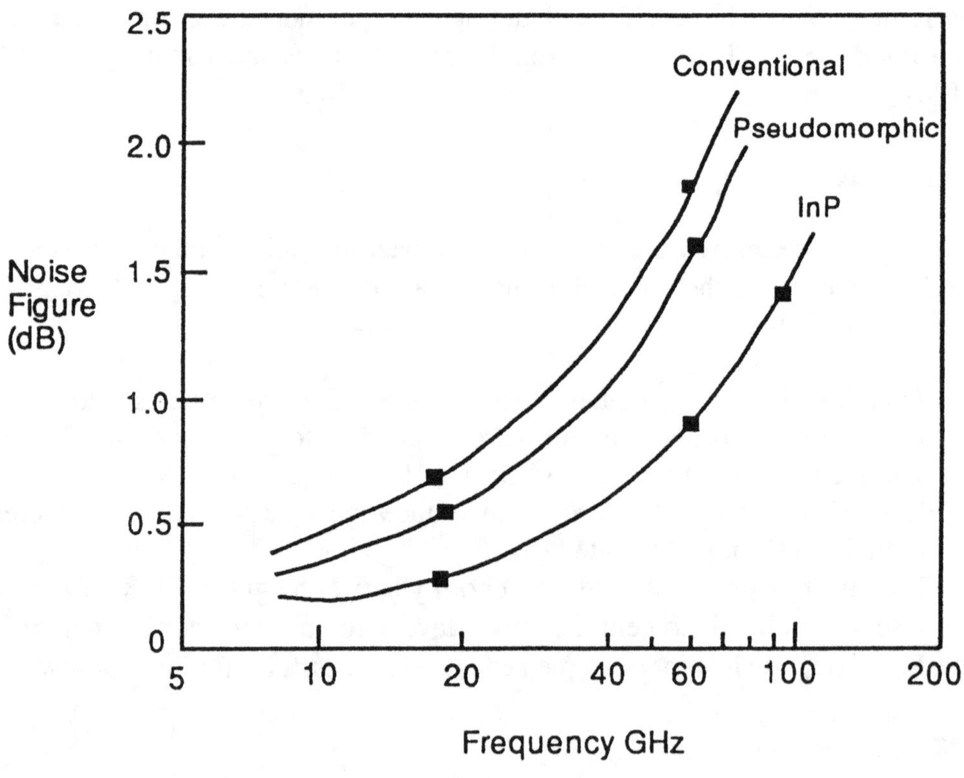

FIGURE 6 Measured noise figure characteristics of 0.15 mm gate-length conventional AlGaAs/GaAs, pseudomorphic AlGaAs/InGaAs and lattice-matched InAlAs/InGaAs HEMTs at room temperature (courtesy of GE).

D2 Strained (Pseudomorphic) FETs

D2.1 Strained (pseudomorphic) HEMTs

As already discussed in Section D1.1 lattice matched InAlAs/InGaAs HEMTs show superior performance over AlGaAs/GaAs devices due to higher 2 DEG density, mobility and velocity in the InGaAs channel. The increase of InAs mole fraction of the InGaAs channel in the strained devices allows further performance improvement. The first demonstration of this approach by Kuo [37] indicated a mobility improvement by 20% using 60% rather than 53% In channels. A systematic study of the effects of strain has been reported by researchers at the University of Michigan [38-40].

The effect of increasing the In content on device performance was studied by Ng et al using 60% and 65% In InGaAs channel [39]. 1 μm gate HEMTs were shown to have an average increase of intrinsic transconductance from 520 mS/mm to 700 mS/mm by increasing the In content.

Microwave measurements made on HEMTs with 1 μm long gates also show improvement in extrinsic f_T from 40 GHz to 45 GHz and intrinsic f_T's from 48 GHz to 56 GHz for the 60% and 65% In samples respectively. The f_T increase is attributed to the enhanced velocity with higher In content channel [39]. Independent studies in the submicron gate-length regime have also shown improvement of device performance with In content [43]. 0.1 mm gate length InAlAs/$In_xGa_{1-x}As$ HEMTs showed g_m increased from 1000 mS/mm to 1160 mS/mm and f_T's increased from 175 GHz to 205 GHz for x = 0.53 and 0.62 respectively.

While g_m increases with the In content, the output resistance (R_{ds}) shows the opposite trend both at DC and microwave frequencies [43]. This is probably due to more electrons acquiring higher energy with the result of increased injection into the buffer. The decrease in R_{ds} resulted in lower gain and f_{max} values; the f_{max} for the 60% In device is 60 GHz whereas it is 46 GHz for the 65% In device despite the higher In content for the latter.

Despite the fact that pseudomorphic InAlAs/InGaAs HEMT technology is rather new, outstanding device performance has been demonstrated by various laboratories and continues to advance during the years. A record high f_T value of 305 GHz was reported by Nguyen et al at Hughes Laboratory [42] using 0.065 μm self-aligned gates and a pseudomorphic $In_{0.80}Ga_{0.20}As$ channel. Tan et al at TRW [43], using a 0.1 μm gate-length pseudomorphic $In_{0.52}Al_{0.48}As/In_{0.60}Ga_{0.40}As$ HEMT, have also obtained a gain of 7.3 dB measured directly at 141.5 GHz; a minimum noise figure of 1.3 dB with an associated gain of 8.2 dB was also measured at 95 GHz for the same device. Kao et al at GE [44] have also obtained excellent millimetre-wave power performance using 0.15 μm gate-length pseudomorphic InAlAs/InGaAs HEMTs. At 94 GHz, the device has a peak power-added efficiency of 26% with an output power density of 0.2 W/mm and a gain of 4.9 dB; when tuned for maximum output power, the device can deliver a maximum output power density of 0.28 W/mm with a power-added efficiency of 24% and a gain of 4.3 dB.

The cryogenic (77 K) rather than room operation of HEMTs showed an improvement of f_T from 22 to 29 GHz, 29 to 35 GHz and 39 to 51 GHz for 0.53, 0.60 and 0.70% In-content in the channel with no noticeable device degradation [45].

Improved transport properties have also been demonstrated by incorporating a thin (~ 40 Å) InAs layer in the channel, sandwiched by undoped InGaAs [46]. A transconductance of 2.1 S/mm, f_T = 264 GHz and f_{max} = 166 GHz have been obtained in this way using 0.08 μm long gates.

Various basic effects such as dispersion of electrical characteristics at low frequencies and low frequency noise have been studied. The evaluated dispersion is smaller in InAlAs/InGaAs HEMTs then in MEDFETs [2]. FIGURE 7 shows the normalised g_m versus frequency plot of 0.7 μm gate $InAlAs/In_xGa_{1-x}As$ ($0.53 \leq x \leq 0.70$) HEMTs biased at saturation (V_{ds} = 1.5 V) [48]. The dispersion for all samples is small; the maximum dispersion (2%) occurs for the 53% In sample and the smallest dispersion (1.7%) is found for the 60% In sample. When the same devices are biased in the ohmic region (V_{ds} = 0.25 V) the 53% In sample showed the largest dispersion (6%). The evaluated dispersion of 2% to 6% is small compared to MESFETs [47,48] where values as high as 20% or more have been observed.

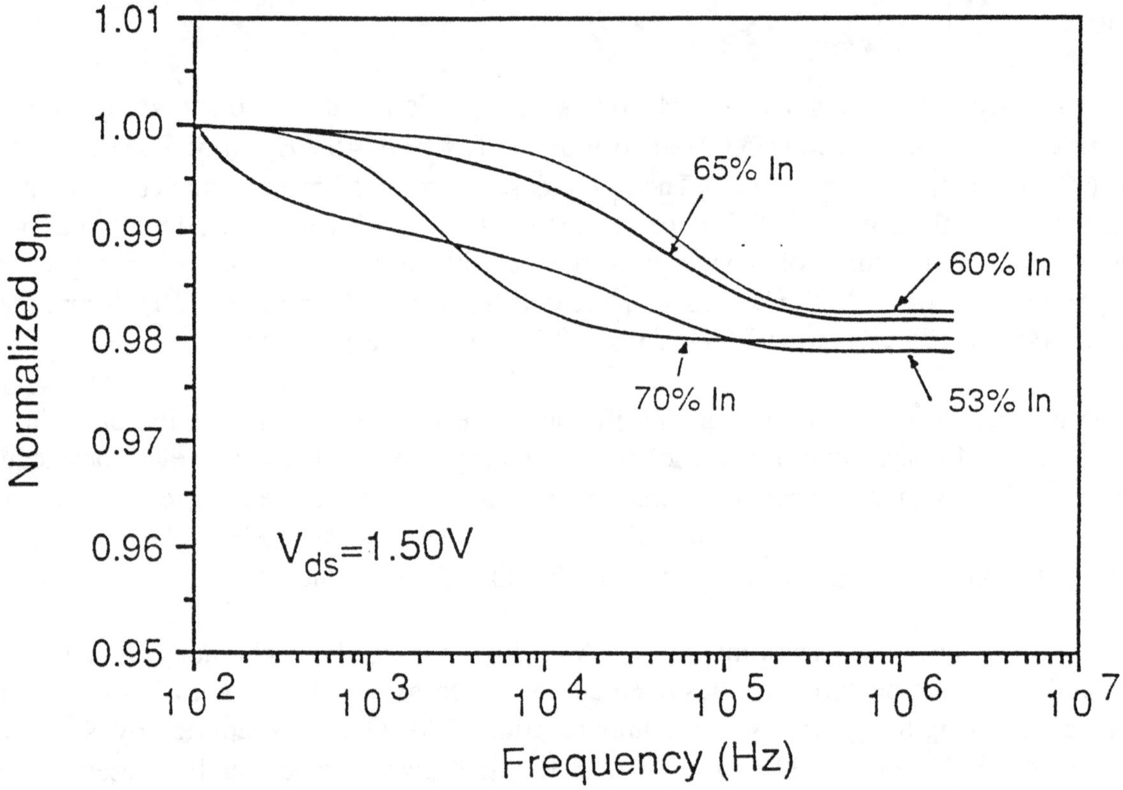

FIGURE 7 Normalised transconductance ($g_m(f)/g_{m0}$) as a function of frequency for InAlAs/InGaAs HEMTs with different channel In contents at V_{ds} = 1.5 V and peak g_m gate-bias conditions.

The g_m dispersion is attributed to the channel region under the gate rather than the access regions between the gate and the ohmic contacts as in the case of MESFETs. This is due to the fact that the 2DEG channel is protected by the n^+ InGaAs cap, i-InAlAs and n-InAlAs donor layers from direct exposure to the device surface.

Studies of R_{ds} dispersion showed the same dependence on In as the trap density [47]. The 53% In channels have the highest trap density and are associated with the largest R_{ds}-f dispersion, followed by the 70% In and 65% In devices. The 60% In channels give the least trap density and smallest R_{ds}-f.

The low frequency noise of InAlAs/InGaAs HEMTs has also been studied [49, 50]. One observes the following features:

(a) On average, there are no significant differences in the LF input noise with strain particularly under saturated bias conditions. The average noise spectra of all devices varies with In content by less than an order of magnitude.

(b) The output-noise voltage spectral density at $V_{DS} = 0.2$ V is slightly higher for the strained HEMTs compared to the lattice matched devices. This reflects a similar trend to that of the input noise with the exception of the 70% In samples which have the highest noise. The increased noise of these samples is due to their high gain compared to the other devices The overall noise level at $V_{DS} = 0.2$ V is smaller than that found at the input because of the HEMT gain being less than one.

(c) The output noise at $V_{DS} = 1.5$ V for all devices. Unlike the characteristics of the input noise, the 65% In samples have the highest noise. This is due to their gain being larger than for the other devices.

(d) The output-noise voltage spectral density is higher at larger V_{DS} due to the higher gain under saturated rather than linear operating conditions.

Although many state-of-the-art device results have been reported for InP-based HEMTs, only a few studies on their reliability issues have been published. FIGURE 8 shows the major degradation mechanisms of the InP-based HEMTs [51]. These can be categorised as follows:

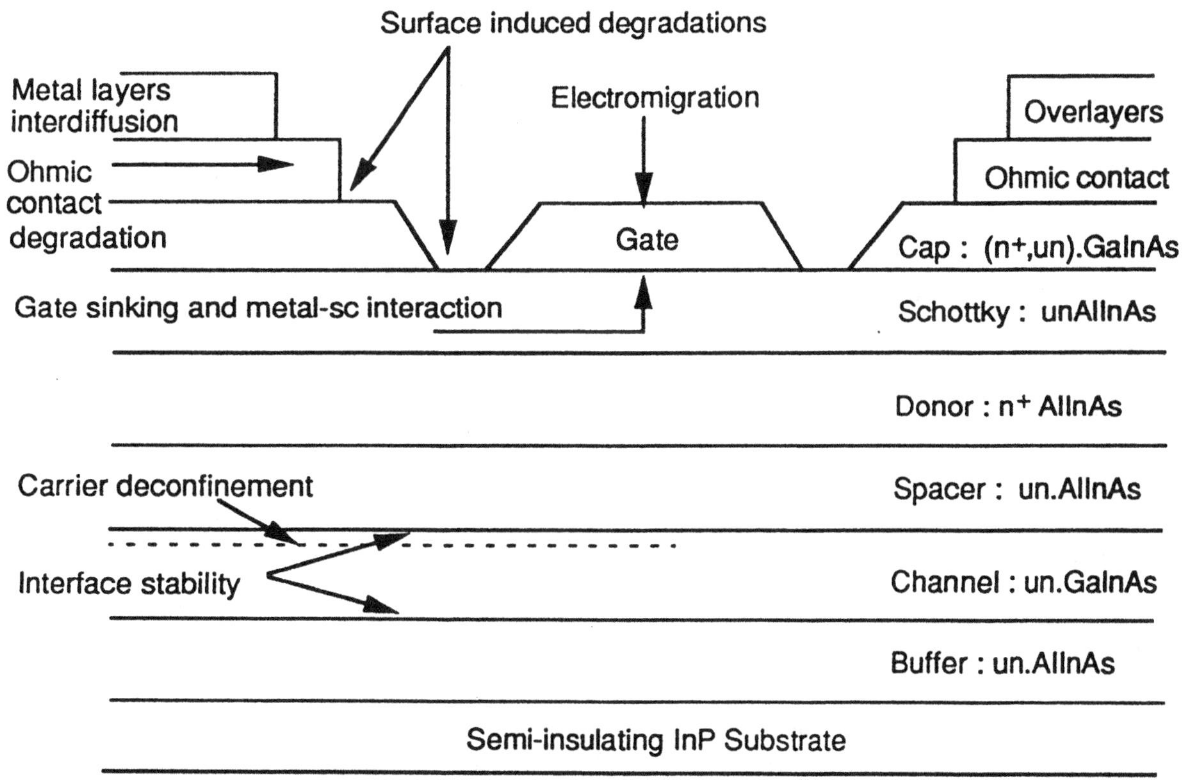

FIGURE 8 Locations of the major degradation mechanisms in InP-based HEMTs [51] (© IEEE 1992).

(1) Ohmic contact related degradation

Typical ohmic metallisation schemes consist of Ge/Au/Ni/Au similar to those used in GaAs devices. As a result, similar degradation mechanisms have been found such as Ga outdiffusion into the ohmic metal. This, however, can be limited by depositing a thick Ti/Pt/Au overlayer on the ohmic metal.

(2) Gate related degradation

Ti/Pt/Au is typically used for Schottky contact and deposited on the undoped InAlAs layer of the HEMT. Despite the fact that this metal system is widely used, Ti on InAlAs exhibits a modest Schottky barrier height of 0.59 eV and poor linearity in the forward diode characteristics [27]. By adding a thin (~ 100 Å) Pt layer between the InAlAs and Ti/Pt/Au, Harada et al [27] have shown improved linearity and a larger Schottky barrier height of 0.82 eV; in addition, better thermal stability can be obtained due to the reduced interaction of the In and Al from the InAlAs layer with the Pt.

(3) Surface induced degradation

This typically refers to the regions between the gate-source and the gate-drain. These areas are normally passivated using plasma enhanced chemical vapour deposited (PECVD) silicon nitride on GaAs FETs for improved reliability. Although the same approach has been successfully implemented for InGaAs/InP photodiodes, no such study has yet been reported for InAlAs/InGaAs HEMTs. Further investigations are therefore needed to understand better the effects of passivation on the InAlAs layer.

(4) Material related degradation

When subjected to thermal stress, InAlAs/InGaAs HEMTs have shown degradation such as decreased drain-source current, transconductance and increased kink effects [52]. These can be attributed to the carrier deconfinement due to the decreased ΔE_c and the channel-buffer interface and layer instability which are manifested by the presence of increased trap density.

D2.2 Strained (pseudomorphic) HIGFETs

As in the case of HEMTs, by introducing excess In in the $In_xGa_{1-x}As$ channel one obtains larger band discontinuities and better carrier confinement.

A systematic study of the DC microwave properties of $In_{0.52}Al_{0.48}As/In_xGa_{1-x}As$ ($0.53 \leq x \leq 0.70$) HIGFETs with a quantum well design showed the merits of the strained approach [53]. DC and microwave transconductances (g_m) are enhanced as the In content is increased in the InGaAs channel. An intrinsic microwave g_m value of 428 mS/mm and K-factor of 1140 mS/mm V have been obtained for 1.0 µm gate length with the 65% In channel devices. The sheet charge density, drift mobility, transconductance, current-gain cutoff frequency (f_T), and maximum oscillation frequency (f_{max}) all show a continuous improvement up to 65% In content (f_T = 22.5 GHz with 53% and f_T = 27 GHz with 65% In; the corresponding f_{max} change is from 6.5 to 8 GHz). The device performance degrades as the In content is increased to 70%. DC and microwave characteristics show the presence of Negative Differential Resistance (NDR) up to 2.7 GHz.

A similar study was performed for p-channel Heterostructure FETs (HFETs) where the channel is p-doped [54]. The study revealed the features of both lattice matched (x = 0.53) and strained (x = 0.65) $In_{0.52}Al_{0.48}As/In_xGa_{1-x}As$ p-HFETs. The role of doping density is studied with the help of two designs (dual-channel with low doping and single-channel with high doping). The strained dual-channel devices demonstrate an improvement of mobility from 108 cm^2/V s (53% In) to 265 cm^2/V s (65% In) at 300 K. The corresponding intrinsic transconductance enhancement is from 23 mS/mm (53% In) to 46.5 mS/mm (65% In) using 1.0-μm-long gates. The cutoff frequency (f_T) also improves from 1.0 to 1.4 GHz. The impact of strain in the highly doped single-channel device is small. The band structure under lattice-matched and strained conditions and the position of the Fermi leve' ... cording to doping seem to be the main factors determining the reported features.

E CONCLUSION

InGaAs channel heterostructure field effect transistors have emerged as high-performance devices for low-noise amplification and for integrated circuits. The DC, microwave and noise performance of several variations of these devices are described in this Datareview and represent the state-of-the-art.

REFERENCES

[1] T.J. Drummond, W. Kopp, R. Fischer, H. Morkoç, R.E. Thorne, A.Y. Cho [*J. Appl. Phys. (USA)* vol.53 (1982) p.1238]

[2] A.J. Valois, G.Y. Robinson, K. Lee, M. Shur [*J. Vac. Sci. Technol. B (USA)* vol.1 (1983) p.190-5]

[3] T.J. Drummond et al [*IEEE Trans. Electron Devices (USA)* vol.30 (1983) p.1806-11]

[4] S.M. Liu, M.B. Das, W. Kopp, H. Morkoç [*IEEE Electron Device Lett. (USA)* vol.6 (1985) p.453-5]

[5] A. Ketterson et al [*IEEE Electron Device Lett. (USA)* vol.6 (1985) p.628-30]

[6] T. Henderson, M.I. Aksun, C.K. Peng, H. Morkoç [*IEEE IEDM Tech. Dig.* (1986) p.464-6]

[7] L.F. Lester et al [*IEEE IEDM Tech. Dig. (USA)* (1988) p.172-5]

[8] G.W. Wang, Y.K. Chen, D.C. Radulescu, L.F. Eastman [*IEEE Electron Device Lett. (USA)* vol.9 (1988) p.4-6]

[9] M.Y. Kao et al [*IEEE Electron Device Lett. (USA)* vol.10 (1989) p.580-2]

[10] R. Fischer et al [*J. Vac. Sci. Technol. B (USA)* vol.2 (1984) p.170-4]

[11] P.M. Smith et al [*IEEE Electron Device Lett. (USA)* vol.10 (1989) p.437-9]

[12] P. Saunier, J.W. Lee [*IEEE Electron Device Lett. (USA)* vol.7 (1986) p.503-5]

[13] D.C. Streit et al [*IEEE Electron Device Lett. (USA)* vol.12 (1991) p.149-50]

[14] P.M. Smith, P.C. Chao, J.M. Ballingall [Presentation at IEEE Princeton Section Sarnoff Symposium, Princeton, NJ, 1992]

[15] K.H.G. Duh et al [*IEEE MTT-S Int. Microw. Symp. Dig. (USA)* (1990) p.595-8]

[16] K.L. Tan et al [*IEEE Electron Device Lett. (USA)* vol.11 (1990) p.585-7]

[17] W.W. Hu, P.C. Chao, M.Y. Kao, P. Ho, R.J. Finke, A.W. Swanson [*IEEE GaAs IC Symposium Tech. Dig. (USA)* (1991) p.191-4]

[18] C.K. Peng, A. Ketterson, H. Morkoç, P.M. Solomon [*J. Appl. Phys. (USA)* vol.60 (1986) p.1709-12]

[19] U.K. Mishra, A.S. Brown, L. Jelloian, L.H. Hackett, M.J. Delaney [*IEEE Electron Device Lett. (USA)* vol.9 (1988) p.41-3]

[20] A.S. Brown, U.K. Mishra, J.A. Henige, M.J. Delaney [*J. Vac. Sci. Technol. B (USA)* vol.6 (1988) p.678-81]

[21] A.S. Brown et al [*IEEE GaAs IC Symposium Tech. Dig. (USA)* (1989) p.143-6]

[22] G.I. Ng, D. Pavlidis, M. Tutt, J. Oh, P.K. Bhattacharya [*IEEE Electron Device Lett. (USA)* vol.10 (1989) p.114-6]

[23] G.I. Ng, R. Lai, D. Pavlidis, J. Pamulapati, P.K. Bhattacharya, K. Studer-Rabeler [*Proc. Int. Conf. on Indium Phosphide and Related Materials*, Denver, Colorado, 1990 (IEEE, 1990) p.424-7]

[24] J.B. Kuang, Y.K. Chen, D. Sivco, A.Y. Cho [*Appl. Phys. Lett. (USA)* vol.57 (1990) p.1784-6]

[25] Y.C. Pao, C.K. Nishimoto, R. Majidi-Ahy, J. Archer, N.G. Bechtel, J.S. Harris Jr. [*IEEE Trans. Electron Devices (USA)* vol.37 (1990) p.2165-70]

[26] M. Matloubian, L.D. Nguyen, A.S. Brown, L.E. Larson, M.A. Melendes, M.A. Thompson [*IEEE MTT-S Int. Microw. Symp. Dig. (USA)* (1991) p.721-4]

[27] N. Harada, S. Kuroda, T. Katakami, K. Hikosaka, T. Mimura, M. Abe [*Proc. 3rd Int. Conf. Indium Phosphide and Related Materials*, Cardiff, Wales, April 1991, p.377-80]

[28] K. Imanishi, T. Shikawa, K. Kondo [*Inst. Phys. Conf. Ser. (UK)* no.106 ch.9 (1989) p.637-41]

[29] W.P. Hong, P.K. Bhattacharya [*IEEE Trans. Electron Devices (USA)* vol.35 (1988) p.8-13]

[30] Y.J. Chan, D. Pavlidis, G.I. Ng [*IEEE Electron Device Lett. (USA)* vol.12 (1991) p.360-2]

[31] U.K. Mishra, A.S. Brown, L.M. Jelloian, M. Thompson, L.D. Nguyen, S.E. Rosenbaum [*IEEE IEDM Tech. Dig. (USA)* (1989) p.101-4]

[32] P.M. Smith et al [*Proc. 2nd Indium Phosphide and Related Compounds Conf.*, Denver, Colorado, April 1990 (IEEE, 1990) p.39-43]

[33] R. Majidi-Ahy et al [*IEEE Trans. Microw. Theory Tech. (USA)* vol.38 (1990) p.1986-93]

[34] Y. Kwon, D. Pavlidis, M. Tutt, G.I. Ng, R. Lai, T. Brock [*Electron. Lett. (UK)* vol.26 (1990) p.1425-6]

[35] Y. Kwon, D. Pavlidis, M. Tutt, G.I. Ng, T. Brock [*GaAs IC Symposium Tech. Dig.* (1990) p.181-4]

[36] Y. Kwon, D. Pavlidis, P. Marsh, M. Tutt, G.I. Ng, T. Brock [*IEEE GaAs IC Symposium Tech. Dig.* (1991) p.165-8]

[37] J.M. Kuo, B. Lalevic, T.Y. Chang [*IEEE IEDM Tech. Dig.* (1986) p.460-3]

[38] M. Jaffee, Y. Sekiguchi, J. Singh, Y.J. Chan, D. Pavlidis, M. Quillec [*Proc. IEEE/Cornell Conf.*, Ithaca, NY, 1987 (IEEE, 1987) p.70-9]

[39] G.I. Ng, D. Pavlidis, M. Jaffee, J. Singh, H.F. Chau [*IEEE Trans. Electron Devices (USA)* vol.36 (1989) p.2249-59]

[40] Y.J. Chan, D. Pavlidis, G.I. Ng, M. Jaffe, J. Singh, M. Quillec [*IEEE IEDM Tech. Dig. (USA)* (1989) p.427-30]

[41] U.K. Mishra, A.S. Brown, S.E. Rosenbaum [*IEEE IEDM Tech. Dig. (USA)* (1988) p.180]

[42] L.D. Nguyen, A.S. Brown, M.A. Thompson, L.M. Jelloian, L.E. Larson, M. Matloubian [*IEEE Electron Device Lett. (USA)* vol.13 (1992) p.143-5]
[43] K.L. Tan et al [*IEEE IEDM Tech. Dig. (USA)* (1991) p.239-42]
[44] M.Y. Kao, P.M. Smith, P.C. Chao, P. Ho [*Proc. IEEE/Cornell Conf.*, Ithaca, NY, 1991 (IEEE, 1991) p.469-77]
[45] R. Lai, P.K. Bhattacharya, S.A. Aterovitz, A.N. Downey, C. Chorey [*IEEE Electron Device Lett. (USA)* vol.11 (1990) p.564-6]
[46] T. Akazaki, T. Enoki, K. Arai, Y. Umeda, Y. Ishii [*Electron. Lett. (UK)* vol.28 (1992) p.1230-1]
[47] P.H. Ladbrooke, S.R. Blight [*IEEE Trans. Electron Devices (USA)* vol.35 (1988) p.257-67]
[48] C. Canali et al [*IEEE Trans. Electron Devices (USA)* vol.37 (1990) p.1163-5]
[49] G.I. Ng et al [*Proc. IEEE/Cornell Conf.*, Ithaca, NY, August 1989, p.73-82]
[50] G.I. Ng, D. Pavlidis, M. Tutt, R.M. Weiss, P. Marsh [*IEEE Trans. Electron Devices (USA)* vol.39 (1992) p.523-32]
[51] J.M. Dumas, D. Lecrosnier [Presented at the *4th Indium Phosphide and Related Materials Conference*, Newport RI, April 1992]
[52] M. Tutt, G.I. Ng, D. Pavlidis, J. Mansfield [*GaAs IC Symposium Tech. Dig. (USA)* (1991) p.101-4]
[53] Y.J. Chan, D. Pavlidis [*IEEE Trans. Electron Devices (USA)* vol.38 (1991) p.1999-2005]
[54] Y.J. Chan, D. Pavlidis [*IEEE Trans. Electron Devices (USA)* vol.39 (1992) p.466-72]

www.ingramcontent.com/pod-product-compliance
Lightning Source LLC
LaVergne TN
LVHW080023210926
848481LV00011B/3091